Assessing Sustainability over Space and Time: The Emerging Roles of GIScience and Remote Sensing

Assessing Sustainability over Space and Time: The Emerging Roles of GIScience and Remote Sensing

Editor

Ronald C. Estoque

Basel • Beijing • Wuhan • Barcelona • Belgrade • Novi Sad • Cluj • Manchester

Editor
Ronald C. Estoque
Center for Biodiversity and
Climate Change, Forestry and
Forest Products Research
Institute (FFPRI)
Tsukuba, Japan

Editorial Office
MDPI
St. Alban-Anlage 66
4052 Basel, Switzerland

This is a reprint of articles from the Special Issue published online in the open access journal *Remote Sensing* (ISSN 2072-4292) (available at: https://www.mdpi.com/journal/remotesensing/special_issues/Sustainability_Space_Time).

For citation purposes, cite each article independently as indicated on the article page online and as indicated below:

Lastname, A.A.; Lastname, B.B. Article Title. *Journal Name* **Year**, *Volume Number*, Page Range.

ISBN 978-3-0365-8498-0 (Hbk)
ISBN 978-3-0365-8499-7 (PDF)
doi.org/10.3390/books978-3-0365-8499-7

Contents

About the Editor

Ronald C. Estoque

Dr. Ronald C. Estoque is a Senior Researcher at the Center for Biodiversity and Climate Change, Forestry and Forest Products Research Institute (FFPRI), Japan. He is a forester and geoenvironmental scientist by training. His research interests are at the interface of nature and society and include sustainability, land use/land cover change, forest cover change, ecosystem services, urbanization, quality of life, climate change mitigation, vulnerability, risk and adaptation, and the nexus approach. He applies various tools and techniques, including GIS, remote sensing, and social–ecological approaches. He has published more than 60 journal articles. Dr. Estoque was also a contributing author (CA) to the Intergovernmental Panel on Climate Change's (IPCC) sixth assessment report (AR6) (WG II, Chapter 10). Currently, he serves as a review editor (RE) for the Intergovernmental Science-Policy Platform for Biodiversity and Ecosystem Services' (IPBES) nexus assessment (Chapter 2), a section Editor-in-Chief for *Sustainability* (section: Sustainable Forestry), an associate editor for *Frontiers in Forests and Global Change*, and a member of the editorial board for *Science of the Total Environment*, *Remote Sensing*, and *Ecological Civilization*.

Preface

Monitoring progress towards the Sustainable Development Goals (SDGs) and sustainability in general is essential, not only at the global and national scales, but also at the subnational and landscape levels. Geographic information science (GIScience) and remote sensing (RS) have made significant advances in this area. Such advances include the increasing availability of geospatial data and the development of more sophisticated approaches or techniques for spatiotemporal analysis, which can play a crucial role in this regard. This Special Issue aims to contribute to the growing body of knowledge in this area, highlighting the crucial roles of GIScience and RS in advancing sustainability assessment and research.

The Guest Editor expresses gratitude to the authors for their valuable contributions, as well as to the editors and reviewers for their expertise, time, and effort. The MDPI team is also acknowledged for their kind assistance and support.

Ronald C. Estoque
Editor

remote sensing 

Editorial

Assessing Sustainability over Space and Time: The Emerging Roles of GIScience and Remote Sensing

Ronald C. Estoque

Center for Biodiversity and Climate Change, Forestry and Forest Products Research Institute, 1 Matsunosato, Tsukuba 305-8687, Japan; estoquerc21@affrc.go.jp

Sustainability is a critical global challenge that requires comprehensive assessments of environmental, social, and economic indicators. The formulation of the 17 Sustainable Development Goals (SDGs) represents a significant leap forward in humanity's pursuit of sustainability. The SDGs now serve as a platform for global development, guiding current actions and shaping visions for a sustainable future. Tracking the spatiotemporal dynamics of progress towards the SDGs and sustainability in general is essential, not only at global and national scales, but also at subnational and landscape levels. Geographic Information Science (GIScience) and Remote Sensing (RS) have made significant advancements, including the increase in availability of geospatial data, which can play a crucial role in this regard.

GIScience focuses on spatial data, information systems, and technologies for managing, analyzing, and visualizing spatial information. GIScience can help identify patterns and trends, assess sustainability-related indicators, evaluate the effectiveness of policies and management strategies, and support decision-making processes. RS, on the other hand, involves the acquisition and interpretation of data about the Earth's surface and atmosphere from RS platforms such as satellites, airplanes, or drones. RS provides a wide range of spatial data, such as land cover, vegetation indices, and atmospheric parameters, which can be used to monitor and assess sustainability-related indicators.

This Special Issue aims to bring together novel contributions on the assessment of sustainability and sustainability-related indicators over space and time using geospatial data, tools, and techniques (GIScience and RS). It consists of eleven peer-reviewed papers, including two review articles and nine research articles.

The first review article focuses on the concept of sustainability and the state of SDG monitoring using RS [1]. It traces the conceptual origins of sustainability and discusses the role of RS in SDG monitoring, as well as the current status, challenges, and opportunities. The review reveals that the pursuit of a sustainable future likely began in the 17th century when declining forest resources in Europe led to proposals for the re-establishment and conservation of forests, embodying the great idea that the current generation bears responsibility for future generations. As of April 2020, preliminary statistical data were available for 21 (70%) of the 30 RS-based SDG indicators, according to the Global SDG Indicators Database, with 10 (33%) also included in the 2019 SDG Index and Dashboards. However, at the time of the review, these statistics may not have necessarily reflected the actual status and availability of raw and processed geospatial data for RS-based indicators, which is an important issue to consider. Nevertheless, the review also identifies various initiatives that have been initiated to address data-related issues, which are crucial for SDG monitoring.

The second review article explores the role of RS in international peace and security [2]. Specifically, it examines the key research concepts and implementation of RS for applications related to international peace and security. The article presents a meta-analysis of how advanced sensor capabilities can support various aspects of peace and security, including relief operations, monitoring armed conflicts, tracking acts of genocide, international peace missions, human rights, and disease control and prevention. The review concludes that

Citation: Estoque, R.C. Assessing Sustainability over Space and Time: The Emerging Roles of GIScience and Remote Sensing. *Remote Sens.* **2023**, *15*, 2764. https://doi.org/10.3390/rs15112764

Received: 25 April 2023

Accepted: 9 May 2023

Published: 26 May 2023

RS, as a surveillance tool, has immense potential in safeguarding the environment, peace, and security, provided it is used actively and transparently. However, there are future challenges that may hinder the application of RS in peace and security, such as discrepancies in image classifications due to varying types of sensors, as well as issues related to cost, resolution, and ground-truth validation in conflict areas. Nevertheless, with emerging technologies and sufficient secondary resources available, the article argues that RS will continue to be an important tool in aiding peacekeeping processes in conflict-affected areas.

The nine research articles can be grouped into two: (1) index and framework development, and (2) social–ecological research applications. The remotely sensed urban surface ecological index (RSUSEI) is proposed as an index to assess urban surface ecological status, which is influenced by surface biophysical, biochemical, and biological properties [3]. The RSUSEI is developed using five RS-derived variables, namely the normalized difference vegetation index, normalized difference soil index, wetness index, land surface temperature, and impervious surface cover, based on principal component analysis. The RSUSEI was successfully applied in six cities across the United States, and the study concludes that it has significant potential for modeling and comparing urban surface ecological status in cities with diverse geographical, climatic, environmental, and biophysical conditions.

Another proposed index, called the improved comprehensive remote sensing ecological index (IRSEI), is designed for evaluating urban ecological quality [4]. The IRSEI incorporates four RS-derived ecological elements, including humidity, greenness, heat, and dryness, and is constructed using a combination of entropy weight and principal component analysis. The index has been applied in Wuhan, China, and the study concludes that IRSEI, along with the urban ecological quality approach in general, can be a valuable tool for ecological management and protection, as well as for assessing progress towards urban sustainable development.

A new framework, based on GIScience and spatial justice (fair allocation), has been proposed with a focus on the sustainable development of life service resources [5]. This framework addresses two key questions: (1) why spatial justice should be considered when studying life service resources, and (2) how spatial justice should be applied and interpreted for life service resources. The framework has been applied in Beijing, China, using multi-source data such as population density and building outlines, and GIScience methods such as nearest neighbour and kernel density analysis. Based on the findings, the study recommends that future planning should aim to narrow the development gap through the optimal spatial allocation of life service resources to improve spatial justice.

Among the research application articles is a study that focuses on the impacts of urbanization on the Muthurajawela Marsh and Negombo Lagoon in Sri Lanka, which is an important wetland ecosystem providing a wide range of ecosystem services [6]. Using multi-temporal RS and other spatial data, the study found that the spatial and socioeconomic elements of rapid urbanization in the area have been the main drivers of the wetland's environmental transformation over the past 20 years (1997–2017). This is evident from the substantial expansion of settlements (+68%) and significant decrease in marshland and mangrove cover (−41% and −21%, respectively). The study concludes that there is an urgent need for forward-looking landscape and urban planning to ensure the sustainability of this valuable wetland ecosystem.

Another contribution focuses on multi-temporal mapping of population distribution using RS data and deep learning techniques [7]. The study argues that understanding the spatiotemporal distribution of population is crucial in various fields, such as resource management, disaster response, public health, and urban planning, all of which are relevant to the sustainability goal. However, the study observes that there is a lack of continuous, multi-temporal gridded population data over a long historical period, and attributes this to the absence of appropriate auxiliary datasets and effective methodological frameworks. The study aims to address this knowledge gap by proposing a methodological framework that integrates deep learning architecture and Landsat data, which has been applied in

China. The study concludes that this proposed framework can be particularly useful in low-income and data-poor regions.

Understanding the relationship between eco-environmental quality and urbanization has been a focal point of another contribution [8]. The study argues that comprehending the interactive coupling mechanism between eco-environmental quality and urbanization is of great significance in achieving urban sustainability. The study focuses on China as a case study and utilizes multi-source RS data and the coupling coordination degree model to facilitate the analysis. The findings of the study reveal that rapid urbanization has resulted in a significant decline in eco-environmental quality in certain areas of the country. The authors emphasize the need to prioritize the protection of the ecological environment while pursuing social and economic development in the future.

A contribution focusing on the neighborhood level in Guilin, China, assessed the spatiotemporal changes of three SDG indicators (11.2.1, 11.3.1, and 11.7.1) using high-resolution RS images and a big data approach [9]. The study found that the proportion of the population with convenient access to public transport gradually improved from 42% in 2013 to 52% in 2020. However, the increase in built-up land was relatively fast, resulting in a decrease in the areal proportion of public open space from 56% in 2013 to 24% in 2020. The authors highlight the role of big data and Earth observation technology in monitoring urban sustainable development. This study provides an example of a neighborhood-level assessment of SDG indicators, which is crucial for local urban governance and planning practices.

Also focusing on SDG 11, a case study in Manila, Philippines, assessed the urban heat island phenomenon using RS data [10]. Consistent with findings from other studies around the world, the study revealed that residential areas, asphalted and concrete roads and walkways, and certain commercial establishments and buildings exhibited higher surface temperatures compared to areas with vegetation and near bodies of water. The study proposed strategies to mitigate the impacts of urban heat islands, including the use of cool materials for pavements and roofs; the conversion of regular walls to green walls; and increased planting in plant boxes, road isles, and indoors. The authors also recommended the establishment of additional meteorological stations in urban heat island hotspots to help improve the current understanding of outdoor thermal characteristics in the city.

Finally, in Mexico, a case study assessed the drivers that influence the dynamics of a tropical dry forest in Ayuquila River watershed using multi-temporal RS and other spatial data [11]. The study estimated a tropical dry forest loss rate of 1.6% per year between 2019 and 2022. The study also identified an inverse relationship between forest loss and slope and distance to roads. This is related to the fact that flat areas are preferred for agricultural and livestock activities due to easy access and lower costs. Careful consideration of these factors is essential when addressing tropical dry forest loss in the region. The loss of tropical dry forests not only contributes to carbon loss, but also leads to biodiversity loss, soil erosion, and increased vulnerability of local communities that rely on these forests for sustenance and shelter.

In summary, GIScience and RS and their integration have emerged as valuable tools for assessing sustainability and sustainability-related indicators over space and time. This Special Issue contributes to the growing body of knowledge in this area, underscoring the crucial roles of GIScience and RS in advancing sustainability assessment and research.

Funding: This work was supported by the Japan Society for the Promotion of Science (JSPS) through its Grants-in-Aid for Scientific Research (KAKENHI) Program: Grant-in-Aid for Scientific Research (C), Number 22K01038 (Principal Investigator: Ronald C. Estoque). The views expressed in this paper are of the author and do not necessarily reflect the position of his institution and of the funder.

Acknowledgments: The author (Special Issue Guest Editor) wishes to thank the authors for their valuable contributions to this Special Issue. Likewise, to the Editors and Reviewers for their expertise, time, and effort.

Conflicts of Interest: The author declares no conflict of interest.

References

1. Estoque, R.C. A Review of the Sustainability Concept and the State of SDG Monitoring Using Remote Sensing. *Remote Sens.* **2020**, *12*, 1770. [CrossRef]
2. Avtar, R.; Kouser, A.; Kumar, A.; Singh, D.; Misra, P.; Gupta, A.; Yunus, A.P.; Kumar, P.; Johnson, B.A.; Dasgupta, R.; et al. Remote Sensing for International Peace and Security: Its Role and Implications. *Remote Sens.* **2021**, *13*, 439. [CrossRef]
3. Firozjaei, M.K.; Fathololoumi, S.; Weng, Q.; Kiavarz, M.; Alavipanah, S.K. Remotely Sensed Urban Surface Ecological Index (RSUSEI): An Analytical Framework for Assessing the Surface Ecological Status in Urban Environments. *Remote Sens.* **2020**, *12*, 2029. [CrossRef]
4. Li, J.; Gong, J.; Guldmann, J.-M.; Yang, J. Assessment of Urban Ecological Quality and Spatial Heterogeneity Based on Remote Sensing: A Case Study of the Rapid Urbanization of Wuhan City. *Remote Sens.* **2021**, *13*, 4440. [CrossRef]
5. Xu, Z.; Niu, L.; Zhang, Z.; Huang, J.; Lu, Z.; Huang, Y.; Wen, Y.; Li, C.; Gu, X. Sustainable Development of Life Service Resources: A New Framework Based on GIScience and Spatial Justice. *Remote Sens.* **2022**, *14*, 2031. [CrossRef]
6. Athukorala, D.; Estoque, R.C.; Murayama, Y.; Matsushita, B. Impacts of Urbanization on the Muthurajawela Marsh and Negombo Lagoon, Sri Lanka: Implications for Landscape Planning towards a Sustainable Urban Wetland Ecosystem. *Remote Sens.* **2021**, *13*, 316. [CrossRef]
7. Zhuang, H.; Liu, X.; Yan, Y.; Ou, J.; He, J.; Wu, C. Mapping Multi-Temporal Population Distribution in China from 1985 to 2010 Using Landsat Images via Deep Learning. *Remote Sens.* **2021**, *13*, 3533. [CrossRef]
8. Xu, D.; Cheng, J.; Xu, S.; Geng, J.; Yang, F.; Fang, H.; Xu, J.; Wang, S.; Wang, Y.; Huang, J.; et al. Understanding the Relationship between China's Eco-Environmental Quality and Urbanization Using Multisource Remote Sensing Data. *Remote Sens.* **2022**, *14*, 198. [CrossRef]
9. Han, L.; Lu, L.; Lu, J.; Liu, X.; Zhang, S.; Luo, K.; He, D.; Wang, P.; Guo, H.; Li, Q. Assessing Spatiotemporal Changes of SDG Indicators at the Neighborhood Level in Guilin, China: A Geospatial Big Data Approach. *Remote Sens.* **2022**, *14*, 4985. [CrossRef]
10. Purio, M.A.; Yoshitake, T.; Cho, M. Assessment of Intra-Urban Heat Island in a Densely Populated City Using Remote Sensing: A Case Study for Manila City. *Remote Sens.* **2022**, *14*, 5573. [CrossRef]
11. Gao, Y.; Solórzano, J.V.; Estoque, R.C.; Tsuyuzaki, S. Tropical Dry Forest Dynamics Explained by Topographic and Anthropogenic Factors: A Case Study in Mexico. *Remote Sens.* **2023**, *15*, 1471. [CrossRef]

 *remote sensing*

Review

A Review of the Sustainability Concept and the State of SDG Monitoring Using Remote Sensing

Ronald C. Estoque

National Institute for Environmental Studies, Tsukuba, Ibaraki 305-8506, Japan; estoque.ronaldcanero@nies.go.jp or rons2k@yahoo.co.uk

Received: 4 April 2020; Accepted: 11 May 2020; Published: 31 May 2020

Abstract: The formulation of the 17 sustainable development goals (SDGs) was a major leap forward in humankind's quest for a sustainable future, which likely began in the 17th century, when declining forest resources in Europe led to proposals for the re-establishment and conservation of forests, a strategy that embodies the great idea that the current generation bears responsibility for future generations. Global progress toward SDG fulfillment is monitored by 231 unique social-ecological indicators spread across 169 targets, and remote sensing (RS) provides Earth observation data, directly or indirectly, for 30 (18%) of these indicators. Unfortunately, the UN Global Sustainable Development Report 2019—The Future is Now: Science for Achieving Sustainable Development concluded that, despite initial efforts, the world is not yet on track for achieving most of the SDG targets. Meanwhile, through the EO4SDG initiative by the Group on Earth Observations, the full potential of RS for SDG monitoring is now being explored at a global scale. As of April 2020, preliminary statistical data were available for 21 (70%) of the 30 RS-based SDG indicators, according to the Global SDG Indicators Database. Ten (33%) of the RS-based SDG indicators have also been included in the SDG Index and Dashboards found in the Sustainable Development Report 2019—Transformations to Achieve the Sustainable Development Goals. These statistics, however, do not necessarily reflect the actual status and availability of raw and processed geospatial data for the RS-based indicators, which remains an important issue. Nevertheless, various initiatives have been started to address the need for open access data. RS data can also help in the development of other potentially relevant complementary indicators or sub-indicators. By doing so, they can help meet one of the current challenges of SDG monitoring, which is how best to operationalize the SDG indicators.

Keywords: sustainable development; Earth observation; SDG indicators; EO4SDG; SDG global indicator framework; global SDG indicators database; social-ecological indicators

---

1. Introduction

In the report "Our Common Future" (also known as the Brundtland Report), the World Commission on Environment and Development has defined sustainable development as "development that meets the needs of the present without compromising the ability of future generations to meet their own needs" [1]. This principle is at the core of the 17 sustainable development goals (SDGs) adopted in 2015 by the United Nations (UN) General Assembly [2]. These SDGs collectively help to guide actions for global development and shape visions for the future. At the global level, the UN Statistical Commission serves as the oversight body for SDG efforts. The commission oversees the UN Statistics Division, which is responsible for the maintenance of the Global SDG Indicators Database.

In 2015, the UN Statistical Commission created the Inter-agency and Expert Group on SDGs (IAEG-SDGs), which was tasked with the development and implementation of a global indicator framework for monitoring global progress toward fulfillment of the 17 SDGs. In 2017, the UN General Assembly adopted the global indicator framework developed by the IAEG-SDGs. In the adopted framework, various social-ecological indicators are identified and assigned to the SDG targets [3,4]. Here, the term "social" includes both the social and economic dimensions of sustainability, whereas the term "ecological" refers to its environmental dimension. Together, these are often referred to as the three pillars of sustainability: people (social), profit (economic), and planet (environmental), respectively [5–7]. In this review, the social-ecological indicators are regarded as a set of outcomes for assessing and monitoring the sustainability of social-ecological systems (also called human–environment systems, coupled human–environment systems, and coupled human and natural systems [8,9]), analyzed in the context of the social-ecological system framework proposed by Elinor Ostrom [10–12].

Among various descriptions of indicators [13–15], one that is often cited states that "desirable indicators are those that summarize or otherwise simplify relevant information, makes [sic] visible or perceptible phenomena of interest, and quantify, measure, and communicate relevant information" [13] (p. 108). Indicators, like goals [16], should also be SMART: specific, measurable, achievable, relevant, and time-bound [17]. The major functions of indicators are to (i) assess conditions and trends, (ii) compare across places and situations, (iii) assess conditions and trends in relation to goals and targets, (iv) provide early warning information, and (v) anticipate future conditions and trends [13]. The SDG global indicator framework, though developed primarily for monitoring global progress toward achievement of the SDGs [2,4], supports all these indicator functions. More generally, sustainable development indicators are "scientific constructs whose principal objective is to inform public policy-making" [18] (p. 45).

The achievement of SDGs relies on the performance of countries with respect to the SDG targets and indicators. Thus, the development of methodologies for monitoring progress toward SDG achievement has become "a new vital science" [19]. Custodian agencies are responsible for the development of the necessary methodologies and the collection and compilation of data related to SDG indicators. The international organizations and agencies designated as custodian agencies by the IAEG-SDGs include the World Health Organization (WHO), the Food and Agriculture Organization (FAO), the United Nations Development Programme (UNDP), and the World Bank [20]. Available data for all SDG indicators are compiled in the Global SDG Indicators Database (https://unstats.un.org/sdgs/indicators/database), and the UN Statistics Division also issues annual SDG progress reports (https://unstats.un.org/sdgs/).

The global indicator framework includes a total of 231 unique indicators spread across 169 SDG targets [3,21]. Twelve indicators are associated with two or three targets [3,21]. SDG 17 (Partnerships for the Goals) has the highest number of targets, with 19, followed by SDG 3 (Good Health and Well-Being) with 13 (Figure 1). By contrast, SDGs 7 (Affordable and Clean Energy) and 13 (Climate Action) have the lowest number of targets, with five each. SDG 3 has the largest number of indicators (28), followed by SDGs 16 (Peace, Justice and Strong Institutions) and 17 (24 each), and SDGs 7 and 13 have the smallest number of indicators (6 and 8, respectively).

The SDG indicators are classified into three tiers according to the level of methodological development and the global availability of data as follows [20]: a Tier I indicator "is conceptually clear, has an internationally established methodology and standards are available, and data are regularly produced by countries for at least 50 per cent of countries and of the population in every region where the indicator is relevant." A Tier II indicator "is conceptually clear, has an internationally established methodology and standards are available, but data are not regularly produced by countries." For a Tier III indicator, "no internationally established methodology or standards are yet available for the indicator, but methodology/standards are being (or will be) developed or tested." Indicators for which data availability is under review are labeled as "pending data availability review" [20].

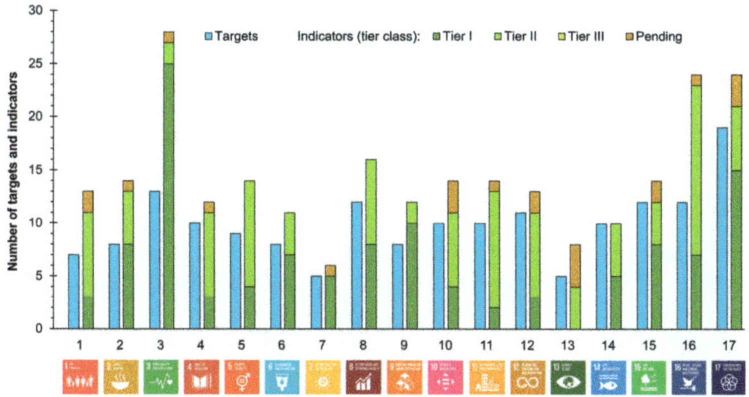

Figure 1. Distribution of targets and indicators across the 17 SDGs. These data were sourced from the March 2020 version of the global indicator framework for the SDGs [3] and the April 2020 version of the tier classification for global SDG indicators [20]. The global indicator framework was adopted by the UN General Assembly in July 2017 and is contained in the resolution designated A/RES/71/313 [4,21]. As of this writing, the latest version of the list of SDG indicators bears the following notations in the upper right corner of the downloadable PDF document: A/RES/71/313, E/CN.3/2018/2, E/CN.3/2019/2, and E/CN.3/2020/2 [3]. These notations mean that the current official list of indicators [3] includes the global indicator framework as contained in A/RES/71/313, the refinements agreed by the UN Statistical Commission at its 49th session in March 2018 (E/CN.3/2018/2, Annex II) and 50th session in March 2019 (E/CN.3/2019/2, Annex II), and the changes from the 2020 Comprehensive Review (E/CN.3/2020/2, Annex II) and annual refinements (E/CN.3/2020/2, Annex III) from the 51st session in March 2020 [21].

Explicit in the 2030 Agenda (Transforming Our World: The 2030 Agenda for Sustainable Development) is a declaration (no. 76) to support developing countries by ensuring access to high-quality, timely, reliable, and disaggregated data, including geospatial and Earth observation data [2]. Recognizing that the "integration of statistical data and geospatial information will be key for the production of a number of indicators," the IAEG-SDGs created the Working Group on Geospatial Information (WGGI) in 2016 [22] (p. 1) under the UN's Committee of Experts on Global Geospatial Information Management (UN-GGIM). The primary objective of the WGGI is "to ensure from a statistical and geospatial perspective that one of the key principles of the 2030 Agenda, to leave no one behind, is reflected in the global indicator framework" [22] (p. 2).

The Group on Earth Observations (GEO), created in 2005, is a global partnership of governments and organizations that "envisions a future where decisions and actions for the benefit of humankind are informed by coordinated, comprehensive and sustained Earth observations" (www.earthobservations. org). In 2016, recognizing the potential of Earth observation data for SDG monitoring, the GEO launched an initiative called "Earth Observations in Service of the 2030 Agenda for Sustainable Development" (EO4SDG) (http://eo4sdg.org/) [23–25]. The purpose of this initiative is to "organize and realize the potential of Earth observations and geospatial information to advance the *2030 Agenda* and enable societal benefits through achievement of the Sustainable Development Goals" [25] (p. 4). The EO4SDG initiative set forth three goals in its Strategic Implementation Plan 2020–2024: "(i) demonstrate how Earth observations, geospatial information, and socio-economic and other data contribute in novel and practical ways to support sustainable development efforts and the SDGs, (ii) increase skills and capabilities in uses of Earth observations for SDG activities and their broader benefits, and (iii) broaden interest, awareness, and understanding of Earth observations support to the SDGs and contributions to social, environmental, and economic benefits" [25] (p. 4).

A couple of years into the implementation of the 2030 Agenda, the GEO has seen an increasing demand for EO data for monitoring progress toward achievement of the SDGs [25]. The GEO has identified 71 (42%) targets and 30 (13%) indicators for the SDGs that can be supported, directly or indirectly, by EO data [25] (see Section 4). The EO4SDG initiative supports the WGGI task stream called "Application of Earth Observations for the SDG Indicators" [25].

The purpose of this review is to describe the conceptualization of sustainability leading to the formulation of the SDGs and then to discuss the current status, challenges, and opportunities in SDG monitoring using remote sensing (RS).

2. Review Approach

This review is intended to be an overview, that is, a survey of the literature and a description of its characteristics [26]. The search and appraisal of reference materials for this review did not follow a set of pre-defined rules as would be done for a systematic review. Potentially relevant reference materials were searched online and appraised according to whether they included information on the conceptualization of sustainability and SDG monitoring with RS. This review thus aims to provide a broad introductory understanding of (a) the conceptualization of sustainability leading to the formulation of the SDGs, (b) the role of RS in SDG monitoring, and (c) the current status, challenges, and opportunities of SDG monitoring with RS.

3. From the Conceptualization of Sustainability to the Formulation of the SDGs

This section provides an overview of the historical origin and development of the sustainability concept and the progression that eventually led to the formulation of the SDGs. In the past three decades, a number of works have addressed this topic. Kidd [27], who focused on the development of the sustainability concept in the 19th and 20th centuries, suggested that the concept has its origins in six ideas (termed "roots"): the ecological/carrying capacity root, the resource/environment root, the biosphere root, the critique of technology root, the no growth-slow growth root, and the ecodevelopment root. According to Grober [28], the sustainability concept probably originated in the 17th and 18th centuries when declining forests in Europe led to the idea that sustained yield (or sustainable use) of forest resources could be achieved through conservation and reforestation. Warde [29], however, thought that the "invention of sustainability" might have occurred earlier, between c. 1500 and 1870.

A timeline of the conceptual development of sustainability (Figure 2) shows that before 1970, the literature on sustainability was dominated by books and essays, whereas after 1970, peer-reviewed articles and global policy initiatives and reports became more dominant. Indeed, environmental issues and initiatives reached the global stage when the UN General Assembly convened in the 1972 Stockholm Conference. Before this conference, environmental governance was apparently not considered to be an international priority. The Stockholm Conference led to the Declaration on the Human Environment (Stockholm Declaration) and resulted in the creation of the UN Environment Programme (UNEP), which is today "the global champion for the environment with programmes focusing on sustainable development, climate, biodiversity and more" (www.unenvironment.org).

Historically, the sustainability concept has its roots in the works of English author John Evelyn (*Sylva*, 1664) and the French statesman Jean Baptist Colbert (*Ordonnance*, 1669), who called for the re-establishment and conservation of forests [28,30] (Figure 2). Although the focus of these works was on Europe, both books included the idea that the current generation has responsibility for future generations [28]. Moreover, they were important sources and models for German nobleman Hanns Carl von Carlowitz [28]. In 1713, von Carlowitz published *Sylvicultura Oeconomica*, in which he introduced the term "sustainable" in its modern sense for the first time: in German, "*nachhaltende Nutzung,*" referring to the "sustainable use" [31] or "sustained use" [28] of forest resources. During the 18th century, *nachhaltend* was modified to *nachhaltig*, and the use of the related noun *nachhaltigkeit*, with reference to "sustained yield" forestry, became widespread [28]. Today, *nachhaltigkeit* is generally translated as "sustainability."

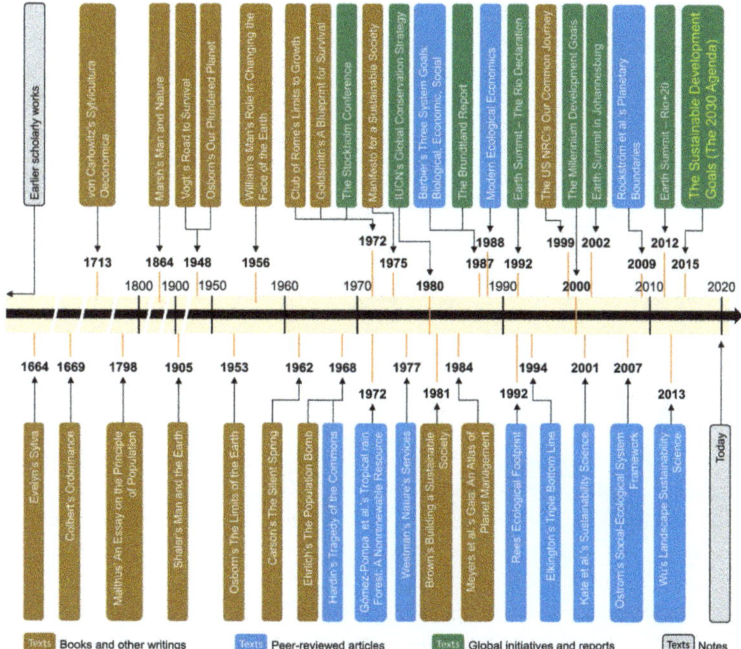

Figure 2. Timeline showing how the sustainability concept has been advanced by scholarly works and global policy initiatives.

Over the past three centuries, many influential books have contributed to the development of the sustainability concept as we understand it today (Figure 2). These include *An Essay on the Principle of Population* (1798) by Thomas Robert Malthus, *Man and Nature* (1864) by George Perkins Marsh, *Man and the Earth* (1905) by Nathaniel Shaler, *Road to Survival* (1948) by William Vogt, *Our Plundered Planet* (1948) and *The Limits of the Earth* (1953) by Fairfield Osborn, *Man's Role in Changing the Face of the Earth* (1956) by William L. Thomas, Jr., *The Silent Spring* (1962) by Rachel Carson, *The Population Bomb* (1968) by Paul R. Ehrlich and Anne H. Ehrlich, *Limits to Growth* (1972) by the Club of Rome, and *A Blueprint for Survival* (1972) by Edward Goldsmith and Robert Allen. Reviews of the sustainability concept and its origin, including the conceptualization of sustainable development, commonly cite these books [27,30,31]. However, the term "sustainable development," as presented in the Brundtland Report (1987) [1], was first used in *World Conservation Strategy*, published in 1980 by the International Union for Conservation of Nature (IUCN) [32]. Other works contributing to the conceptualization of sustainable development include *Building a Sustainable Society* (1981) by Lester R. Brown and *Gaia: An Atlas of Planet Management* (1984) by Norman Meyers and colleagues [31].

Among articles in peer-reviewed journals that have helped shape and advance the sustainability concept, "The tragedy of the commons" (1968) [33] was one of the earliest and most influential. The "tragedy of the commons" is that, in the absence of proper regulation, self-interest can lead to the over-exploitation and destruction of non-renewable common resources, such as the atmosphere, the ocean, and biodiversity, threatening their sustainability. "Tropical rain forest: A nonrenewable resource" (1972) [34] emphasized the importance of tropical rainforests as a non-renewable resource. "How much are nature's services worth?" [35] introduced the concept of nature's services, which, together with some other works [36,37], led to the "ecosystem services" concept [38]. Interestingly, the Intergovernmental Science-Policy Platform on Biodiversity and Ecosystem Services (IPBES) recently proposed, without referring to these earlier works, "nature's contribution to people" as an umbrella term that also includes ecosystem services [39,40]. The ecosystem services concept gained popularity with the publication of the book *Nature's Services: Societal Dependence on Natural Ecosystems* (1997) [41], the article "The value of the world's ecosystem services and natural capital", which appeared in *Nature* in 1997 [42], and the Millennium Ecosystem Assessment report in 2005 (www.millenniumassessment.org/).

Several important ideas have helped advance the sustainability concept. In 1987, Barbier [5] introduced the three system goals of sustainability (biological and resource, economic, and social), which might be the origin of the idea that there are three pillars, dimensions, components, or aspects of sustainability [30,43]. Along with Barbier's work [5], the notion of the triple bottom line [6,7] also contributed to the idea that there are three pillars of sustainability. Ecological economics, a transdisciplinary field of study [44,45], has helped advance the concept of natural capital, a term first used by Ernst F. Schumacher in his book *Small is Beautiful* (1973). The ecological footprint concept [46] eventually led to the establishment of the global footprint network, which aims to advance the science of sustainability. The term "sustainability science" [47] was introduced in *Our Common Journey: A Transition Toward Sustainability* (1999) by the US National Research Council (NRC) [48]. The field of sustainability science has since been advanced by the development of the social-ecological system framework [10] which couples human and environmental systems. In this context, the planetary boundaries concept [49] aims to define a safe operating space for humanity, and landscape sustainability science [50] focuses on the dynamic relationship between ecosystem services and human well-being.

The development of the idea of the human-environment system, together with global initiatives such as the Earth Summit and the Millennium Development Goals (MDGs) (see Figure 2), have helped advance the field of sustainability science and influenced the formulation of the current SDGs. In particular, the three pillars of sustainability are explicitly embedded in the formulation of the SDGs [30,51]. However, the sustainability (or sustainable development) concept is frequently criticized as vague and ambiguous [52–56]. These critiques ask, the sustainability of what and for whom? This reviewer argues that, in the context of the SDGs, sustainability means the sustainability of Earth's resources—its life-support system—for the benefit, or at least for the survival, of the current and future generations of humankind.

In their comprehensive review on the conceptual evolution of sustainability, Purvis et al. [30] suggest that the concepts of the three pillars of sustainability (social, economic, and environmental) do not have theoretically rigorous support, and they conclude that "the absence of such a theoretically solid conception frustrates approaches towards a theoretically rigorous operationalisation of 'sustainability'" [30] (p. 681). This reviewer recognizes this issue. Nevertheless, the viewpoint of this reviewer is that the idea of sustainability is of great importance to humanity, because, like the concepts of freedom, justice, and democracy, which are also dialectically vague, it expresses a fundamental principle that can guide our actions and shape our visions for the future [50,57].

4. The Role of Remote Sensing for SDG Monitoring

The development of a "human capability to observe regions of the electromagnetic spectrum outside the range of wavelengths discernable by the human eye" was fundamental to the evolution of remote sensing technology [58] (p. 685). The term "remote sensing" was coined in the 1950s by Evelyn Pruitt, a geographer and oceanographer formerly with the Office of Naval Research (https://earthobservatory.nasa.gov/features/RemoteSensing). According to Gerald K. Moore [59] (p. 478), remote sensing is "the use of reflected and emitted energy to measure the physical properties of distant objects and their surroundings". The Encyclopedia of Remote Sensing defines remote sensing as "the technique of obtaining information about objects through the analysis of data collected by special instruments that are not in physical contact with the objects of investigation" [58] (p. 684). Nicholas M. Short, in his Remote Sensing Tutorial (An Online Handbook), has defined remote sensing as "the acquisition and measurement of data/information on some property(ies) of a phenomenon, object, or material by a recording device not in physical, intimate contact with the feature(s) under surveillance; techniques involve amassing knowledge pertinent to environments by measuring force fields, electromagnetic radiation, or acoustic energy employing cameras, radiometers and scanners, lasers, radio frequency receivers, radar systems, sonar, thermal devices, seismographs, magnetometers, gravimeters, scintillometers, and other instruments."

Remote sensing (RS) is multi-functional because it is (1) a source of basic data, (2) a science, and (3) a tool. RS is a source of basic data because the measurements of physical properties of distant objects and their surroundings with the use of reflected and emitted energy are themselves data, regardless of where they are recorded [59]. RS is a science because it utilizes a scientific process: measurements, data processing, interpretation of the results, and scientific inference [59]. Finally, RS is a tool because results obtained from this scientific process can be used for various purposes, from the making of inventories of resources to the solving of ecological problems [59].

Monitoring of the SDG indicators is vital, and Earth observation technologies such as RS have an important role to play in indicator monitoring. RS data are particularly useful because they can be used for both temporal and spatial monitoring. As early as two decades ago, before the formulation of the SDGs [2,51] and launch of the EO4SDG initiative [23–25], and also before the development of concepts such as "seeing sustainability from space" [60] and "remote sensing for sustainability" [61], Rao [62] foresaw the potential of RS technology for sustainability-related research and for helping to achieve sustainable development. Through the EO4SDG initiative, the potential of RS technology is now being explored at a global scale. RS-derived data have been shown to be useful across many fields, such as in the field of land cover monitoring and ecosystem assessment [63–71], hydrological studies [72,73], meteorological/climatological and climate change studies [74,75], thermal and urban remote sensing [76–85], air quality monitoring [86,87], health geographics [88–91], and disaster risk management [92,93].

The importance of EO technologies such as RS has become more apparent with the formulation of the SDG framework, compared with their importance to its predecessor, the Millennium Development Goals (MDG) framework (www.un.org/millenniumgoals/). Although the MDG framework helped narrow the data gap with regard to the social dimension of sustainability, the inclusion of various environmental indicators under the SDG targets has increased the need for accurate, timely, and reliable environmental data. RS is an important environmental monitoring tool that can help fill gaps in environmental data [24,25,94].

Some essential characteristics of RS data have important advantages for SDG monitoring: spatial resolutions ranging from coarse to very high; temporal resolutions ranging from ca. bimonthly to daily; various spectral resolutions; spatial scales from local to global; long-period time series (starting from 1972 for Landsat); consistency (in terms of data capture or measurement); and complementarity (ability to be validated) [95] (Tables A1 and A2). These features mean that RS data are useful for the development of policy-relevant environmental SDG indicators [94,95] that can be monitored over space and time.

The 30 social-ecological indicators that can be directly or indirectly supported by EO data [25] are related to 13 SDGs (Figures 3 and A1). Among these 13 SDGs, SDGs 6 (Clean Water and Sanitation), 11 (Sustainable Cities and Communities), 14 (Life Below Water), and 15 (Life on Land) offer "the greatest opportunities for the application of EO data" [25] (p. 8). Information on the current status of these RS-based indicators is given in Section 5 and Figure 3.

In general, the monitoring of progress toward achieving the SDGs by way of the SDG global indicator framework increases the demand for various statistical data from countries all over the world. This increased demand for these data necessitates an increase in investments of money, manpower, and time in building the capacity of national statistical offices–investments that might be better made in research and substantive development projects with clear impacts on meeting SDG targets [96]. In this regard, the EO4SDG initiative is especially important for developing regions that have low capacity for database development. Moreover, the potential use of EO data for more indicators, in addition to the 30 indicated currently identified as EO-supported, presents valuable opportunities in research and development (details are discussed in Section 5).

Goal	Indicators				
No Poverty	1.4.2 Tier II				
Zero Hunger	2.4.1 Tier II				
Good Health and Well-Being	3.9.1 Tier I				
Quality Education					
Gender Equality	5.a.1 Tier II				
Clean Water and Sanitation	6.3.1 Tier II	6.3.2 Tier II	6.4.2 Tier I	6.5.1 Tier I	6.6.1 Tier I
Affordable and Clean Energy	7.1.1 Tier I				
Decent Work and Economic Growth					
Industry, Innovation and Infrastructure	9.1.1 Tier II	9.4.1 Tier I			
Reduced Inequalities					
Sustainable Cities and Communities	11.1.1 Tier I	11.2.1 Tier II	11.3.1 Tier II	11.6.2 Tier I	11.7.1 Tier II
Responsible Consumption and Production	12.a.1 Tier Pending				
Climate Action	13.1.1 Tier II				
Life Below Water	14.1.1 Tier II	14.3.1 Tier II	14.4.1 Tier I	14.5.1 Tier I	
Life on Land	15.1.1 Tier I	15.2.1 Tier I	15.3.1 Tier I	15.4.1 Tier I	15.4.2 Tier I
Peace, Justice and Strong Institutions					
Partnerships for the Goals	17.6.1 Tier I	17.18.1 Tier Pending			

Data availability (as of April 2020)

☐ Available ☐ Not available

Inclusion in the 2019 SDG Index and Dashboards

○ Included (exact indicator) ○ Included (proxy indicator) ○ Not included

Figure 3. Current status of the RS-based SDG indicators. This list of indicators is based on GEO [25]. Data availability is based on the Global SDG Indicators Database as of April 2020 (https://unstats.un. org/sdgs/indicators/database/). The 2019 SDG Index and Dashboards are available in the Sustainable Development Report 2019 [97]. The tier classification of the indicators are those as of April 2020 [20].

5. Current Status, Challenges, and Opportunities

5.1. Status of the 2030 Agenda

Just under 10 years remain to achieve the 2030 Agenda. The UN Global Sustainable Development Report 2019 (UN-GSDR 2019) [98] has concluded that (i) despite initial efforts, the world is not on track for achieving most of the 169 targets that comprise the SDGs, (ii) recent trends along several dimensions with cross-cutting impacts across the entire 2030 Agenda (rising inequalities, climate change, biodiversity loss, and increasing amounts of waste from human activity) are not even moving in the right direction, (iii) under current trends, the world's social and natural biophysical systems cannot support the aspirations for universal human well-being embedded in the SDGs, and (iv) no country is yet convincingly able to meet a set of basic human needs at a globally sustainable level of resource use.

To achieve the desired transformations at the necessary scale and speed, the UN-GSDR 2019 identifies six entry points: human well-being and capabilities, sustainable and just outcomes, food systems and nutrition patterns, energy decarbonization with universal access, urban and peri-urban development, and global environmental commons [98]. It also identifies four levers that "can be coherently deployed through each entry point to bring about the necessary transformations": governance, economy and finance, individual and collective action, and science and technology [98] (p. xxi).

The report also highlights the importance of sustainability science [47] to "help tackle the trade-offs and contested issues involved in implementing the 2030 Agenda." Accordingly, "new initiatives are needed that bring together science communities, policymakers, funders, representatives of lay, practical and indigenous knowledge and other stakeholders to scale up sustainability science and transform scientific institutions towards engaged knowledge production for sustainable development." To achieve this, the UN "should launch a globally coordinated knowledge platform to synthesize existing international and country-by-country expertise on transformation pathways from scientific and nonscientific sources, including lay, practical and indigenous knowledge," and, at the same time, "educational institutions at every level, especially universities, should incorporate high-quality theoretical and practically oriented courses of study on sustainable development" [98] (p. 120).

One feedback on the UN-GSDR 2019 relates to the effort paid to the improvement of the SDG global indicator framework, because although there are frequent opportunities for input, clarity and transparency on the dynamics of decision-making is less frequent [96]. The fear is that more is being invested in the development of the SDG global indicator framework, including its databases, than is being invested in actual projects that can deliver desirable outcomes and bring about progress, and, as a result, "many of the targets will not only not be met, but unless things change radically, will never be met" [96] (p. 5).

The standpoint of this reviewer is that, although both the refinement of the SDG global indicator framework and the development of relevant databases are necessary to enable the monitoring of progress toward the SDGs, these activities should not overshadow the need to implement projects needed to help achieve the SDGs and which can be expected to have actual and positive impacts on people's lives, society, and the environment.

5.2. Status of the RS-Based Indicators and Their Inclusion in the SDG Index 2019

As of April 2020, 21 (70%) of the 30 RS-based SDG indicators have at least some preliminary statistical data, according to the Global SDG Indicators Database (Figure 3; Table 1). Among those indicators with statistical data, 16 are classified as Tier 1 and five as Tier II indicators. Among the nine indicators still without statistical data, seven are Tier II indicators and the tier classification of the other two is pending. None of the 30 RS-based indicators is currently classified as a Tier III indicator [20].

The SDG indicators are also being monitored indirectly by the UN Sustainable Development Solutions Network (SDSN), in partnership with the Bertelsmann Stiftung (BS), which produces annual reports that assess each country's performance on the 17 SDGs [97]. The mission of the SDSN, which was set up in 2012 at the direction of the UN Secretary-General, is to "mobilize global scientific and technological expertise to promote practical solutions for sustainable development, including the implementation of the SDGs and the Paris Climate Agreement" (www.unsdsn.org). The SDG Index and Dashboards (SDR-ID 2019) includes 10 (33%) of the RS-based indicators (Figure 3) [97]; six of these exactly match official SDG indicators, and the other four are proxy indicators that are related to or closely aligned with official indicators.

Table 1. Status of remote sensing (RS)-based sustainable development goals (SDG) indicators. Only those indicators with data as of April 2020, according to the Global SDG Indicators Database (https://unstats.un.org/sdgs/indicators/database/), are included. The complete list of RS-based SDG indicators is given in Figure 3. The custodian and tier classification information is current as of April 2020 [20]. The number of countries/territories/regions is for the described variable only.

RS-based SDG Indicator	Custodian	Tier	Available Data		
			No. and Name of Data Variables	Year	No. of Countries/Territories/ Regions
3.9.1 Mortality rate attributed to household and ambient air pollution	WHO	I	6 The 6th variable is the crude death rate attributed to household and ambient air pollution	2016	219
5.a.1 (a) Proportion of total agricultural population with ownership or secure rights over agricultural land, by sex; and (b) share of women among owners or rights-bearers of agricultural land, by type of tenure	FAO	II	2 The 1st variable is proportion of people with ownership or secure rights over agricultural land, by sex	2009–2019 (varies by country/ territory)	10 (total for all data years)
6.3.1 Proportion of wastewater safely treated	WHO, UN-Habitat, UNSD	II	1 Proportion of safely treated domestic wastewater flows	2018	79
6.3.2 Proportion of bodies of water with good ambient water quality	UNEP	II	4 The 1st variable is proportion of bodies of water with good ambient water quality	2017	52
6.4.2 Level of water stress: freshwater withdrawal as a proportion of available freshwater resources	FAO	I	1 Level of water stress: freshwater withdrawal as a proportion of available freshwater resources	2000, 2005, 2010, 2015	269 (2015)
6.5.1 Degree of integrated water resources management implementation (0–100)	UNEP	I	2 The 1st variable is degree of integrated water resources management implementation	2018	182
6.6.1 Change in the extent of water-related ecosystems over time	UNEP, Ramsar	I	16 The 4th variable is nationally derived proportion of water bodies with good quality	2017	28
7.1.1 Proportion of population with access to electricity	World Bank	I	1 Proportion of population with access to electricity, by urban/rural	2000–2017 (annual)	236 (2017)
9.4.1 CO_2 emission per unit of value added	UNIDO, IEA	I	3 The 3rd variable is CO_2 emissions per unit of manufacturing value added	2000–2017 (annual)	182 (2017)
11.1.1 Proportion of urban population living in slums, informal settlements or inadequate housing	UN-Habitat	I	1 Proportion of urban population living in slums	2000, 2005, 2010, 2014, 2016	126 (2016)
11.6.2 Annual mean levels of fine particulate matter (e.g. PM2.5 and PM10) in cities (population weighted)	WHO	I	1 Annual mean levels of fine particulate matter in cities, urban population	2016	215
13.1.1 Number of deaths, missing persons and directly affected persons attributed to disasters per 100,000 population	UNDRR	II	10 The 2nd variable is number of deaths and missing persons attributed to disasters per 100,000 population	2005–2018 (annual)	43 (2018)
14.3.1 Average marine acidity (pH) measured at agreed suite of representative sampling stations	IOC-UNESCO	II	1 Average marine acidity (pH) measured at agreed suite of representative sampling stations	2010–2019 (annual)	3 (2019)

Table 1. *Cont.*

RS-based SDG Indicator	Custodian	Tier	Available Data		
			No. and Name of Data Variables	Year	No. of Countries/Territories/Regions
14.4.1 Proportion of fish stocks within biologically sustainable levels	FAO	I	1 Proportion of fish stocks within biologically sustainable levels (not overexploited)	2000–2017 (varied interval)	1 (global)
14.5.1 Coverage of protected areas in relation to marine areas	UNEP-WCMC, UNEP, IUCN	I	3 The 2nd variable is coverage of protected areas in relation to marine areas (Exclusive Economic Zones)	2018	192
15.1.1 Forest area as a proportion of total land area	FAO	I	3 The 2nd variable is forest area as a proportion of total land area	2000, 2005, 2010, 2015	292 (2015)
15.2.1 Progress towards sustainable forest management	FAO	I	5 The 4th variable is proportion of forest area with a long-term management plan	2000, 2005, 2010	292 (2010)
15.3.1 Proportion of land that is degraded over total land area	UNCCD	I	1 Proportion of land that is degraded over total land area	2015	294
15.4.1 Coverage by protected areas of important sites for mountain biodiversity	UNEP-WCMC, UNEP, IUCN	I	1 Average proportion of Mountain Key Biodiversity Areas covered by protected areas	2000–2019 (annual)	197 (2019)
15.4.2 Mountain Green Cover Index	FAO	I	3 The 1st variable is Mountain Green Cover Index	2017	276
17.6.1 Fixed Internet broadband subscriptions per 100 inhabitants, by speed	ITU	I	2 The 1st variable is fixed Internet broadband subscriptions per 100 inhabitants, by speed	2000–2018 (annual)	179 (2018)

Twelve RS-based indicators that have statistical data according to the Global SDG Indicators Database are not included in the SDR-ID 2019. These include nine Tier I indicators. Recent updates to the Global SDG Indicators Database may partly explain this inconsistency between these two monitoring platforms. However, one indicator (i.e., 11.2.1) that, according to the Global SDG Indicators Database, had no data as of April 2020 is included in the SDR-ID 2019 [97]. Overall, the information in these two monitoring platforms considered in combination (Figure 3; Table 1) is indicative of the current status of EO contributions to SDG progress monitoring at the global level. Clearly, there is more that needs to be done.

The set of criteria used to select indicators for inclusion in the SDR-ID 2019 might also account for some of the inconsistency between the two monitoring platforms. These criteria were (i) global relevance and applicability to a broad range of country settings, (ii) statistical adequacy, (iii) timeliness, (iv) data quality, and (v) coverage [97]. Given these criteria, it is surprising that indicator 15.1.1 (forest area as a proportion of total land area) was not included in the SDR-ID 2019 (Figure 3) [97], because forest cover data are available in Forest Resources Assessment (FRA) reports (www.fao.org/forest-resources-assessment) published by the FAO–the custodian for indicator 15.1.1 (Table 1) [20]. Unfortunately, the SDR-ID 2019 does not provide any explanation.

Here, some important issues regarding SDG indicator 15.1.1 are highlighted. It is important to note that the forest cover data available in the FRA reports are based on statistics consolidated from country reports, which are not backed up by publicly available geospatial data. For the upcoming 2020 FRA report, FAO is "conducting a participatory global remote sensing survey (FRA 2020 RSS) with the scope of improving estimates of forest area change at global and regional scales." Accordingly, "the FRA secretariat, in collaboration with the Joint Research Center of the European Commission (JRC) and the FAO working group on remote sensing, has developed a worldwide methodology for the FRA 2020 RSS, which is also scalable to national assessments" (http:

//www.fao.org/forest-resources-assessment/remote-sensing/fra-2020-remote-sensing-survey/en/). A switch to a collaborative remote sensing survey approach with the use of a harmonized method for forest cover mapping is necessary so that forest cover change statistics reported in the future will be comparable across countries, landscapes, and forest monitoring studies that employ RS technologies. More importantly, so that such statistics will be backed up by publicly available geospatial data. At present, forest cover change statistics from the FRA reports are neither comparable nor backed up by publicly available geospatial data.

One major challenge to the implementation of the new approach is the harmonization of the definition of forest. The FAO defines a forest in terms of both tree cover and land use; a forest may include bare areas where trees are expected to regenerate, but areas with tree cover in agricultural or urban land use classes are excluded [65,99]. In contrast, RS-based data (e.g., Global Forest Watch) define forest only in terms of tree cover [63,99]. Harmonization of the definition of forest would help clarify important but conflicting records. For instance, a recent global land change study found that global tree cover increased by 2.24 million km^2 from 1982 to 2016 as a result of a net gain in the extratropics [66], but this finding contradicts FRA reports of a global decline in forest area [99]. This reported increase in global tree cover is supported, however, by another study that reported a net increase of 5.4 million km^2 of new leaf area from 2000 to 2017, two-thirds of which was attributed to the greening of croplands and forests [67].

Of these two forest definitions, land use plus tree cover and tree cover alone, the latter has greater potential to harmonize methods for global forest cover monitoring with the application of RS technologies. The use of remotely sensed tree cover as the basis for mapping forest cover and monitoring changes is also timely because, as reported by the GEO in a 2017 press release, full satellite coverage of the world's forests has now been achieved, so that all countries have the data necessary for annual forest cover monitoring (https://www.earthobservations.org/article.php?id=250). Nevertheless, if the FAO continues to use the land use plus tree cover concept of forest, the view of this reviewer is that the FAO should ensure the use of a harmonized method for identifying and classifying land use from RS data, not only tree cover. It should also make sure that gross forest cover losses and gains are also reported and backed up by publicly available geospatial data (both raw and processed).

Among databases relevant to the 2030 Agenda for Sustainable Development, the Global SDG Indicators Database (https://unstats.un.org/sdgs/indicators/database/) is of primary importance. In terms of geospatial data, the data for the EO4SDG initiative are stored in the Global Earth Observation System of Systems (GEOSS) portal (www.geoportal.org). The UN's Open SDG Data Hub also stores available geospatially referenced data for each SDG (http://unstats-undesa.opendata.arcgis.com). Other sources of geospatial data include NASA's Open Data Portal (https://data.nasa.gov) and the Center for International Earth Science Information Network (CIESIN) (www.ciesin.org). Other initiatives that make SDG-relevant data available include the Global Partnership for Sustainable Development Data (www.data4sdgs.org) and the Open Data Watch (https://opendatawatch.com).

5.3. Challenges, Opportunities, and Insights

The production of data for the other RS-based indicators and the subsequent inclusion of more RS-based indicators in the annual updates of the SDR-ID 2019 are among the current challenges to the realization of the EO4SDG initiative. Other challenges relate to the identification and development of relevant sub-indicators or complementary indicators, because some SDG indicators are not specific enough to be properly addressed at present. For example, indicator 15.3.1, the "proportion of land that is degraded over total land area" has been identified as one of the indicators that can be supported by EO data (Figure 3; Table 1). However, land degradation itself is a broad concept, as indicated by target 15.3: "by 2030, combat desertification, restore degraded land and soil, including land affected by desertification, drought, and floods, and strive to achieve a land degradation-neutral world" [2,3].

To address this issue, three sub-indicators for indicator 15.3.1 have been identified: land cover and land cover change, land productivity, and above- and belowground carbon stocks [24,100]. With respect to land productivity as a sub-indicator of 15.3.1, a recent study [100] (p. 1) argues that "current use of vegetation indices alone to remotely sense degradation of ecosystem services does not provide an adequate productivity indicator"; thus a more robust methodology is needed. Another study has developed an integrated approach to the operationalization of the three proposed sub-indicators for indicator 15.3.1 [95]. In this approach, all three sub-indicators are taken into account by a modeling process that results in an overall indicator of land degradation [95].

RS data can also be used to examine and identify priority areas for sustainable management to realize targets under SDG 6 (Clean Water and Sanitation) [101]. Furthermore, RS data can also be used to provide evidence for some of the currently identified non-RS-based indicators. For instance, RS data can be used to derive relevant landscape-based indicators (e.g., landscape fragmentation and connectivity metrics) as complementary indicators for assessing the effectiveness of financial investments in the conservation and sustainable use of biodiversity and ecosystems (target 15.a), as well as in sustainable forest management, including conservation and reforestation (target 15.b). RS data can also help operationalize social and economic SDG targets. For instance, RS data can be used to advance the study and monitoring of household poverty [102], which is relevant to targets 1.1 and 1.4. An idea for assessing slavery from space, which is relevant to target 8.7, has also been proposed [103].

According to the Global SDG Indicators Database, some indicators have available data in >200 countries or territories, but others have available data in fewer than 50 (Table 1). Moreover, for some indicators, data are available at multiple spatial scales (regions, countries, and territories), whereas for others, data are available only for countries. To ensure consistency, geographical units for data production, collection, and reporting for global monitoring need to be harmonized across all SDG indicators. Likewise, the baseline year for all SDG indicators may also need to be defined and harmonized. Simply harmonizing the baseline year would help not only the groups that are in charge of data production and collection but also researchers interested in SDG monitoring at the global level.

It is clear that the SDG global indicator framework in general and the RS-based indicators in particular need to be improved. Other current challenges, as well as future research directions, related to SDGs, targets, and indicators include addressing the potential pitfalls (ethical, legal, and reputational) in the compilation and use of big data [104] and the analysis of synergies and trade-offs [101,105–108]. Other important issues include the development of other frameworks for assessing the suitability of EO-derived data for SDG indicators [60] and of another aggregation method for the SDG Index [108], as well as the regionalization (sub-national) of SDG progress monitoring [109]. In particular, for the possible improvement of the SDG Index as presented in the SDR-ID 2019 [97] and in its earlier versions (www.sdgindex.org), an aggregation method based on a multidimensional synthesis of indicators and that takes into account the trade-offs and synergies between goals and targets and across the three pillars of sustainable development (social, economic, and environmental) has been proposed [108]. Regionalized or localized SDG progress monitoring is also particularly important because it allows individual countries to assess their own progress in space and time toward sustainable development [109].

RS-based indicator status as presented here (Figure 3; Table 1) is based solely on the Global SDG Indicators Database. This database only consolidates available statistics at the country or territory and regional levels; it does not include information on the actual availability of the geospatial data (both raw and processed) from which the statistical data were supposedly derived. This is important to note because some of the RS-based indicators identified as already having at least preliminary statistical data (Table 1) may not yet have geospatial data. For example, data for indicator 15.1.1 (forest area as a proportion of total land area) recorded in the database are based on FRA reports, but, as mentioned earlier, these recorded statistical data are not backed up by geospatial data. Some information on the availability of geospatial data are available from the UN Open SDG Data Hub, but it would be better if the GEOSS portal provided information related to the EO4SDG initiative. As of this writing, however, no specific page is dedicated to the currently identified RS-based SDG indicators and for tracking the actual status and availability of raw and processed geospatial data for each of these indicators.

Perhaps future updates of the EO4SDG initiative and the GEOSS portal could address this issue, in collaboration with the concerned designated SDG indicator custodians.

A limitation of this review is that it is only an overview of the conceptualization of sustainability leading to the formulation of the SDGs and of the current status, challenges, and opportunities in SDG monitoring with RS. Some important issues need to be covered at greater depth, such as (i) the applications of RS, including data availability, across ecosystems but with particular focus on the SDGs, (ii) how the number of SDG indicators might be different from one ecosystem to another, (iii) challenges and opportunities of RS data in terms spectral and spatial resolutions as applied across ecosystems, (iv) how the SDG indicators were decided (e.g., how much is based on concerns in terms of the economic, social, and environmental pillars of sustainability and how much is based on logistics, e.g., data availability), and (v) the specific issues about the SDG global indicator framework, e.g., inappropriateness of some of the (initial) indicators.

6. Conclusions

Although the concept of sustainability, the idea that the current generation has responsibility for future generations, originated at least as far back as the 17th century, according to the UN-GSDR 2019, the world is not on track for achieving most of the targets that comprise the SDGs by the target date of 2030. Meanwhile, through the EO4SDG initiative of the GEO, the full potential of RS for SDG monitoring is now being explored at a global scale. As of April 2020, 21 (70%) of the RS-based SDG indicators already have at least some preliminary statistical data according to the Global SDG Indicators Database, and 10 (33%) of the RS-based SDG indicators are included in the SDR-ID 2019. These statistics, however, do not necessarily reflect the actual status and availability of raw and processed geospatial data for the RS-based indicators, which remains an important issue. Nevertheless, various initiatives have also been started to address the need for open access data. RS data can also help in the development of potentially relevant complementary indicators or sub-indicators, which will help address one of the current challenges in SDG monitoring, which is how to operationalize the SDG indicators.

Author Contributions: R.C.E. conceptualized this review. He performed the review, prepared all the tables and figures, and wrote the paper. The author has read and agreed to the published version of the manuscript.

Funding: This research was supported by the Japan Society for the Promotion of Science (JSPS), through Grant-in-Aid for Young Scientists (B): 20K13262 (PI: Ronald C. Estoque).

Acknowledgments: This work was also supported by the Climate Change Adaptation Program of the National Institute for Environmental Studies, Japan. The author also acknowledges the four anonymous reviewers whose comments and suggestions helped improve this manuscript.

Conflicts of Interest: The author declares no conflict of interest.

Appendix A

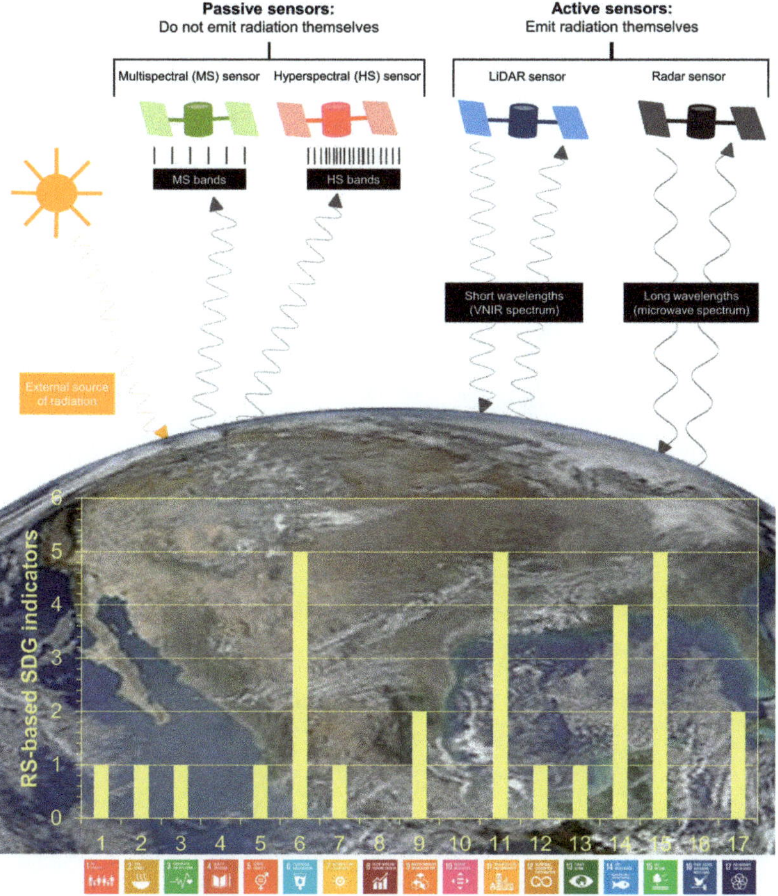

Figure A1. Contributions of RS to SDG monitoring. The diagrams of passive and active remote sensors are from [110]. Passive sensors (e.g., multispectral and hyperspectral sensors) rely on an external source of energy (the sun) and record the radiation reflected by Earth's surface to produce an image, whereas active sensors (e.g., LiDAR and Radar) emit energy in the microwave part of the electromagnetic spectrum and measure the amount of energy reflected back at them [110]. The column graph shows the number of indicators for each of the 17 SDGs to which Earth Observation data can contribute, directly or indirectly [25].

Table A1. Widely used multispectral satellite RS data.

Satellite	Sensor and Number of Bands	Spatial Resolution	Revisit Interval	Reference
Landsat 4/5 TM	MS 6; T 1	MS &T: 30 m	16 days	[111]
Landsat 7 ET+/ 8 OLI/TIRS	MS 6/8; T 1/2; Pan 1	MS & T: 30 m Pan: 15 m	16 days	[111]
SPOT 1/2/3 (2 HRVs)	MS 3; Pan 1	MS: 20 m Pan: 10 m	1 to 3 days	[112]
SPOT 4 (2 HRVIRs)	MS 4; Pan 1	MS: 20 m Pan: 10 m	2 to 3 days	[112]
SPOT 5 (2 HRGs)	MS 4; Pan 2	VNIR: 10 m; SWIR: 20 m Pan: 5 m (2.5 m)	2 to 3 days	[112]
SPOT 6/7 (2 NAOMI)	MS 4; Pan 1	MS: 8 m Pan: 2 m	Daily	[113]
NOAA AVHRR	MS & T 4–5	1.1 km	Daily (Vis), 2×/day (IR)	[114]
OrbView 2 (SeaWiFS)	MS 8	1 km	Daily	[113]
IKONOS 2	MS 4; Pan 1	MS: 3.20 m Pan: 0.82 m	1 to 3 days	[113]
Terra ASTER	MS up to 10; T 5	VNIR: 15 m; SWIR: 30 m, T: 90 m	All bands: at least 1×/16 days, VNIR: 5 days	[115]
MODIS Terra/Aqua	MS & T 36	Bands 1–2: 250 m, Bands 3–7: 500 m, Bands 8–36: 1 km	1 to 2 days	[116]
Envisat MERIS	MS 15	Ocean: 1040 m × 1200 m; Land: 260 m × 300 m	3 days	[117]
QuickBird	MS 4; Pan 1	MS: 2.40 m Pan: 0.60 m	1.5 to 2.8 days	[117]
GeoEye	MS 4; Pan 1	MS: 1.64 m Pan: 0.41 m	≤ 3 days	[113]
RapidEye	MS 5	MS: ~6.5 m	1 to 5.5 days	[117]
WorldView 2	MS 8; Pan 1	MS: 1.80 m, Pan: 0.46 m	1.1 days	[113]
Sentinel 2A/2B	MS 13	MS: 10 m, 20 m, 60 m	A or B: 10 days A & B: 5 days	[118]

Abbreviations: MS, multispectral band; T, thermal band; Pan, panchromatic band; SWIR, short-wave infrared; VNIR, visible and near infrared; Vis, visible; IR, infrared.

Table A2. Widely used radar satellite RS data.

Satellite	Sensor	Spatial Resolution	Revisit Interval	Reference
ERS 1/2	C-band SAR	30 m to 50 km	35 days	[113]
JERS 1	L-band SAR	18 m	44 days	[113]
RADARSAT 1/2	C-band SAR	10–100 m/ 3–100 m	24 days	[117]
Envisat ASAR	C-band SAR	28–980 m	35 days	[117]
ALOS PALSAR	L-band SAR	7–100 m	46 days	[119]
TerraSAR-X	X-band SAR	1–16 m	11 days	[117]
TanDEM-X	X-band SAR	12 m	11 days	[117]
ALOS-2	L-band SAR	3–100 m	14 days	[119]
Sentinel 1/2 SAR	C-band SAR	5–100 m	1/2: 6 days; 1 or 2: 12 days	[118]

References

1. WCED (World Commission on Environment and Development). *Our Common Future*; Oxford University Press: New York, NY, USA, 1987; ISBN 978-0-19-282080-8.
2. UN General Assembly. Transforming Our World: The 2030 Agenda for Sustainable Development. Resolution Adopted by the General Assembly on 25 September 2015. Available online: https://undocs.org/A/RES/70/1 (accessed on 19 March 2020).
3. UN IAEG-SDGs. Global Indicator Framework for the Sustainable Development Goals and Targets of the 2030 Agenda for Sustainable Development. Available online: https://unstats.un.org/sdgs/indicators/Global%20Indicator%20Framework%20after%202020%20review_Eng.pdf (accessed on 19 March 2020).
4. UN General Assembly. Work of the Statistical Commission Pertaining to the 2030 Agenda for Sustainable Development. Resolution Adopted by the General Assembly on 6 July 2017. Available online: https://undocs.org/A/RES/71/313 (accessed on 19 March 2020).
5. Barbier, E.B. The concept of sustainable economic development. *Environ. Conserv.* **1987**, *14*, 101–110. [CrossRef]
6. Elkington, J. Towards the sustainable corporation: Win-win-win business strategies for sustainable development. *Calif. Manage. Rev.* **1994**, *36*, 90–100. [CrossRef]
7. Elkington, J. *Cannibals with Forks. The Triple Bottom Line of 21st Century Business*; Capston, Publishing Ltd.: Oxford, UK, 1997; ISBN 1-900961-27-X.
8. Estoque, R.C.; Murayama, Y. Social-ecological status index: A preliminary study of its structural composition and application. *Ecol. Indic.* **2014**, *43*, 183–194. [CrossRef]
9. Estoque, R.C.; Murayama, Y. A worldwide country-based assessment of social-ecological status (c. 2010) using the social-ecological status index. *Ecol. Indic.* **2017**, *72*, 605–614. [CrossRef]
10. Ostrom, E. A diagnostic approach for going beyond panaceas. *Proc. Natl. Acad. Sci. USA* **2007**, *104*, 15181–15187. [CrossRef]
11. Ostrom, E. A general framework for analyzing sustainability of social-ecological systems. *Science* **2009**, *325*, 419–422. [CrossRef]
12. McGinnis, M.D.; Ostrom, E. Social-ecological system framework: Initial changes and continuing. *Ecol. Soc.* **2014**, *19*, 30. [CrossRef]
13. Gallopin, G.C. Environmental and sustainability indicators and the concept of situational indicators. A systems approach. *Environ. Model. Assess.* **1996**, *1*, 101–117. [CrossRef]
14. Heink, U.; Kowarik, I. What are indicators? On the definition of indicators in ecology and environmental planning. *Ecol. Indic.* **2010**, *10*, 584–593. [CrossRef]
15. Lehtonen, M. Indicators: Tools for informing, monitoring or controlling? In *The Tools of Policy Formulation: Actors, Capacities, Venues and Effects. New Horizons in Public Policy*; Jordan, A.J., Turnpenny, J.R., Eds.; Edward Elgar: Cheltenham, UK; Northampton, MA, USA, 2015; pp. 76–99. ISBN 9781783477036.
16. Estoque, R.C. *GIS-Based Multi-Criteria Decision Analysis in Natural Resource Management*; University of Tsukuba: Tsukuba, Japan, 2011.
17. Essex, B.; Koop, S.H.A.; Van Leeuwen, C.J. Proposal for a national blueprint framework to monitor progress on water-related Sustainable Development Goals in Europe. *Environ. Manag.* **2020**, *65*, 1–18. [CrossRef]
18. Boulanger, P.-M. Sustainable development indicators: A scientific challenge, a democratic issue. *Surv. Perspect. Integrating Environ. Soc.* **2008**, *1*, 45–59. [CrossRef]
19. Maurice, J. Measuring progress towards the SDGs — A new vital science. *Lancet* **2016**, *388*, 1455–1458. [CrossRef]
20. UN IAEG-SDGs. Tier Classification for Global SDG Indicators (as of 17 April 2020). Available online: https://unstats.un.org/sdgs/iaeg-sdgs/tier-classification/ (accessed on 2 May 2020).
21. UN Statistics Division. SDG Indicators. Available online: https://unstats.un.org/sdgs/indicators/indicators-list/ (accessed on 19 March 2020).
22. WGGI (Working Group on Geospatial Information). *Terms of Reference*; Inter-Agency and Expert Group on SDG Indicators, United Nations: New York, NY, USA, 2019.
23. Anderson, K.; Ryan, B.; Sonntag, W.; Kavvada, A.; Friedl, L. Earth observation in service of the 2030 Agenda for Sustainable Development. *Geo-Spat. Inf. Sci.* **2017**, *20*, 77–96. [CrossRef]

24. Paganini, M.; Petiteville, I.; Ward, S.; Dyke, G.; Steventon, M.; Harry, J.; Kerblat, F. *Satellite Earth Observations in Support of the Sustainable Development Goals: The CEOS Earth Observation Handbook*; Special 2018 Edition; The Committee on Earth Observation Satellites and the European Space Agency: Paris, France, 2018.

25. GEO. *EO4SDG: Earth Observations in Service of the 2030 Agenda for Sustainable Development. Strategic Implementation Plan 2020–2024*; Group on Earth Observations: Geneva, Switzerland, 2019.

26. Grant, M.J.; Booth, A. A typology of reviews: An analysis of 14 review types and associated methodologies. *Health Inf. Libr. J.* **2009**, *26*, 91–108. [CrossRef]

27. Kidd, C.V. The evolution of sustainability. *J. Agric. Environ. Ethics* **1992**, *5*, 1–26. [CrossRef]

28. Grober, U. *Deep Roots: A Conceptual History of "Sustainable Development" (Nachhaltigkeit)*; WZB Discussion Paper, No. P 2007-002; Wissenschaftszentrum Berlin für Sozialforschung (WZB): Berlin, Germany, 2007.

29. Warde, P. *The Invention of Sustainability: Nature and Destiny*; Cambridge University Press: Cambridge, UK, 2018; ISBN 1-107-15114-7.

30. Purvis, B.; Mao, Y.; Robinson, D. Three pillars of sustainability: In search of conceptual origins. *Sustain. Sci.* **2019**, *14*, 681–695. [CrossRef]

31. Du Pisani, J.A. Sustainable development – historical roots of the concept. *Environ. Sci.* **2006**, *3*, 83–96. [CrossRef]

32. IUCN. *World Conservation Strategy: Living Resource Conservation for Sustainable Development*; IUCN-UNEP-WWF: Gland, Switzerland, 1980.

33. Hardin, G. The tragedy of the commons. *Science* **1968**, *162*, 1243–1248.

34. Gómez-Pompa, A.; Vázquez-Yanes, C.; Guevara, S. The tropical rain forest: A nonrenewable resource. *Science* **1972**, *177*, 762–765. [CrossRef]

35. Westman, W.E. How much are nature's services worth? *Science* **1977**, *197*, 960–964. [CrossRef]

36. Ehrlich, P.R.; Ehrlich, A.H. *Extinction: The Causes and Consequences of the Disappearance of Species*; Random House: New York, NY, USA, 1981; ISBN 0-394-51312-6.

37. de Groot, R.S. Environmental functions as a unifying concept for ecology and economics. *Environmentalist* **1987**, *7*, 105–109. [CrossRef]

38. Gómez-Baggethun, E.; de Groot, R.; Lomas, P.L.; Montes, C. The history of ecosystem services in economic theory and practice: From early notions to markets and payment schemes. *Ecol. Econ.* **2010**, *69*, 1209–1218. [CrossRef]

39. Diaz, S.; Demissew, S.; Carabias, J.; Joly, C.; Lonsdale, M.; Ash, N.; Larigauderie, A.; Adhikari, J.R.; Arico, S.; Baldi, A.; et al. The IPBES conceptual framework - connecting nature and people. *Curr. Opin. Environ. Sustain.* **2015**, *14*, 1–16. [CrossRef]

40. Diaz, S.; Pascual, U.; Stenseke, M.; Martín-López, B.; Watson, R.T.; Molnár, Z.; Hill, R.; Chan, K.M.A.; Baste, I.A.; Brauman, K.A.; et al. Assessing nature's contributions to people. *Science* **2018**, *359*, 270–272. [CrossRef]

41. Daily, G.C. (Ed.) *Nature's Services. Societal Dependence on Natural Ecosystems*; Island Press: Washington, DC, USA, 1997; ISBN 1-55963-476-6.

42. Costanza, R.; D'Arge, R.; de Groot, R.; Farber, S.; Grasso, M.; Hannon, B.; Limburg, K.; Naeem, S.; O'Neill, R.V.; Paruelo, J.; et al. The value of the world's ecosystem services and natural capital. *Nature* **1997**, *387*, 253–260. [CrossRef]

43. Estoque, R.C.; Murayama, Y. Measuring sustainability based upon various perspectives: A case study of a hill station in Southeast Asia. *AMBIO* **2014**, *43*, 943–956. [CrossRef]

44. Costanza, R. (Ed.) *Ecological Economics: The Science and Management of Sustainability*; Columbia University Press: New York, NY, USA, 1991; ISBN 0-231-07562-6.

45. Røpke, I. The early history of modern ecological economics. *Ecol. Econ.* **2004**, *50*, 293–314. [CrossRef]

46. Rees, W.E. Ecological footprints and appropriated carrying capacity: What urban economics leaves out. *Environ. Urban.* **1992**, *4*, 121–130. [CrossRef]

47. Kates, R.W.; Clark, W.C.; Corell, R.; Hall, J.M.; Jaeger, C.C.; Lowe, I.; McCarthy, J.J.; Schellnhuber, H.J.; Bolin, B.; Dickson, N.M.; et al. Sustainability science. *Science* **2001**, *292*, 641–642. [CrossRef]

48. NRC USA. *Our Common Journey: A Transition Toward Sustainability*; Athens Center of Ekistics: New York, NY, USA, 1999.

49. Rockström, J.; Steffen, W.; Noone, K.; Persson, A.; Chapin, F.S.; Lambin, E.F.; Lenton, T.M.; Scheffer, M.; Folke, C.; Schellnhuber, H.J.; et al. Planetary boundaries: Exploring the safe operating space for humanity. *Ecol. Soc.* **2009**, *14*, 32. [CrossRef]

50. Wu, J. Landscape sustainability science: Ecosystem services and human well-being in changing landscapes. *Landsc. Ecol.* **2013**, *28*, 999–1023. [CrossRef]

51. UN General Assembly. The Future We Want. Resolution Adopted by the General Assembly on 27 July 2012. Available online: https://sustainabledevelopment.un.org/futurewewant.html (accessed on 19 March 2020).

52. Beckerman, W. "Sustainable development": Is it a useful concept? *Environ. Values* **2013**, *3*, 191–209. [CrossRef]

53. Connelly, S. Mapping sustainable development as a contested concept. *Local Environ.* **2007**, *12*, 259–278. [CrossRef]

54. Bell, S.; Morse, S. *Sustainability Indicators: Measuring the Immeasurable*; Earthscan: London, UK, 2008; ISBN 1-84407-299-1.

55. Pesqueux, Y. Sustainable development: A vague and ambiguous "theory". *Soc. Bus. Rev.* **2009**, *4*, 231–245. [CrossRef]

56. Salas-Zapata, W.A.; Ortiz-Muñoz, S.M. Analysis of meanings of the concept of sustainability. *Sustain. Dev.* **2019**, *27*, 153–161. [CrossRef]

57. Daly, H.E. On Wilfred Beckerman's critique of sustainable development. *Environ. Values* **1995**, *4*, 49–55. [CrossRef]

58. Salomonson, V.V. Remote sensing, historical perspective. In *Encyclopedia of Remote Sensing*; Njoku, E.G., Ed.; Springer Science+Business Media: New York, NY, USA, 2014; pp. 684–691.

59. Moore, G.K. What is a picture worth? A history of remote sensing/ Quelle est la valeur d'une image? Un tour d'horizon de télédétection. *Hydrol. Sci. Bull.* **1979**, *24*, 177–485. [CrossRef]

60. Andries, A.; Morse, S.; Murphy, R.J.; Lynch, J.; Woolliams, E.R. Seeing sustainability from space: Using Earth observation data to populate the UN Sustainable Development Goal indicators. *Sustainability* **2019**, *11*, 5062. [CrossRef]

61. Weng, Q. (Ed.) *Remote Sensing for Sustainability*; CRS Press: Boca Raton, FL, USA, 2016; ISBN 1-4987-0071-3.

62. Rao, U.R. Remote sensing for sustainable development. *J. Indian Soc. Remote Sens.* **1991**, *19*, 217–235. [CrossRef]

63. Hansen, M.C.; Potapov, P.V.; Moore, R.; Hancher, M.; Turubanova, S.A.; Tyukavina, A.; Thau, D.; Stehman, S.V.; Goetz, S.J.; Loveland, T.R.; et al. High-resolution global maps of 21st-century forest cover change. *Science* **2013**, *342*, 850–853. [CrossRef]

64. de Araujo Barbosa, C.C.; Atkinson, P.M.; Dearing, J.A. Remote sensing of ecosystem services: A systematic review. *Ecol. Indic.* **2015**, *52*, 430–443. [CrossRef]

65. Estoque, R.C.; Ooba, M.; Avitabile, V.; Hijioka, Y.; DasGupta, R.; Togawa, T.; Murayama, Y. The future of Southeast Asia's forests. *Nat. Commun.* **2019**, *10*, 1829. [CrossRef] [PubMed]

66. Song, X.P.; Hansen, M.C.; Stehman, S.V.; Potapov, P.V.; Tyukavina, A.; Vermote, E.F.; Townshend, J.R. Global land change from 1982 to 2016. *Nature* **2018**, *560*, 639–643. [CrossRef] [PubMed]

67. Chen, C.; Park, T.; Wang, X.; Piao, S.; Xu, B.; Chaturvedi, R.K.; Fuchs, R.; Brovkin, V.; Ciais, P.; Fensholt, R.; et al. China and India lead in greening of the world through land-use management. *Nat. Sustain.* **2019**, *2*, 122–129. [CrossRef]

68. Kerr, J.T.; Ostrovsky, M. From space to species: Ecological applications for remote sensing. *Trends Ecol. Evol.* **2003**, *18*, 299–305. [CrossRef]

69. Wang, K.; Franklin, S.E.; Guo, X.; Cattet, M. Remote sensing of ecology, biodiversity and conservation: A review from the perspective of remote sensing specialists. *Sensors* **2010**, *10*, 9647–9667. [CrossRef] [PubMed]

70. Kwok, R. Ecology's remote-sensing evolution. *Nature* **2018**, *556*, 137–138. [CrossRef]

71. Yu, H.; Liu, X.; Kong, B.; Li, R.; Wang, G. Landscape ecology development supported by geospatial technologies: A review. *Ecol. Inform.* **2019**, *51*, 185–192. [CrossRef]

72. Schmugge, T.J.; Kustas, W.P.; Ritchie, J.C.; Jackson, T.J.; Rango, A. Remote sensing in hydrology. *Adv. Water Resour.* **2002**, *25*, 1367–1385. [CrossRef]

73. Crow, W.T.; Chen, F.; Reichle, R.H.; Liu, Q. L band microwave remote sensing and land data assimilation improve the representation of prestorm soil moisture conditions for hydrologic forecasting. *Geophys. Res. Lett.* **2017**, *44*, 5495–5503. [CrossRef]

74. Thies, B.; Bendix, J. Satellite based remote sensing of weather and climate: Recent achievements and future perspectives. *Meteorol. Appl.* **2011**, *18*, 262–295. [CrossRef]

75. Yang, J.; Gong, P.; Fu, R.; Zhang, M.; Chen, J.; Liang, S.; Xu, B.; Shi, J.; Dickinson, R. The role of satellite remote sensing in climate change studies. *Nat. Clim. Change* **2013**, *3*, 875–883. [CrossRef]

76. Voogt, J.A.; Oke, T.R. Thermal remote sensing of urban climates. *Remote Sens. Environ.* **2003**, *86*, 370–384. [CrossRef]

77. Weng, Q.; Lu, D.; Schubring, J. Estimation of land surface temperature-vegetation abundance relationship for urban heat island studies. *Remote Sens. Environ.* **2004**, *89*, 467–483. [CrossRef]

78. Weng, Q. Thermal infrared remote sensing for urban climate and environmental studies: Methods, applications, and trends. *ISPRS J. Photogramm. Remote Sens.* **2009**, *64*, 335–344. [CrossRef]

79. Estoque, R.C.; Murayama, Y. Monitoring surface urban heat island formation in a tropical mountain city using Landsat data (1987–2015). *ISPRS J. Photogramm. Remote Sens.* **2017**, *133*, 18–29. [CrossRef]

80. Estoque, R.C.; Murayama, Y.; Myint, S.W. Effects of landscape composition and pattern on land surface temperature: An urban heat island study in the megacities of Southeast Asia. *Sci. Total Environ.* **2017**, *577*, 349–359. [CrossRef]

81. Blaschke, T. Object based image analysis for remote sensing. *ISPRS J. Photogramm. Remote Sens.* **2010**, *65*, 2–16. [CrossRef]

82. Myint, S.W.; Gober, P.; Brazel, A.; Grossman-Clarke, S.; Weng, Q. Per-pixel vs. object-based classification of urban land cover extraction using high spatial resolution imagery. *Remote Sens. Environ.* **2011**, *115*, 1145–1161. [CrossRef]

83. Weng, Q. Remote sensing of impervious surfaces in the urban areas: Requirements, methods, and trends. *Remote Sens. Environ.* **2012**, *117*, 34–49. [CrossRef]

84. Blaschke, T.; Hay, G.J.; Kelly, M.; Lang, S.; Hofmann, P.; Addink, E.; Queiroz, F.; van der Meer, F.; van der Werff, H.; van Coillie, F.; et al. Geographic Object-Based Image Analysis—Towards a new paradigm. *ISPRS J. Photogramm. Remote Sens.* **2014**, *87*, 180–191. [CrossRef]

85. Estoque, R.C.; Murayama, Y. Classification and change detection of built-up lands from Landsat-7 ETM+ and Landsat- 8 OLI/TIRS imageries: A comparative assessment of various spectral indices. *Ecol. Indic.* **2015**, *56*, 205–217. [CrossRef]

86. Gupta, P.; Christopher, S.A.; Wang, J.; Gehrig, R.; Lee, Y.; Kumar, N. Satellite remote sensing of particulate matter and air quality assessment over global cities. *Atmos. Environ.* **2006**, *40*, 5880–5892. [CrossRef]

87. Martin, R.V. Satellite remote sensing of surface air quality. *Atmos. Environ.* **2008**, *42*, 7823–7843. [CrossRef]

88. Beck, L.R.; Lobitz, B.M.; Wood, B.L. Remote sensing and human health: New sensors and new opportunities. *Emerg. Infect. Dis.* **2000**, *6*, 217–227. [CrossRef] [PubMed]

89. Maxwell, S.K.; Meliker, J.R.; Goovaerts, P. Use of land surface remotely sensed satellite and airborne data for environmental exposure assessment in cancer research. *J. Expo. Sci. Environ. Epidemiol.* **2010**, *20*, 176–185. [CrossRef] [PubMed]

90. Seltenrich, N. Remote-sensing applications for environmental health. *Environ. Health Perspect.* **2014**, *122*, A268–A275. [CrossRef] [PubMed]

91. Estoque, R.C.; Ooba, M.; Seposo, X.T.; Togawa, T.; Hijioka, Y.; Takahashi, K.; Nakamura, S. Heat health risk assessment in Philippine cities using remotely sensed data and social-ecological indicators. *Nat. Commun.* **2020**, *11*, 1581. [CrossRef] [PubMed]

92. Joyce, K.E.; Belliss, S.E.; Samsonov, S.V.; McNeill, S.J.; Glassey, P.J. A review of the status of satellite remote sensing and image processing techniques for mapping natural hazards and disasters. *Prog. Phys. Geogr. Earth Environ.* **2009**, *33*, 183–207. [CrossRef]

93. Kaku, K. Satellite remote sensing for disaster management support: A holistic and staged approach based on case studies in Sentinel Asia. *Int. J. Disaster Risk Reduct.* **2019**, *33*, 417–432. [CrossRef]

94. de Sherbinin, A.; Levy, M.A.; Zell, E.; Weber, S.; Jaiteh, M. Using satellite data to develop environmental indicators. *Environ. Res. Lett.* **2014**, *9*, 084013. [CrossRef]

95. Giuliani, G.; Mazzetti, P.; Santoro, M.; Nativi, S.; Van Bemmelen, J.; Colangeli, G.; Lehmann, A. Knowledge generation using satellite earth observations to support sustainable development goals (SDG): A use case on Land degradation. *Int. J. Appl. Earth Obs. Geoinf.* **2020**, *88*, 102068. [CrossRef]

96. Adams, B.B.; Judd, K. *Global Indicator Framework for SDGs: Value Added or Time to Start Over?* Global Policy Forum: New York, NY, USA, 2019.

97. Sachs, J.; Schmidt-Traub, G.; Kroll, C.; Lafortune, G.; Fuller, G. *Sustainable Development Report 2019. Transformations to Achieve the Sustainable Development Goals: Includes the SDG Index and Dashboards*; Bertelsmann Stiftung and Sustainable Development Solutions Network (SDSN): New York, NY, USA, 2019.

98. Independent Group of Scientists appointed by the Secretary-General. *Global Sustainable Development Report 2019: The Future is Now—Science for Achieving Sustainable Development*; United Nations: New York, NY, USA, 2019.

99. FAO. *Global Forest Resources Assessment 2015: How are the World's Forests Changing?* second ed.; UN Food and Agriculture Organization: Rome, Italy, 2015.

100. Prince, S.D. Challenges for remote sensing of the Sustainable Development Goal SDG 15.3.1 productivity indicator. *Remote Sens. Environ.* **2019**, *234*, 111428. [CrossRef]

101. Mulligan, M.; van Soesbergen, A.; Hole, D.G.; Brooks, T.M.; Burke, S.; Hutton, J. Mapping nature's contribution to SDG 6 and implications for other SDGs at policy relevant scales. *Remote Sens. Environ.* **2020**, *239*, 111671. [CrossRef]

102. Watmough, G.R.; Marcinko, C.L.J.; Sullivan, C.; Tschirhart, K.; Mutuo, P.K.; Palm, C.A.; Svenning, J.C. Socioecologically informed use of remote sensing data to predict rural household poverty. *Proc. Natl. Acad. Sci. USA* **2019**, *116*, 1213–1218. [CrossRef] [PubMed]

103. Boyd, D.S.; Jackson, B.; Wardlaw, J.; Foody, G.M.; Marsh, S.; Bales, K. Slavery from Space: Demonstrating the role for satellite remote sensing to inform evidence-based action related to UN SDG number 8. *ISPRS J. Photogramm. Remote Sens.* **2018**, *142*, 380–388. [CrossRef]

104. Macfeely, S. The big (data) bang: Opportunities and challenges for compiling SDG indicators. *Glob. Policy* **2019**, *10*, 121–133. [CrossRef]

105. Pradhan, P.; Costa, L.; Rybski, D.; Lucht, W.; Kropp, J.P. A systematic study of Sustainable Development Goal (SDG) interactions. *Earths Future* **2017**, *5*, 1169–1179. [CrossRef]

106. Singh, G.G.; Cisneros-Montemayor, A.M.; Swartz, W.; Cheung, W.; Guy, J.A.; Kenny, T.A.; McOwen, C.J.; Asch, R.; Geffert, J.L.; Wabnitz, C.C.C.; et al. A rapid assessment of co-benefits and trade-offs among Sustainable Development Goals. *Mar. Policy* **2018**, *93*, 223–231. [CrossRef]

107. Kroll, C.; Warchold, A.; Pradhan, P. Sustainable Development Goals (SDGs): Are we successful in turning trade-offs into synergies? *Palgrave Commun.* **2019**, *5*, 140. [CrossRef]

108. Biggeri, M.; Clark, D.A.; Ferrannini, A.; Mauro, V. Tracking the SDGs in an 'integrated' manner: A proposal for a new index to capture synergies and trade-offs between and within goals. *World Dev.* **2019**, *122*, 628–647. [CrossRef]

109. Xu, Z.; Chau, S.N.; Chen, X.; Zhang, J.; Li, Y.; Dietz, T.; Wang, J.; Winkler, J.A.; Fan, F.; Huang, B.; et al. Assessing progress towards sustainable development over space and time. *Nature* **2020**, *577*, 74–78. [CrossRef]

110. Pettorelli, N.; Schulte toBühne, H.; Shapiro, A.C.; Glover-Kapfer, P. *Satellite Remote Sensing for Conservation*; WWF Conservation Technology Series 1(4); World Wide Fund for Nature: Gland, Switzerland, 2018.

111. USGS. What Are the Band Designations for the Landsat Satellites? Available online: https://www.usgs.gov/faqs/what-are-band-designations-landsat-satellites?qt-news_science_products=0#qt-news_science_products (accessed on 19 March 2020).

112. Airbus Intelligence. SPOT Image: Spot Satellite Technical Data. Available online: https://www.intelligence-airbusds.com/files/pmedia/public/r329_9_spotsatellitetechnicaldata_en_sept2010.pdf (accessed on 19 March 2020).

113. EO Portal Directory. Satellite Missions Database. Available online: https://directory.eoportal.org/web/eoportal/satellite-missions (accessed on 19 March 2020).

114. NOAA. AVHRR. Available online: https://www.avl.class.noaa.gov/release/data_available/avhrr/index.htm (accessed on 19 March 2020).

115. USGS. AST_L1T Product User's Guide. Available online: https://lpdaac.usgs.gov/documents/71/AST_L1T_User_Guide_V3.pdf (accessed on 19 March 2020).

116. NASA MODIS. Specifications. Available online: https://modis.gsfc.nasa.gov/about/specifications.php (accessed on 19 March 2020).

117. ESA (European Space Agency) Earth Online. Earth Observation Information Discovery Platform. Available online: https://earth.esa.int/eogateway (accessed on 19 March 2020).

118. Sentinel Online. User Guides Introduction. Available online: https://sentinel.esa.int/web/sentinel/user-guides (accessed on 19 March 2020).
119. JAXA (Japan Aerospace Exploration Agency) Earth Observation Research Center. Available online: https://www.eorc.jaxa.jp/en/index.html (accessed on 19 March 2020).

 *remote sensing*

Review

Remote Sensing for International Peace and Security: Its Role and Implications

Ram Avtar [1,*], Asma Kouser [2], Ashwani Kumar [3] Deepak Singh [1,4], Prakhar Misra [5], Ankita Gupta [6], Ali P. Yunus [7], Pankaj Kumar [8], Brian Alan Johnson [8], Rajarshi Dasgupta [8], Netrananda Sahu [9] and Andi Besse Rimba [10,11]

[1]   Faculty of Environmental Earth Science, Hokkaido University, Sapporo 060-0810, Japan; deepak84_sse@jnu.ac.in
[2]   Department of Economics, Bengaluru City University, Post Office Road, Ambedkar Veedhi, Bengaluru, Karnataka 560001, India; asma11_isg@jnu.ac.in
[3]   Electrical and Instrumentation Engineering Department, Sant Longowal Institute of Engineering and Technology, Longowal 148106, India; ashwani.ist@sliet.ac.in
[4]   Department of Geography and Resource Management, The Chinese University of Hong Kong (CUHK), Sha Tin, New Territories, Hong Kong 999077, China
[5]   Research Institute for Humanity and Nature (RIHN), Kamigamo, Kita-ku, Kyoto 603-8047, Japan; mprakhar@chikyu.ac.jp
[6]   Graduate School of Information Science and Technology, Hokkaido University, Sapporo 060-0814, Japan; anki@m-icl.ist.hokudai.ac.jp
[7]   State Key Laboratory of Geohazard Prevention and Geoenvironment Protection, Chengdu University of Technology, Chengdu 610059, China; yunusp@cdut.edu.cn
[8]   Natural Resources and Ecosystem Services, Institute for Global Environmental Strategies, Hayama, Kanagawa 240-0115, Japan; kumar@iges.or.jp (P.K.); johnson@iges.or.jp (B.A.J.); dasgupta@iges.or.jp (R.D.)
[9]   Department of Geography, Delhi School of Economics, University of Delhi, New Delhi 110007, India; nsahu@geography.du.ac.in
[10] Department of Civil Engineering, Shibaura Institute of Technology, Tokyo 135-8548, Japan; rimba@unu.edu
[11] United Nations University Institute for the Advanced Study of Sustainability (UNU-IAS), Shibuya-ku, Tokyo 150-8925, Japan
*   Correspondence: ram@ees.hokudai.ac.jp; Tel.: +81-011-706-2261

Citation: Avtar, R.; Kouser, A.; Kumar, A.; Singh, D.; Misra, P.; Gupta, A.; Yunus, A.P; Kumar, P.; Johnson, B.A.; Dasgupta, R.; et al. Remote Sensing for International Peace and Security: Its Role and Implications. *Remote Sens.* **2021**, *13*, 439. https://doi.org/10.3390/rs13030439

Academic Editor: Ronald Estoque
Received: 24 December 2020
Accepted: 20 January 2021
Published: 27 January 2021

Publisher's Note: MDPI stays neutral with regard to jurisdictional claims in published maps and institutional affiliations.

Abstract: Remote sensing technology has seen a massive rise in popularity over the last two decades, becoming an integral part of our lives. Space-based satellite technologies facilitated access to the inaccessible terrains, helped humanitarian teams, support complex emergencies, and contributed to monitoring and verifying conflict zones. The scoping phase of this review investigated the utility of the role of remote sensing application to complement international peace and security activities owing to their ability to provide objective near real-time insights at the ground level. The first part of this review looks into the major research concepts and implementation of remote sensing-based techniques for international peace and security applications and presented a meta-analysis on how advanced sensor capabilities can support various aspects of peace and security. With key examples, we demonstrated how this technology assemblage enacts multiple versions of peace and security: for refugee relief operations, in armed conflicts monitoring, tracking acts of genocide, providing evidence in courts of law, and assessing contravention in human rights. The second part of this review anticipates future challenges that can hinder the applicative capabilities of remote sensing in peace and security. Varying types of sensors pose discrepancies in image classifications and issues like cost, resolution, and difficulty of ground-truth in conflict areas. With emerging technologies and sufficient secondary resources available, remote sensing plays a vital operational tool in conflict-affected areas by supporting an extensive diversity in public policy actions for peacekeeping processes.

Keywords: conflict resources monitoring; disease control and prevention; human rights; genocide tracking; human rights violation; geopolitics

---

1. Introduction

Assuring the individual and collective well-being are the quintessential goals of any society. Different civilizations and societies develop certain principles and ethos that are binding as rules and regulations to their citizens. In return, the state (through its administrative machinery) strives to ensure its residents safety and welfare as a social contract [1]. The technology-oriented industrial revolution propelled the advent of the nation-state system and democratic thought. The Government's mandate and tools to deliver its functions continue evolving with time, demography, and technological innovations. The knowledge frontiers explore more scientific methods with considerable precision and accuracy to testify and ensure compliance with rules [2]. In the initial years, the power to master the earth's natural resources was the fundamental principle for economic expansion [3]. However, this uncontrolled growth became the reason for conflict (strategic control over oil and key mineral resources) among participating powers [4].

The rise of a scientific-industrial-military complex before and during the Cold War era had a lasting effect on the peace and tranquility of certain resource-rich and strategic regions. Although the techno-economic prowess aided development, the armed conflicts affected individual and community rights. For centuries, military commanders have sought out positions with a high elevation, such as mountains and ridges, to gain visual information about their enemies' locations and movements [5]. The First World War is widely regarded as the turning point in history that led to the wide popularity of many advanced techniques and weaponry systems. In particular, remote sensing gained prominence due to the application of high-altitude airplanes for aerial reconnaissance [6]. The aerial photographs were specifically used to locate enemy trenches and hidden positions, troop movements, supply routes, and depots, as well as to verify the effectiveness of artillery attacks against the enemy [7,8].

The development of the man-made satellite was considered one of the largest technological breakthroughs in the military field [9]. The Cold War and Post-Cold War phase saw a vast expansion in such satellite deployments for international peace and security [10]. The United States Air Force's CORONA satellite program operated during the Cold War and collected over 800,000 aerial images of the Union of Soviet Social Republics (USSR), the People's Republic of China (PRC), and other countries and regions. As the platforms for remote sensing applications advanced by leaps and bounds, the sensors themselves also improved drastically. Initially, the CORONA satellite was only able to capture images from orbit with a spatial resolution up to roughly 12 m [11]. However, the spatial resolution of satellite images has shown drastic improvement to below one meter in recent years [12]. Multi spatio-temporal satellite data with local to global data acquisition can be applied in international peace and security in conflict zones. Several applications are shown in Figure 1.

In particular for military and conflict management, the application of remote sensing was initially limited to the technologically advanced nations, like the United States of America (USA) and the former USSR, as well as other countries with significant defense budgets. Satellite data have been used by the forces to identify terrains, rivers, ridges, populated areas, strategic installations, communication networks, etc. [13,14]. With time and technical advancements, remote sensing has also made significant contributions for less developed countries, such as Vietnam, Indonesia, Thailand, India, Cambodia, etc. The type of information accessible from remote sensing for peace and security depends on the sensor's specific properties and platform. Recently, with the availability of high spatio-temporal data, remote sensing technology was actively used in the detection of genocide in Darfur and human and drug trafficking in Afghanistan [15,16]. In addition to the applications in military purposes, aerial and satellite remote sensing have been significantly utilized for international peace through their role in preventing resource conflicts [17], disease control and prevention [18], human rights protection [19], and tracking genocide [20].

Figure 1. Applications of remote sensing for international peace and security.

Earth observation satellites and communication technologies offer precise and accurate means for remote monitoring of conflict zones. Some of the worst forms of human rights violations have been deeply rooted in either war zones (in resource-rich regions) or regions lacking basic resources like water and food [21]. It can be difficult to monitor these dangerous zones using ground information. However, remote sensing techniques can help to monitor such remote and dangerous zones without physical contact. Remote sensing has also been used for verifying international laws, treaties, and resolutions, e.g., for monitoring oil pollution sources [22], exploring renewable energy resources [23]. The technological development and rise in using sensors have led to the surge in remote sensing companies, aiding in the usage of data for the larger social and environmental safety. Geospatial techniques can provide useful information for the implementation of The United Nations Sustainable Development Goals (SDGs#16, i.e., to promote peaceful and inclusive societies for sustainable development, and provide access to justice for all and build effective, accountable and inclusive institutions at all levels) [24,25].

Several approaches of remote sensing for military and civilian applications have been investigated [26]. These studies displayed remote sensing's utility for international peace and security both from a macro-perspective and micro-perspective, respectively. At the macro-level, the application of Geographical Information System (GIS) techniques in identifying the role of historical precedents in territorial disputes has shown valid results. For instance, in the European context, the application of GIS helped in finding the relationship between historical boundaries and conflicts [27]. For instance, studying the micro-level effect on issues like migration led to violent situations in the Goma City (the Democratic Republic of Congo) [28] and city-level consequences of Arab Spring in Jordan [29,30]. Furthermore, along with GIS techniques, other scientific tools like big data have been utilized to understand the intensity of such conflicts [31]. Remote sensing can also assist in understanding the issues emanating due to state classifications like ethnic fractionalization [32]. This, in return, can aid in the consolidation of socio-cultural theoretical frameworks of other humanities disciplines.

This paper builds on the previously mentioned potential importance of space-borne technologies in peace and security missions by highlighting the role of image data in near real-time and archived sources. Based on the available literature, we classified various studies and performed its meta-analysis for understanding the trends, potential, and impediments in applying remote sensing to the issues of peace and security. The selective analysis was conducted for the academic papers, research reports, and handbooks from a wide variety of sources, tracing the development of remote sensing applications in peace, security, and allied areas. We explored conflict resolution and monitoring as a theme by looking at disturbed and volatile regions and investigating instances related to a refugee relief operation, armed conflicts, genocide, human rights, peace-building, and issues crucial for preventing and controlling human disease. In each of these cases, we highlighted how remote sensing was utilized on the ground.

2. Methodology

This scoping review is intended to cover an overlaying body of literature on earth observation technology in supporting emerging issues like sustainable development, human development, international peace, and security [33–35]. It presents a brief description of each application of remote sensing to these developmental and security issues at a global perspective [33–35]. Potentially relevant reference materials were obtained from research databases and appraised according to whether they included information on the previously mentioned themes. Thus, this review provides a broad introductory understanding structured across three nodes. First is the concept of peace and security. Second, the relevance of remote sensing to the issues of peace and security is introduced. Third, the contemporary status, challenges, and opportunities for peace and security through remote sensing are discussed. Figure 2 illustrates the flow of literature review for this study.

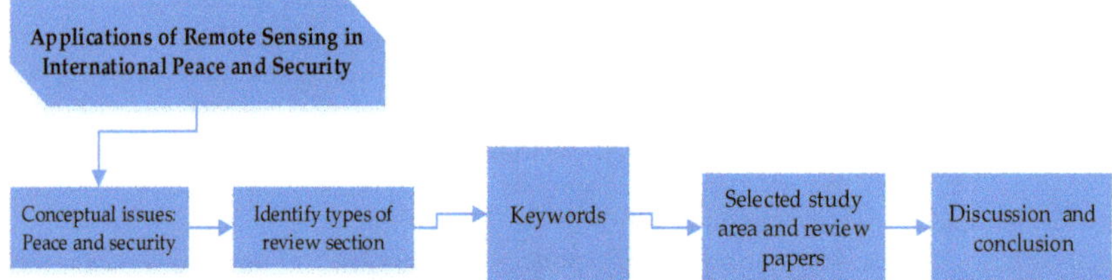

Figure 2. Flowchart of literature review.

2.1. Conceptual Issues

A conceptual analysis of the remote sensing approaches for peace and security has been carried out in Reference [36]. The conceptual issues can be mainly classified into two strands viz. definitions and cross-disciplinary interpretations.

'Peace' can be defined as a state or a period in which there is no war, or a war has ended [37,38]. However, there are many types of social settings where society does not experience peace despite the absence of war or conflict [39]. For instance, Thomas et al. [40] have argued that poverty is one of society's major threats. Countries with a high level of poverty, prohibition, intimidation, repression, terror, and other deterrents cannot be categorized as peaceful countries. Based on these observations, analysts working in peace and conflict management have varying definitions for peace from different perspectives. 'Security' is a state of ensuring protection from the direct/indirect notions and actions threatening an individual or, collectively, a group [37,38]. The international and regional organizations like the United Nations (UN), European Union, and other multi-lateral groups strive for attaining the standards of safety and security of individuals, countries, and regions.

Contextualizing the disciplinary connections becomes germane to broaden the understanding of the applications of remote sensing technology for peace and security. Branch [35] has identified the issues of measurement validity and selection bias with approaches in GIS technologies to the domains of peace and security in the studies of international relations. The measurement validity issues arise when the institutional structures are built over novel propositions with behavioral and pragmatic approaches that are not sufficiently absorbed by technical operational parameters (raster, vector, and other files) of the GIS interfaces [35]. During the analysis stage, the selection bias leads the discrepancies and prejudices between spatial and non-spatial information to creep into the study/system [35].

The successful application of remote sensing for peace and security depends on different kinds of datasets and how it is coded and analyzed within the framework of measurement validity and selection bias [35]. For instance, the large datasets about ethnic conflicts have the potential to overshadow the reasons contributing to their origins. The quantitative study of ethnic conflicts with a focus on micro-level studies has elucidated the essence of individuality during a conflict [41].

2.2. Selection of Related-Articles and Keywords

This study has approached the literature review extensively, exploring the application of remote sensing in international peace and security. Due consideration was given to the geographic distribution of the origin of articles, the discipline of the journal in which the article is published, online as well as offline, and the publishing type as open access or not, etc., using the Latent Dirichlet allocation (LDA) scheme. With an understanding that the topic should equally relate to science as well as the humanities disciplines, we leveraged our search to all the related disciplines. Among many literature database search engines, we primarily used the most popular ones *viz.*—ScienceDirect and Google Scholar, to explore scholarly articles from journals, conference proceedings, book chapters, etc. We searched articles with keywords such as "remote sensing" + "peace and security", bringing in 3240 results, and "remote sensing" + "natural resources monitoring" bringing in 854 results. Apart from that, we decided to use many online journals and research databases viz. ScienceOpen, Directory of Open Access Journals, Education Resources Information Center (ERIC), CIA World Factbook, Web of Science, Social Science Research Network (SSRN), ResearchGate, and Public Library of Science (PLoS) for exploring the scholarly articles.

While choosing an article for the study, we pondered with great attention that the document is relevant to remote sensing applications for international peace and security in domains such as conflict resources monitoring, disease control and prevention, human rights, and track of genocide, etc., as illustrated in Figure 1. After reviewing articles, we noted salient points of each of the articles and finalized a subset of articles most relevant to this review paper. A large number of papers were collected from a database of published articles. Several case studies were referred from the regions witnessing international conflict and bringing peace and security in the context of social dimensions. The literature identified that several troubled regions in Asia, Latin America, and Africa were studied through the use of satellite imagery. Two group discussions were conducted for three months among all authors of this paper. The first meeting resulted in deciding the types of keywords to be used while searching for relevant literature and the second discussion was focused on selecting the relevant examples.

The keywords identified were "role of remote sensing in international peace and security," "international peace and security with remote sensing," "geopolitics and remote sensing," "international conflicts resolution with remote sensing," and "remote sensing-based conflict resolution" were also chosen as search terms to gather more articles related to this study. To know the trend, the search was carried out using the same keyword for different periods. For example, a keyword phrase of "remote sensing for international peace and security" produced a total of 47, 79, 126, 169, 227, and 252 search results for the period 1997–2000, 2001–2004, 2005–2008, 2009–2012, 2013–2016, and 2017–2020, respectively, while keeping the other search filters the same. Figure 3 shows the trend in searched papers. The rising numbers of published studies indicates the growing role of remote-sensing approaches toward international peace and security.

We also split the search results obtained in each of the four categories of remote sensing applications in peace and security, as mentioned in Figure 1. The total number of articles found on Google Scholar were 17,900, 13,700, 11,377, and 2510 for the categories of remote sensing applications in epidemics control and prevention, human rights, conflict resources monitoring, and genocide, respectively. However, category-overlapping was encountered among the searched papers since papers were often connected to more than one category.

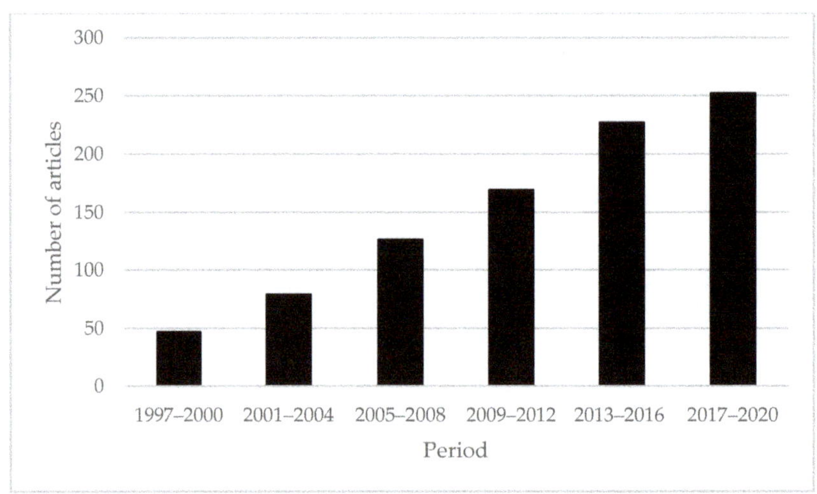

Figure 3. An increasing trend of the role of remote sensing in international peace and security.

A special focus was made for policy-oriented studies vis-à-vis key objectives of this study. To obtain a wide range of related articles, we added papers relating to conflict management with remote sensing, geopolitics and remote sensing, and other related reports. The unrelated papers were omitted after detailed reading. The final selected set of articles were examined carefully and the main findings were compiled with their shortcomings. Many papers did not meet our criteria for inclusion in the study. These papers were generally similar to other papers, or lacked statistically significant analysis, had shaky research methodology, deficient references, and lack of supporting evidence while drawing a conclusion, etc. Such papers were removed from our study's database.

3. Results

3.1. Remote Sensing for Refugee Relief Operations

Mapping displaced zones and refugee camps are vital to relief operations. The satellite data to map and monitor inaccessible conflict-riddled areas for relief work activities has been used by the United Nations High Commissioner for Refugees (UNHCR). It allows for tracking the affected communities and physical as well as environmental impact on the infrastructure at the refugee camps. For instance, the Sudan conflict left more than 2.5 million people displaced internally and approximately 600,000 refugees outside the country. Similarly, 5.5 million displaced Syrians were registered by UNHCR in Lebanon, Turkey, Jordan, Egypt, and Iraq. Furthermore, about 33,000 Syrian refugees were related in North Africa. According to UNHCR, to provide data on refugees-related numbers, food, and logistics, precise land use and individual refugee tents and buildings in the camp have been mapped.

Dalen et al. [42] and Bjorgo et al. [43] have employed European Remote Sensing satellite (ERS) and commercial very high spatial resolution (VHSR) satellite data from the Russian KVR-1000 sensor to demonstrate the pre-operational use of satellite remote sensing techniques. It helped in appropriately planning the strategies at the refugee camps. High-resolution is used for automatic mapping of refugee tents and camps by applying object-based image analysis [44]. Figure 4 shows the mapped tents of Al Zaatari refugee camps in Mafraq Governorate, Jordan using satellite images. As of 3 January 2013, a total of 11,966 shelters were detected within the 314 ha of the camp (Figure 4a). By 4 May 2013, a total of 28,243 shelters were detected within the 530.95 ha of the camp (Figure 4b), indicating that the number of shelters increased by about 16,000. Previously, United Nations Institute for Training and Research (UNITAR)/UNITAR's Operational

Satellite Applications Programme (UNOSAT) found that the number of shelters is about 5138 on 15 November 2012. Thus, the shelters in the camp increased by 450% within less than five months. Besides, the total area of the camp grew from 216 hectares to 531 hectares in the same period, which is a 146% increase. This information, together with the data on the number of people living in each tent, is especially critical for assessing the food and medical aid. The number of refugees in each shelter can be inferred by the size of the structures and total population by multiplying the number of shelters in each camp [45].

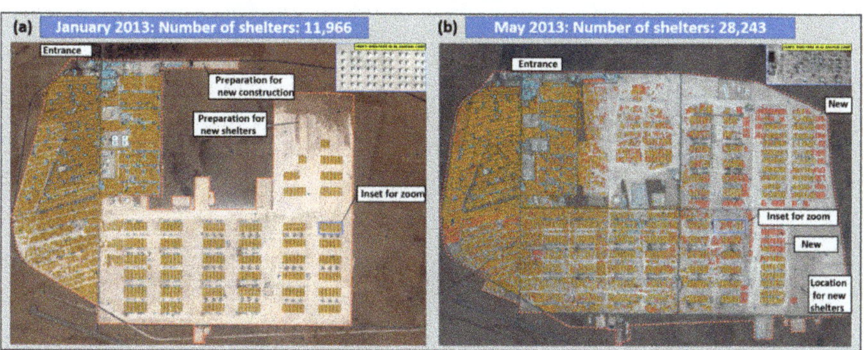

Figure 4. Satellite-detected shelters at the Al Zaatari refugee camp, Jordan (**a**) 3 January 2013 (WorldView-2 image), and (**b**) 4 May 2013 (WorldView-1 image) (Source: UNITAR/UNOSAT).

Nevertheless, there are difficulties in practice on object detection from medium-coarse resolution data. According to Quinn et al. [46], the following challenges are typical in refugee shelter mapping: (i) higher range of variation among refugees locations, (ii) small and clustered close-together shelters, and (iii) a higher log of accuracy is required owing to the critical decision support demanded. Automated machine learning models, such as Mask-Region Based Convolutional Neural Network (RCNN), can detect refugee settlements from high-resolution satellite images with an average Area Under the Curve (AUC) of 0.78 [46]. Apart from the relief operations, the satellites also aid site selection by providing baseline information of the terrain for refugee camps and also by identifying favorable areas with access to drinking water and firewood.

Images acquired from high-resolution satellite data can help in monitoring environmental degradation due to the movement of displaced persons from the conflict area. For instance, PlanetScope data shows an increase in the built-up area near the Kutupalong refugee camp in Cox's Bazar, Bangladesh. Figure 5a,b show the PlanetScope images acquired on 25 November 2016 and 15 November 2020, respectively. The development of the 32 Km^2 of Rohingya refugee camps destroyed also harmed biodiversity. These images acquired before and after the incident provide clear evidence of forest disturbance due to the development of the refugee campsite (Figure 5a,b). Satellite data can be useful to provide information about environmental degradation due to international conflicts [47]. Knowing the size of refugee camps can help international agencies to plan and manage relief operations by providing logistic support, financial aids, and medical supports for these camps.

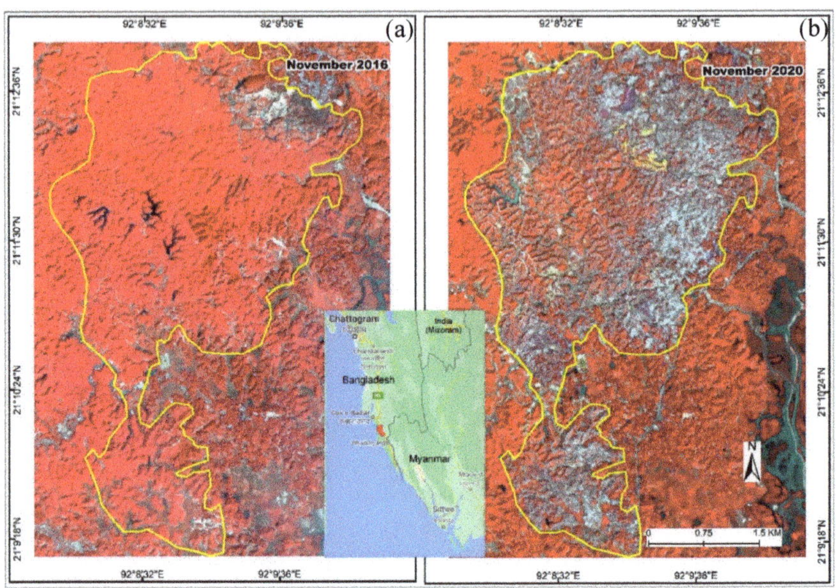

Figure 5. PlanetScope data of Kutupalong refugee campsite: (**a**,**b**) show the destruction of forest area due to the establishment of Rohingya refugee camps as observed from images taken on 25 November 2016 and 15 November 2020, respectively. [Source: prepared by authors].

3.2. Remote Sensing in Armed Conflicts

This technology has also been applied to quantify the causes of armed conflict and how such conflict can impact the environment. An example of this was shown in Brown et al. [16], where authors examined whether remote sensing could be used to confirm that the conflict in Darfur was a conflict across the resources among communities. Their study illustrated that remote sensing could be useful in testing claims, which are difficult to validate with more traditional information. These studies can be further extended with more granularity in studying the role of individuals and communities in the ignition of ethnic conflicts through local and regional databases [34]. Brown et al. [16] also examined how this technology can assist in reducing the devastating impacts on the environment during and after an armed conflict. War can cause extensive water and air pollution as well as the degradation of land and biodiversity. Such long-term environmental impacts might again contribute to future conflict if not addressed. Therefore, understanding environmental impacts exacerbated by conflict is fundamental in both conflict prevention and reconstruction. However, the impacts of the armed disputes and violence across the environmental domain, tending to be multifarious and, hence, extremely difficult to assess, especially during wartime. Nevertheless, remote sensing has augmented environmental impact assessment in conflict scenarios. This has greatly enhanced post-conflict reconstruction and rehabilitation [48].

The use of night-time light data to study armed conflicts has increased considerably over the last decade. It helps in identifying and analyzing the impact of different conflicts. For instance, the Suomi National Polar-Orbiting Partnership Visible Infrared Imaging Radiometer Suite sensor (NPP-VIIRS) has been used to evaluate the crisis in Sana, Yemen by utilizing the time series night-time light images (Figure 6). Jiang et al. [49] observed that between February 2015 and June 2015, the total night-time light of Yemen has decreased by 71.60% because of the severe conflict across the country. Levin et al. [31] utilized Visible Infrared Imaging Radiometer Suite (VIIRS) and Flickr photos to monitor the crisis development and refugee flow in Arabian countries. Change in economic and human capital can be observed using time-series VIIRS and Flickr photos. Their study concludes that the big data with remote sensing has the potential to provide unstructured but timely data on a conflicting situation in case of lacking a conducive approach for changes in environmental and geopolitical variations. Figure 7 shows a false-color composite (FCC) of VIIRS night-time brightness data acquired in October 2012 (in the red band), October 2015 (in the green band), and October 2018 (in the blue band) in Arabian countries. The areas with a decrease in night-time light after 2012 appear to have a red color. Most of the decrease of the night-time light is noticed in the Syrian region, which is related to the intensification of conflicts in the region. The areas with higher intensity of brightness during night-time on the observation dates emerge in a white color. The white color region shows the sites were not affected by the conflicts. VIIRS time-series data can be useful to provide essential information about economic changes in the region.

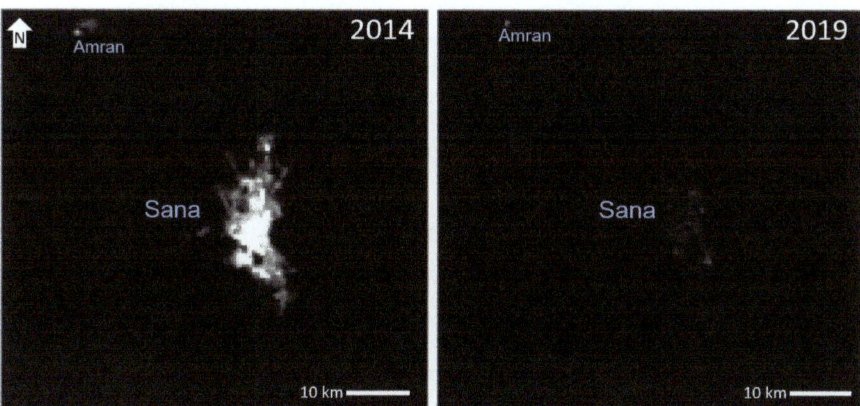

Figure 6. The Suomi National Polar-Orbiting Partnership Visible Infrared Imaging Radiometer Suite sensor (NPP-VIIRS) night-time light images of Sana, Yemen, during the years 2014 and 2019.

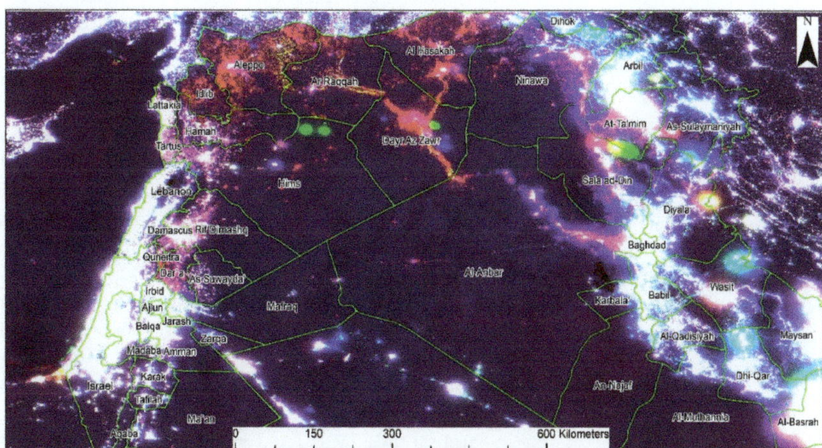

Figure 7. False-color composite (FCC) of Visible Infrared Imaging Radiometer Suite (VIIRS) night-time data acquired in October 2012 (red band), October 2015 (green band), and October 2018 (blue band) in Arabian countries.

Remote sensing has also been utilized in assessing forest cover change aggravated by conflict. Forests are crucial in conflict situations because they are used as safe havens by both combatants and civilians. Poaching and deforestation in unprotected forests tend to increase during the conflict and, thus, threaten terrestrial biodiversity [50]. High spatial resolution satellite data can help determine forest cover change, a change in biodiversity, and species distribution. For instance, Georsevrski et al. [48] used multi-temporal Landsat imagery to determine forest cover change as a result of the Darfur crisis and Uganda clash. The study developed a disturbance index to monitor changes in South Sudan-based Imatong Central Forest Reserve (ICFR) located around Dongotana hills, and Uganda-based Agoro-Agu Forest Reserve (AFR) during the conflict (the 1980s–2001) and post-conflict (2003–2010). The study discovered there was more forest cover loss during the conflict period and that unprotected forests (Dongotana hills) experienced a huge loss compared to reserved forests [48].

Severe conflicts often result in attacks with bombs and firearms. Fires detected with the help of images from the satellite. It can be potentially utilized as an early warning signal, indicating humanitarian crisis like human-rights violations [46]. Armed conflicts disrupt human settlement and economic activities, as many people flee their homes to seek refuge. The refugees and internally displaced persons (IDPs) reduce environmental pressure in places of origin and increase pressure in places of destination. Remote sensing is a valuable tool in assessing these environmental pressures instigated by an influx of IDPs and refugees. In Sudan, Landsat data were used to locate burned structures among habitations around the ethnic violence periods [51–53]. In the Rift Valley province of Kenya, the United Nations Institute for Training and Research (UNITAR) used Moderate Resolution Imaging Spectroradiometer (MODIS) images to identify the areas showing signs of violence occurrences following Kenyan National Elections [54].

Remote sensing has also been used to detect macro-level transformations like land cover changes due to human displacement as well as to setup an association among the land tenure and land use land cover (LULC) classification [55]. If remote sensing techniques are cautiously applied within the limits of measurement validity and selection biases, then it is also possible to estimate the cultural loss, agricultural losses, land degradation and rehabilitation, disruption of water resources, and change in vegetation [33,35,56]. Using multiple satellite images and other explanatory variables, Levin et al. [57] indicated that more than 5% of 1073 World Heritage Sites (WHS) were in danger because of severe internal conflicts in which 25 are caused by armed conflicts.

Previously, in the context of monitoring of damage to the cultural heritage sites, manual satellite image interpretation techniques were used to assess the changes in the affected sites/area. These manual interpretation techniques were laborious and time-consuming. However, the automatic classification of satellite imagery and the change detection methods are promising to monitor the changes recently [58]. In other studies, high-resolution multi-temporal satellite imagery was used to understand the impact on environment at the Sudanese Zam Zam IDP camp, which was established as a result of the Darfur crisis [59]. Figure 8 shows the satellite images of burned sites damaged in the 2011 Sudan conflicts. Figure 8e clearly shows the damaged and non-damaged structures in the Tajalei region of Sudan.

Figure 9 shows another case study in the conflict zones of Syria, where an archaeological site was damaged during the Syrian civil war. Figure 9a,b show the encroachment around the Abbasid Palace and its peripheral area using satellite data before (8 April 2011) and after (1 July 2016) the conflict, respectively. The former Abbasid Palace compound, Syria, which was one of the archaeological sites, is now covered by a modern building visible in satellite data [60]. The temporal satellite data shows the change in the built-up area due to the Islamic State of Iraq and Syria (ISIS) activities. On the city scale, the high-resolution satellite data have been used for mapping ambient light during the night in urban environments for security [61].

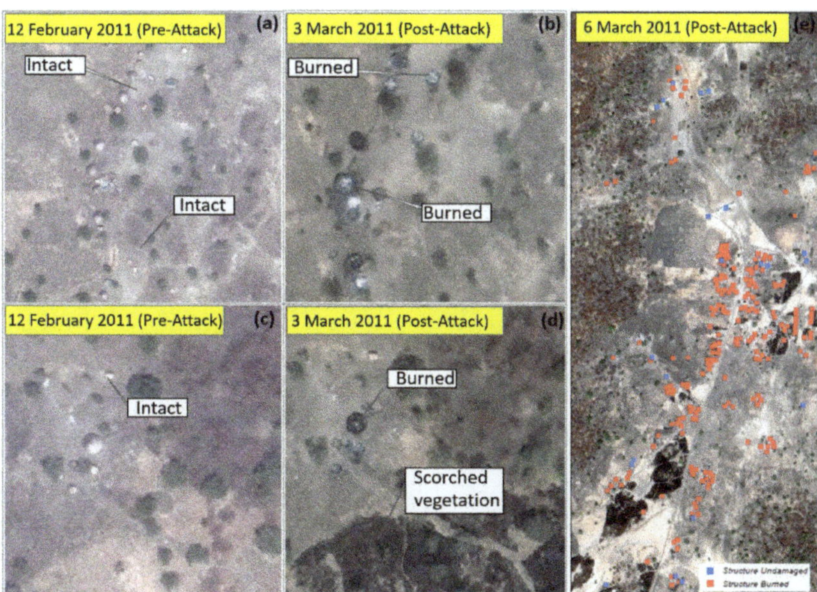

Figure 8. Satellite images of conflict zones in Sudan showing evidence of arson attacks: (**a,c**) pre-conflict images of 12 February 2011 showing intact buildings (**b,d**) are post-conflict images on 3 March 2011 showing burned structures in the Maker Awat region. (**e**) show the damaged and non-damaged structures of Tajalei region mapped from Worldview 2 imagery on 6 March 2011. [Source: UNITAR/UNOSAT].

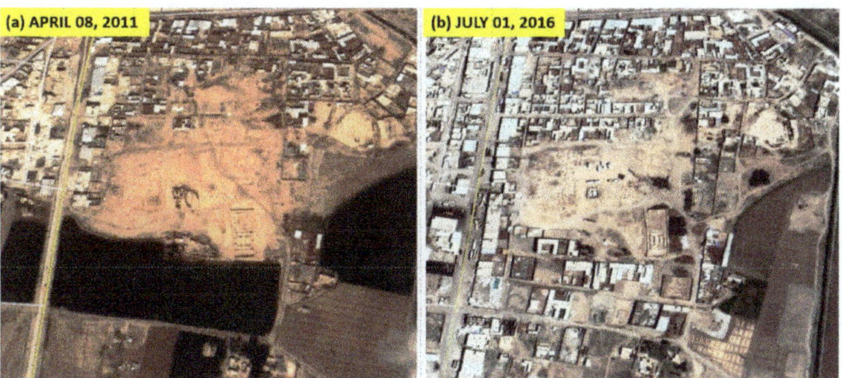

Figure 9. Satellite images of conflict zones in Syria: (**a,b**) the Abbasid Palace area encroachment as observed from images taken in April 2011 and July 2016, respectively [Images are reproduced from Casana and Laugier et al. studies [62] Image sources: GoogleEarth images date of access: 28 November 2020].

In the domain of conflict in resource-rich regions, the oil production of ISIS has been estimated using remote sensing techniques for geopolitics and energy security [62]. The politics of carbon sequestration to awareness about the potentials of negative emissions have been discussed in Reference [63]. A security assessment model for geo-energy that quantifies geo-strategic oil energy security by China for oil pipeline construction in Russian pacific (from 1995 to 2010) has been rebuilt [64]. With inspiration from previous work on methodologies in political geography, a linkage between politics and language has also been examined in Reference [65]. The converging applications of emerging technologies like artificial intelligence and remote sensing through autonomous drones for international peace and security have been discussed in Reference [66].

The contemporary geopolitics in terms of the digital representations of space such as Google Maps, OpenStreetMap, geopolitical properties, and geopolitical subjectivity have been discussed in Reference [39]. The geopolitical remote sensing has been discussed with an emphasis on aerial data collection rather than ground-based data collection [67]. On a large scale, country-specific applications, the imagery captured using commercial satellites have been used for the security of Canada [68] and use of the field spectroscopy for security in Cyprus [69]. The geopolitical approaches to outer space activities and the effect of launching several satellites into space have been discussed in Reference [70].

A survey on the scope of remote sensing technology in assessing the influence of conflict has been carried out by Witmer [71]. A case study on the Arab Spring was carried out to quantify the conflict level using remote sensing techniques [31]. A survey on the scope of remote sensing approach in assessing the influence of conflict has been carried out by Witmer [71]. Another use of remote sensing technology in the region for monitoring of conflict and damage to the heritage sites using satellite data. Previously, manual satellite image interpretation techniques were used to assess damage or changes in the area. These manual interpretation techniques were laborious and time-consuming. However, the recent automatic classification of satellite imagery and change detection methods are promising to monitor the changes [58].

According to Galtung et al. [39], there are many types of social settings that are not considered as war or conflict. Remote sensing techniques have played an instrumental role in supporting complex emergency cases from time to time. Zhang et al. [72] used a high-frequency mixed-mode surface wave radar to detect ships for international peace and security purposes. Unmanned Aerial vehicles (UAVs) equipped with a camera and a ground-based sensor has been used for landmine surveying during humanitarian demining [73]. The recommendations based on an extensive review of global peace, security, and development have been made in Ibrahim et al. studies [74].

Another potential utilization of satellite remote sensing in border armed conflicts is tracking compliance with peacekeeping processes. Due to the recent incidents of violence across the international border of South Sudan and Sudan, the UN security council enacted a significant resolution (Resolution-2046). This resolution requires the governments of these two countries to meet several conditions to ensure peaceful settlements of international disputes, including an outright withdrawal of the respective armed forces from their sides of the border or consequences of sanctions under U.N. Charter's Article 41. However, the satellite-based tracking of border areas shows that both nations did not comply with the UN resolution even after the deadline, thus, allowing for the implementation of strict sanctions against both countries [75].

Furthermore, high-resolution satellite images showing aircraft, tanks, trucks, and troops in strategic locations indicate compelling evidence of soon-to-occur attacks. If monitored in near-real-time, such activities can be recognized as an early warning system for preparations and alert civilians about the danger of the attack. A classic example is from Sudan, where satellite-based tracking of artillery troops in a striking distance of Kurmuk area averted a potential disaster by issuing a human security warning that enables civilian people to flee away before the massive bombs strike [76].

On a similar note, in the peace talks between China and India after the conflict in Galwan valley (in 2020), the tensions cease to end after multiple negotiations from both sides about an agreement to withdraw troops from border areas. The data from satellite and aerial images confirmed the disengagement of troops from the border zones, which helped prevent a war-like situation amid COVID-19 peaks [77].

3.3. Remote Sensing in Tracking the Acts of Genocide

Genocide is the intentional mass murder or destruction of a people group and is among the most severe acts of human rights violation. Genocides have occurred in several regions, but, in most cases, these mass murders are conducted secretly and are only discovered afterward [78]. Remote sensing appears promising vis-à-vis international security from genocidal issues. An analysis of two case studies by the Genocide Studies Program at Yale University demonstrates the usefulness of remote sensing in tracking genocidal acts, corroborating eyewitness testimony, and supplementing reports by independent organizations. In addition, when the world is facing challenges in countries and regions like Syria, Myanmar, and the Democratic Republic of Congo (DRC), it can help to assess the severity of acts of genocide and population displacement arising from the armed conflicts [78].

Schimmer Russell [79] reported how a single Landsat image (taken four days after the announcement of results of the East Timorese independence referendum in 1999) was used to identify the concentrations of ground fires to assess the extent of destruction by pro-Indonesian militant forces. Schimmer's work found correlations between eyewitness reports and the location and pattern of fires identified from satellite images. The data also located other areas around the image-based boundaries, which were most likely subjected to related demolition but with missing reports. Using satellite imagery, the researchers showed that the fires were mainly concentrated around the capital city of Dili. A United Nations Security Council delegation that visited East Timor after the referendum, and, other eyewitnesses, reported that much of Dili had been burned and ransacked [79]. Expanding the analysis to the rest of East Timor, the researchers were able to positively identify locations of fires across the country and subsequently corroborate this with eyewitness and other independent reports. This study demonstrated how remote sensing data can be used alongside eyewitness accounts to ascertain the extent of destruction in these types of situations. It should, however, be noted that there was some fortune in the timing of the capture of the image. As the author notes—"It is rare that an image both temporally and spatially coincides with a place and moment in time under research" [79]. In another study, trends in vegetation cover change were directly related to variations among land-use caused by the genocidal activities in Darfur. The researchers used rainfall data and vegetation indices (NDVI) obtained from MODIS Terra and Satellite Pour l'Observation de la Terre (SPOT) satellites [80]. In this particular case, the remote sensing data were applied to illustrate the temporal changes in vegetation cover (2000–2003).

The availability of satellite-clusters with short revisit periods are providing rapid alert of human right violations associated with genocide. In the Rakhine state, the crisis is between Rakhine Buddhists and ethnic Muslims and several villages belonging to the latter have undergone arson while those belonging to the former are usually standing intact. Satellite analysis was associated with the information on the habitations in Myanmar, which indicates that approximately 392 towns and villages from a total number of 993 were affected (Figure 10). A UNITAR study reported the use of WorldView-3 data to identify destroyed villages in the Northern Myanmar's Rakhine state.

Marx et al. [81,82] developed a methodology for detecting burned villages in the Rakhine state of Myanmar. Burned or damaged villages can be identified by sudden drops in near-infrared reflectance between two observation dates. Village huts with dried plantation roofs have a low NIR reflectance when damaged or destroyed. The researchers utilized the short repeat cycle of Planet's Dove satellites to compare and detect changes in the near-infrared band over rural household structures belonging to adjacent Buddhist and Muslim communities. This enabled them to identify the potential destruction of villages and inform agencies like Human Rights Watch. The information can then also be used for corroborating eyewitness accounts for the destruction by refugees. Madden et al. [83] combined qualitative data collected from personal narratives with the data obtained using the Geographical Information System (GIS) technologies to study mass atrocities in the northern part of Uganda. This study shows the role of GIS for cartographic functions and geo-visualization. The previous case studies demonstrate the usefulness of remote sensing data to track cases of population displacement and genocide, supplementing independent eyewitnesses and other reports.

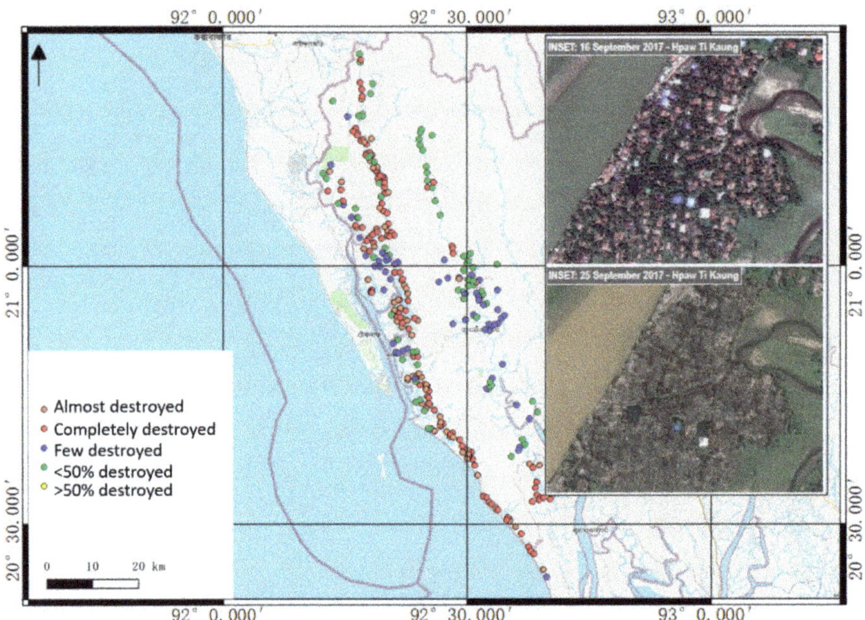

Figure 10. Areas of satellite-detected destroyed or otherwise damaged settlements in Northern Rakhine State in Myanmar. Inset figures showing destroyed villages (Source: UNITAR/UNOSAT).

3.4. Remote Sensing in International Peace Missions

Remote sensing technologies can contribute to various components of the fulfillment of the peace process. GIS can help to provide spatial information about homicides, conflict-related deaths, violence, etc., which can help the implementation of SDGs# 16 i.e., to promote peaceful and inclusive societies for sustainable development, and provide access to justice for all and build effective, accountable and inclusive institutions at all levels [24]. Satellite data can provide valuable evidence in remote or conflict zones where field visits are difficult [71]. Remote sensing data can be an effective tool for resolving running disputes at international borders. By addressing the discerning factors causing the conflict in monitoring the border situation remotely can be done through the deployment of remote sensing techniques. It can be used as the identifiable indicators, with little warning signs before a major crisis erupts [71]. Verifying the agreements put forth in conflict zones is an important aspect of handling international disputes of this kind [22]. Satellites have always been identified with national technical means of verification with increased cooperative missions at an international level that are associated with it. For instance, satellite images have been widely used as the source of verification in decommissioning of facilities, disengagements, storage, and destruction that are vital in the peace process—from negotiating to overseeing implementation to post-conflict peacebuilding processes.

3.5. Applications in Peace and Conflict Areas

Remote sensing can also aid in peace and conflict policy formulation with the identification of exploited natural resources [23]. For example, in the context of the DRC, the extraction of natural minerals, such as concentrated columbite-tantalite and cassiterite, has been discussed [52]. Both these minerals are used to build components of electronic products and have been in demand because of their application-specific competitive pricing in the international market. This work has looked into the possibility of detecting resource-based conflict before they inflate into a violent confrontation.

Gorsevski et al. [48] estimated 2000-armed conflicts that occurred about the same time when the market value of these minerals was at its peak. To monitor the extraction of concentrated columbite-tantalite and cassiterite, a workflow based on object detection and GIS analysis was adopted. Site-activity indicators such as roads, settlements, or rivers were used to narrow down relevant areas for analysis [84]. The object-based multiscale image analysis approach was also able to locate mining areas in North and South Kivu. However, the researchers did not discount the possibility of overlooking small scale mining areas. Through this process, the researchers argued that remote sensing could aid in the development of early warning systems for resource-based conflict by monitoring the extraction of conflict resources that can result in the prevention of conflict escalation and adoption of peace and security policies that comprehensively address the roots of resource-based conflict. The study demonstrated that remote sensing could be highly useful in monitoring resource extraction activities in conflict-affected areas.

Another case where remote sensing was applied for monitoring resource-related conflicts was in the Afghan drug industry. The researchers used remote sensing and population survey data to develop a virtual, rural Afghan population, where poppy is cultivated [15]. The virtual population developed by the researchers was used to model illicit economic activities in the country torn by complex conflicts. A multi-agent model helped to identify the spatial distribution of pertinent agents/actors involved in the Afghan drug industry. This information benefits the monitoring of the trade by relevant peace and security actors. This example shows how qualitative data collected from the narrative description from natives was combined with remote sensing data to build a virtual rural map and, hence, was successfully utilized to accomplish conflict resolution in Afghanistan.

Besides these applications, radar satellites with Synthetic Aperture Radar (SAR) technology and VIIRS nighttime images can identify shipping vessels and the possible nature of their activity. Detection from SAR is based on the principle that the backscatter from the 3-dimensional planar surfaces of a ship along with the overall size can be used to distinguish them into oil-tankers, shipping boats, patrol vessels, and armed naval ships. Large ships or clusters of small-boats also emit light, which makes their detection possible in the day/night band of VIIRS. Recently Elvidge et al. [85] have shown detection of low light emission from small fishing boats (smaller than 19 m in length) by using a combination of nighttime light imagery. By cross-matching these detections with tracking data from Automatic Identification System (AIS) and Vessel Monitoring Systems (VMS), the "dark fishing" vessels can be identified, which have intentionally shut down their tracking devices for a considerable time. This system is used by the Global Fishing Watch to identify illegal fishing in international waters or foreign maritime zones.

The use of chemicals and explosive weapons can have devastating impacts on land, water, and air. It is hard to monitor these effects with the naked eye. Nevertheless, remote sensing helps detect both the short-term and long-term impacts of these weapons. It can help to determine the effects of a crisis on aquatic life and water quality. Besides, it is very pertinent in assessing the geomorphologic impact of the usage of explosive weapons. Remote sensing can also identify changes in soil texture and content as well as changes in the landscape. It can also be used to identify landmines during the post-conflict phase to minimize further damage. Moreover, remote sensing can be used in monitoring air pollution during armed conflict, which has been trivialized for a long time. After the devastating Gulf war in Kuwait, remote sensing was utilized to monitor sand encroachment and land deformation due to the usage of explosive weapons [86].

A case study of forestry conflict in Sumatra, which is a large island in Indonesia, has been presented in Yasmi et al. [87]. The authors discussed how remote sensing techniques have been helpful in conflict crisis resolution between the communities and the associated organizations. The challenges in conflict management have also been presented exhaustively. The management of cross-border conflicts using remote sensing data collected with Google Earth layers is a good example highlighting the importance of remote sensing toward international peace and security [88].

3.6. Remote Sensing and Human Rights

From 2006 onward, the American Association for the Advancement of Science (AAAS) Geospatial Technologies and Human Rights Project has globally aligned with human rights organizations. The remote sensing data has been used for human rights in two distinct ways, namely documentation and analysis, and as scientific evidence in the court of law [89]. In courts, scientific proof has been used to supplement other, more conventional evidence. In the context of the former Yugoslavia's case, the satellite imagery was used to validate the witness accounts of civilian killings by the International Criminal Tribunal (ICT). The Amnesty International [89] has suggested using remote sensing as evidence in courts for bringing the perpetrators to the book [89]. Still, there is a lack of consensus on the admissibility of such evidence in different courts [90]. In addition, this has made it difficult to make generalized conclusions about the consistent use of remote sensing in the field of jurisprudence.

Various remote sensing techniques viz. radiometric, SAR, aerial photography, Light Detection and Ranging (LiDAR), and thermal imaging, etc. play a vital role in various aspects of human rights, such as equality, freedom, education, dignity, prosperity, justice, and speech. Figure 11 depicts the interconnection of remote sensing techniques with the various aspects of human rights. Although remote sensing platforms provide useful data, it can be interpreted differently by different users and algorithms. During a dispute between Nigeria and Cameroon before the International Court of Justice (ICJ), the remote sensing evidence presented before the court had conflicting interpretations [90]. To establish the legitimacy of using satellite data in legal proceedings, it is important to create universal standards for the interpretation and validation of data across different courts worldwide. To date, it has been used on an ad hoc basis by the European Court of Human Rights for the conflict between Russia and Georgia over South Ossetia, the African Commission, and the African Court of Human and People's Rights in the case of forced evictions in Zimbabwe, and the International Criminal Court in the case of human rights violations in Darfur [91].

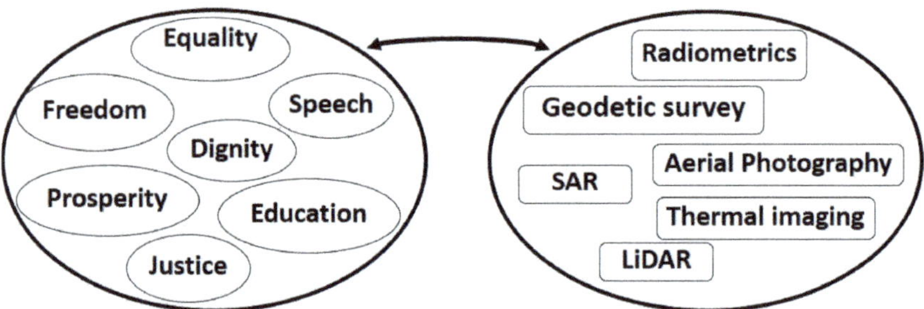

Figure 11. Remote sensing and human rights [15].

Different tools are being used to document and understand human rights violations. For instance, the imagery processed from Afghanistan in 2006–2007 supported the findings of Physicians for Human Rights on the suspicions of mass graves in Northern Afghanistan [92]. On the other hand, the review of different images in Myanmar compared to the reports on attacks against civilians in Karen State by the government in 2006–2007 and 2009 provided proof of destruction and the construction of new military occupation camps [93]. In collaboration with the Amnesty International and the US Holocaust Memorial Museum, AAAS has also looked into evidence of attacks against civilians in Chad and Darfur, Sudan. In 2008, AAAS and Amnesty International in Georgia assessed the damage of the region of Tskhinvali in conflict with Russia. In 2005, the Zimbabwe Lawyers for

Human Rights used remote sensing techniques in assessing the damage caused by the demolition of houses by the government [89].

In the human rights field, remote sensing has opened a new window of opportunity to reach into areas of the world that are not easily accessible either for geographical or political reasons. In these circumstances, remote sensing data serves to supplement and validate reports of human rights abuse and violations coming from the ground. For instance, satellite images confirmed the increased number of prison camps in North Korea over the years [94]. It has also been ingeniously used to create a web resource serving as an open-source of information for human rights violations in places like Pakistan and to identify the location of individuals at risk [95]. These events and attributes suggest that non-governmental actors have been active in using remote sensing techniques to identify and report on human rights violations. There are often three common components of humanitarian remote sensing projects: (1) they usually involve a combination of different actors, including the provider of images, (2) they consist of imagery analysis experts partnering with an advocacy organization, and, (3) typically, the funder is directly involved with the project [96].

Looking at the collaboration networks among various actors and agencies, the Satellite Sentinel Project (SSP) is an important example [97]. The project is funded by the American actor George Clooney's organization Not On Our Watch [98]. The goal of the project is to prevent the return of all-out civil war between South Sudan and Sudan. In the SSP program, a non-governmental organization is working with a private donor with aid through remote sensing for assessing, deterring, and documenting human rights violations [97]. For SSP to accomplish this goal, they have partnered with the imagery provider DigitalGlobe and their imagery analysis experts. DigitalGlobe's high-resolution commercial satellites, known as QuickBird, WorldView-1, and WorldView-2, pass over Sudan and South Sudan in order to understand the impact on civilians. After collecting the satellite images, analysts at DigitalGlobe work with the rights-based group called the Enough Project to analyze imagery and ground sourcing information. If experts detect human rights abuses, then the project releases a report to the press and policymakers to generate a rapid response when it comes to human rights abuses and human security concerns [96].

An example of the SSP using remote sensing to document atrocities is the Mass Atrocity Alert report called *Cover-Up: New Evidence of Three Mass Graves in South Kordofan*. With the analysis of the imagery and eyewitnesses who admitted to seeing the government of Sudan's militia groups dumping white body bags in the area, SSP analysts were able to conclude that the suspected areas were the locations of mass graves. After concluding that these areas were, in fact, efforts by the government of Sudan to cover up large-scale burial operations, SSP immediately sent out *Mass Atrocity Alerts* in hopes of alerting the press and policymakers of the human rights abuses happening in the area [97]. According to Wang et al. [96], the Satellite Sentinel Project [97] is one of the few non-governmental organizations that can afford the high cost of near-real-time imagery from high-resolution commercial satellites. It is not surprising that most human rights organizations and advocacy groups are unable to use remote sensing techniques to uncover and report on mass atrocities in locations where international crimes like genocide and other gross human rights violations are taking place [96].

3.7. Remote Sensing for Disease Control and Prevention

With accelerating global warming and climate change combined with human-animal conflict, there is a rise in communicable and non-communicable diseases. Infectious diseases such as HIV/AIDS, Ebola, SARS, Zika, and COVID-19 posed a challenge to the global peace order [99]. The spread of a disease, whether it is within a developed or a developing country, can have severe security repercussions. A widespread outbreak can shut down a country, causing both fear and an economic downturn. Consequently, improvements in methods by which the government can identify and stem the outbreak of disease early on are necessary. Remote sensing has been suggested as such a method, using different characteristics of diseases to identify either the outbreak or possibilities of risk. It can be used for monitoring, surveying, or conducting a risk analysis of vector-borne diseases [100]. Disease mapping once believed to be a final frontier for remote sensing [101] has come a long way with better spatio-temporal availability of remote sensing platforms and socially sensed datasets [102,103].

Remote sensing has been used during studies done on malaria [102], Lyme disease [100], Rift Valley fever, and the COVID-19 pandemic [101,104–109], among others [101,104,107–109]. However, it should be noted that these studies have only been done post facto, as a way to verify whether different remote sensing indicators matched the actual occurrence on the ground. Subsequently, this means that these methods have never been tested ex-ante, either as an outbreak was occurring or to prevent an outbreak from happening. An example of such a study concerned the spread of Lyme disease in the Northeastern United States [110]. Remote sensing was used in several ways to examine the "map relative tick abundance on residential properties by using Landsat TM-derived indices of vegetation greenness and wetness" [110]. Essentially, this study aims to identify risk factors for Lyme disease exposure in different populated areas. The study extracted information about contiguous forest patches using Landsat TM data. Contiguous forest patches are a favorite location of white-tailed deer, and its illegal transfer from the region. Furthermore, it was examined whether these were areas for human host exposure. Using these factors, as well as examining the suitable "tasseled cap greened and wetness," which measures the greenness and wetness of an area, residential areas harboring abundant tick populations were identified. The researchers could then determine the areas that were most at risk of exposure to the ticks and, therefore, Lyme disease [110].

Another example is the case of cholera in Bangladesh. Using remote sensing, the researchers looked for sediment loads within the Bay of Bengal, which contain nutrients that could support the plankton that could contain *Vibrio cholerae* [110]. The researchers also examined the height, temperature of the sea surface, and the amount of chlorophyll within the water to determine the possibility of there being *Vibrio cholerae* in the Bay (Ibid.). Since this study was done after the event, the different factors showed a positive correlation in the appearance of cholera in the water. By completing this research, it is hoped that remote sensing could be used to examine the existence of cholera within the water before an outbreak happens or within its beginning stages. Both the above studies mentioned above were done after a disease outbreak. Currently, the Japanese Aerospace Exploration Agency (JAXA) operates the Public-health Monitor and Analysis Platform (JMAP) for surveillance and prediction of malaria and cholera in African countries [111]. However, the reliability of remote sensing-based predictions is still unclear for this type of investigation. Furthermore, the hope is that this type of analysis can be expanded to include other water-borne and soil-borne diseases through model analysis to identify warning signs of a disease outbreak.

4. Discussion

Remote sensing applications have come a long way, ever since the testing of the *Transit* (1960), which is the United States Navy's first military navigation satellite. The defense applications have expanded leaps and bounds from intelligence gathering, positioning, and navigation to communications. This technique has become essential in many aspects of international peace and security [112,113]. The costs of using satellite and other imagery data are very high, and, therefore, such techniques are not accessible to all actors playing a role in peacemaking and peacebuilding processes.

Applications of remote sensing techniques have been improved with the advancement of sensor technology and processing algorithms over time. Table 1 shows the major case studies related to remote sensing applications in peace and security in the last two decades. Landsat data has played a vital role and is popular among most of the major case studies presented in Table 1 because of free access since 2008 [71]. There were fewer case studies before 2010 and most of these studies used conventional visual image interpretation techniques using common satellite data. The number of published papers increased after

2010 and used advanced image processing algorithms with advanced sensors. Machine learning algorithms have recently become popular to process satellite data with better resolutions. The use of advanced sensor data with advanced spatio-temporal strengths can provide more accurate information about incidents. This information can help to develop an early warning system to prevent conflicts. However, there is still a need to establish a standard methodology and code of ethics [96] to use these scientific data as a source for international cooperation and international courts of law.

For the operations of law enforcement agencies, the satellite data can be used as a piece of credible evidence in a court of law. However, since there is a paucity of legal criteria for interpretation and admissibility of remote sensing imaging, the legality of its use as uncontested evidence requires refinements in country-specific bye-laws. As indicated, the issues of measurement validity and selection biases require assessment through a multidisciplinary lens. Where the local and standard definitions, classifications, and constructs regarding socio-cultural and historical aspects need to be a combiner. This will strengthen the capacity of legal institutions to respond to human rights violations and bring perpetrators to justice. Furthermore, the field of remote sensing technology must look beyond the courts and tribunals by educating members from other disciplines like sociologists, demographers, and members from law enforcement and judicial agencies, such as judges, prosecutors, and paralegal professions. The value of satellite imaging for human rights and its use as evidence in legal proceedings need to be strengthened across the multidisciplinary dimensions.

Much of the research on remote sensing and its applications in counter-terrorism initiatives are based on data that was mainly derived by the experts in the field. Several computer algorithms are available, which determine what objects are present in satellite imagery. Effective data interpretation is very important in such cases where information is dynamic and dependent on many variables. Still, information gathered through remote sensing can increase transparency in the cases of counter-terrorism. However, many measures should be taken to understand and contextualize data as well as ensure its protection because of the sensitivity of the remote sensing-based information. The existing literature indicates a paucity of papers in which a deeper contextual analysis based on the socio-economic and historical data is conducted.

The use of visual images post facto as evidence of genocide, population displacement, environmental damage, the supplementation of the results with eyewitness and independent reports, and the versatility of the use of data are all examples of how important remote sensing has become in the domain of global peace and security. Though it is not novel for use as military means, it has only recently been incorporated into other areas, such as quantifying conflict zones. More research is needed regarding how to use remote sensing imagery as an early warning tool for conflict prevention. The credibility of remote sensing information and its application in court or by policymakers is still a grey region. In the field of peace and security, the need for strong communication between scientific applications and policymakers is crucial. The *Eyes on Pakistan* project is one of the successful examples of how remote sensing data documenting human rights violations have been used and publicized through an open-source platform accessible to public and government officials [95,114].

Table 1. Major case studies associated with applications of remote sensing data in peace and security.

No.	Authors	Publication Year	Study Area	Remote Sensing Sensors	Methodologies
1	Koch and El-Baz, [86]	1998	Kuwait	Landsat, SPOT	Visual image interpretation
2	Bjorgo [43]	2000	Thailand	Russian KVR-1000 sensor	Visual image interpretation

Table 1. *Cont.*

No.	Authors	Publication Year	Study Area	Remote Sensing Sensors	Methodologies
3	Giada et al. [45]	2003	Tanzania	IKONOS	Supervised, unsupervised image classification
4	Schimmer R. [79]	2006	East Timor	Landsat	Visual image interpretation
5	Schimmer R. [80]	2008	Darfur, Sudan	MODIS, SPOT-vegetation, Climate data	Temporal change in vegetation phenology
6	Prins [53]	2008	Darfur, Sudan	Landsat ETM+	Normalized burn ratio (NBR)
7	Anderson et al. [54]	2008	Rift Valley province, Kenya	MODIS	Active fire detection
8	Madden et al. [83]	2009	Uganda	Landsat, Google Earth	Visual interpretation
9	Schoepfer et al. [84]	2010	The Democratic Republic of the Congo	Rapideye, Geoeye-1	Object-based image classification
10	Gorsevski et al. [48]	2012	South Sudan and Uganda border	Landsat, MODIS, Aerial photographs	Image classification, TCA, disturbance index (DI), NDVI
11	Hagenlocher et al. [59]	2012	Northern Darfur, Sudan	QuickBird	LULC, Object-based image analysis (OBIA)
12	Marx and Loboda [52]	2013	Darfur, Sudan	Landsat	Reflectance, TCA
13	Jiang et al. [49]	2017	Yemen	NPP-VIIRS	Theil-Sen Median Trend Method, Nighttime Light Indexes
14	Casana et al. [60]	2017	Southern Turkey, Syria, and Northern Iraq	High-resolution satellite (DigitalGlobe)	Image interpretation
15	Pech et al. [28]	2017	Goma city, the Democratic Republic of the Congo	Landsat, Worldview-2, topographic maps	Image processing and visual interpretation
16	Sawalhah et al. [39]	2018	Jordan	Landsat 8	Maximum likelihood classification
17	Levin et al. [31]	2018	Arab countries	VIIRS, Flickr photos	Temporal trends in monthly time-series
18	Quinn et al. [46]	2018	NA	NA	Machine learning
19	Hassan et al. [47]	2018	Bangladesh	Sentinel-2A and Sentinel-2B	Random forest classification
20	Marx et al. [82]	2019	Rakhine, Myanmar	PlanetScope	Pixel-based value extraction
21	Levin et al. [57]	2019	World heritage sites	VIIRS, MODIS, Global Terrorism Database	Statistical analysis
22	Prem et al. [50]	2020	Colombia	Landsat	Empirical model
23	Shantnawi et al. [30]	2020	North Jordan	Landsat	Supervised classification and change analysis

Remote sensing techniques have been instrumental in supporting international treaties, monitoring crises, and predicting natural catastrophes. Various United Nations, governmental, and non-governmental organizations have shown a substantial interest in remote sensing technologies for international peace and security. The main purpose of these organizations is to provide scientific information toward international peace and security. The cooperation between various space agencies can provide useful information to decision-makers, helping to establish global peace and security. For an example, the "International Charter Space and Major Disasters" platform provides satellite-based information for relief during humanitarian disasters. Satellite data are made available for rapid response to reduce disaster losses and damages. Advanced near real-time geospatial data can provide useful information to control the conflicts if combined with mobile data [115]. Big data initiatives can be helpful in conflict prevention and uncovering the relationship between conflict dynamics and development goals [116].

This review indicated that there has been significant progress in several technologies (like electronics, sensors, data analysis, and machine learning) being subsumed together with remote sensing applications for better understanding and realization of the project objective. However, the role of disciplines from social sciences and humanities along with remote sensing is still an underexplored territory. The absence of humanities somewhere also limits in understanding the key issues behind major conflicts and violence within a society. The future applications of remote sensing can be made more comprehensive by enriching the foundations with qualitative information from the ground level like issues with race, ethnicity, social structures, economic conditions, historical and anthropological foundations, and so on.

5. Future Perspectives

5.1. Technology Development Perspective

Advances in remote sensing technology rely on the development of improved sensors, imaging devices, and communication networks, among others. Some future perspectives in these areas are listed below.

5.1.1. Use of New Sensors

The remote sensing applications for international peace and security rely on the availability of accurate, reliable, fast, and high-resolution data from sensors. Advancements in sensor technology include the development of miniaturized sensors that reliably and accurately capture data with high spatial and temporal resolution. The terrestrial LiDAR sensors, which have been widely used in 3D city modeling, transport infrastructure mapping, vegetation mapping, and utility surveying, etc. will be effective tools for monitoring international peace and security. Such sensors are capable of collecting a 3D point cloud, which can be processed by state-of-the-art registration methods to reconstruct 3D surfaces. The 3D models are developed from 3D surfaces using computer graphics techniques. The 3D models help trace the location and extent of territories, avoid international border conflicts, and, hence, contribute toward peace and security [117]. The micro-electromechanical system (MEMS) technology [72], 3D sensors, and sensor fusion [118] are some of the notable areas, which are augmenting the remote sensing technology toward its role in international peace and security.

5.1.2. UAV-Based Survey

Unmanned Aerial vehicles (UAVs) have become more sophisticated during the last few decades [119]. The use of UAVs in the world's major military services for international security has been growing enormously. UAVs are being used to survey the battlefield remotely with a bundle of sensors mounted onto the aerial vehicle, giving adequate image resolution. An ensemble of tiny UAVs working with each other in collaboration is another technological advancement that can be used for remote sensing-based surveys for international peace and security applications. UAVs can be used to improve border control by monitoring large areas of border quickly and more efficiently than conventional means of border patrolling [120,121].

5.1.3. Internet of Things (IoT) Based Survey

The emergence of distributed computing has revolutionized the world across various domains. The use of the Internet of Things (IoT) in remote sensing applications for international peace and security is gaining popularity, enabled by the facility of cost-effective and reliable network services. The mitigation strategies, challenges, and risks of using IoT-based remote sensing for international peace and security have been disheveled [122].

5.1.4. Visual Inertial System

The role of the machine vision in inertial navigation systems for robotic vehicles dates back to the 1960s where cameras mounted on robotic vehicles were used for industrial inspection. The conventional cameras were replaced by 3D cameras, which were capable of capturing depth information along with intensity variation in the scene. The depth information is beneficial for the localization of objects in a scene and also overcomes the limitations of variation, in contrast to ambient light and parallax. However, in such systems, inertial drift is an issue as it makes the registration of consecutive scans erroneous. Leica Geosystems came up with a solution to overcome the problem of inertial drift with Visual Inertial System (VIS) technology. The technology uses five cameras along with an inertial measurement unit (IMU) to introduce a delta pose between two consecutive scans in real-time. It helps in the fusion of huge amounts of visual information very accurately, which can be used by various remote sensing applications for international peace and security [123].

5.1.5. Simultaneous Localization and Mapping

Simultaneous Localization and Mapping, commonly known as SLAM, is a computational technique that builds a map of an unknown environment along with tracking the position or movement of the camera in the environment. SLAM uses various kinds of sensors such as grayscale cameras, RGB cameras, radar, LiDAR, time-of-flight cameras, and monocular as l as stereo cameras to capture the scene. The feature points are extracted from the captured images using feature extraction and description algorithms, which are used to estimate the movement of features from one frame to another, and, hence, the movement of the camera. Using SLAM technology, the movement of objects can be accurately articulated in an environment, which finds enormous applications in monitoring international peace and security [124].

5.2. Conflict Management Perspective

By combining local socio-economic and historical information with remote sensing techniques can improve measurement validity and selection bias. This can potentially result in making it a major tool for conflict prevention, peacekeeping, peacemaking, and peace enforcement. It can provide the policymakers with better foresight or the future perspectives in some and/or the areas listed below.

5.2.1. Conflict Prevention

In the future, remote sensing should more actively be used to identify and monitor resources that might be the source of conflicts. Letouzé et al. [116] discussed the role of big data including geospatial information in conflict prevention. However, specific international cooperation and policy are required to avoid breaching national boundaries during such applications.

More research is needed on how to effectively use remote sensing techniques to prevent conflicts from happening. This may be an analysis of the regions known for recurring conflicts and identification of regions most likely to be hampered by future challenges of climate change. Furthermore, geographical indicators of possible sites with militant activity should be established [125].

5.2.2. Peacekeeping

A cluster of remote sensing satellites launched by the USA and Russia keeps the planet on a constant mode of surveillance. The sovereign nations often question this as espionage. However, the utility of remote sensing satellites for international peacekeeping efforts is commendable. To the extent of which international security norms and regulations permit, remote sensing research in areas with known militant activity should be enhanced through the improvement of data available for such regions and the creation of databases of similar research. In the future, remote sensing should also aid strategic positioning and monitoring of peace support operations (PSO) and disarmament, demobilization, and reintegration (DDR) [126].

5.2.3. Peacemaking and Peace Enforcement

The role of remote sensing in peacemaking and peace enforcement in many parts of the world has been pivotal for decades. Further case studies and examples of systematic exchange and sharing of conflict-related remote sensing data among both developing and the developed world can be helpful in the resolution of ongoing conflicts. Moreover, remote sensing can support and strengthen civil-military cooperation. With an increase in the number of satellites orbiting the earth, a huge amount of earth data has a significant potential for its use in peacemaking and peace enforcement. The synergistic approach of civilian science and remote sensing can help consolidate the environmental safety standards among the troubled regions [115].

5.2.4. Peacebuilding

Remote sensing will useful in post-conflict community rehabilitation like zoning, resource accessibility, and community planning. With the advancement in image processing algorithms, satellite imagery can be used in a better way than it was used a few decades ago for peacebuilding. Besides that, community project mapping by remote sensing should be considered to avoid project fatigue and duplication. In conflict zones, security aspects along with the political and social aspects influences the data collection. Here, remote sensing technology can be an enabler for exploring and assessing the risks to the ecosystem during uncertain and violent situations [115].

6. Conclusions

This review presents the applications of remote sensing technologies in supporting international peace and security missions and provided a detailed overview of each application with various case study examples from literature. In particular, we focused on earth observation for monitoring armed conflicts, human rights, tracking genocides, impact on environmental aspects and disease control preventions, and identified different interactions among these 4-level processes. This study noticed that the current research on peace and security issues is in the initial phase. Due to this reason, it is affected with incommensurate data for correspondence limiting granular information essential for characterization. Nevertheless, some recent studies have demonstrated the usefulness of high-resolution satellites in the area of refugee rehabilitation, quantifying conflict zones, and border security controls. With Kurmuk, Sudan as a case example, we also showed that the synergistic use of multi-source data from satellites, secondary sources, and ground-data could help in developing early warning systems to minimize civilian casualties in war zones. With the increased availability of higher spatial resolution images in the future, there is significant potential in space-borne technologies for early-warning of attacks, particularly for knowing the potential affected sites of targets.

On the other hand, near-real-time data provides useful information to rapid-response in relief activities and crisis management. Additionally, recent improvements in sensor capabilities make it an essential and relevant tool for documenting and preventing armed conflicts. The salient understanding from this study is that there is an ample requirement for disseminating information for ensuring prompt and efficient humanitarian responses at a global level. In particular, this is evident from Yemen civil conflicts. The Syrian crisis has the Sudan-South Sudan border issues and Rohingya crisis, etc. However, the initiatives on such works remain elusive with a lack of co-ordinate ion and setting between the geospatial community and relief organizations. It is hoped that initiatives such as the UNITAR/UNOSAT by United Nation, 'Not on Our Watch' funded the Satellite Sentinel Project (SSP), as a partnership among the Enough Project and DigitalGlobe, and the continuing upward trend in scientific work at the regional level (e.g., Witmer, [71]) will help to minimize the impact of crisis and disasters. Considerable capacity building for data processing, knowledge dissemination, and resource sharing is, therefore, required. Automated object-based analysis, data fusion algorithms, and tools to process Big Data in remote sensing are the way forward in terms of building a peaceful and secure world. This study concludes that remote sensing as a surveillance tool has immense potential to guard our planet's environment, peace, and security, provided it is used actively and transparently. The trajectory, intensity, and orientational control of those applications will depend on how the major stakeholders from academia, governance, and civil society converge to advance the foundational principles for conflict prevention, detection, and related human rights issues.

Author Contributions: Conceptualization, R.A., A.K. (Asma Kouser), D.S., and A.K. (Ashwani Kumar); Methodology, R.A., A.K. (Asma Kouser), D.S., A.P.Y. and A.K. (Ashwani Kumar); Formal analysis, R.A., A.K. (Ashwani Kumar), D.S., and P.M. Resources, R.A., A.K. (Asma Kouser), A.K. (Ashwani Kumar), D.S., P.M., A.G., A.P.Y., P.K., B.A.J., R.D., N.S., and A.B.R. Writing—original draft preparation, R.A., A.K. (Asma Kouser), D.S., A.P.Y. and A.K. (Ashwani Kumar); Writing—review and editing, R.A., A.K. (Asma Kouser), A.K. (Ashwani Kumar), D.S., P.M., A.G., A.P.Y., P.K., B.A.J., R.D., N.S., and A.B.R. All authors have read and agreed to the published version of the manuscript.

Funding: The publication fee was supported by the Publication Support Grants of Hokkaido University, Japan.

Data Availability Statement: The datasets generated in the current study are available from the corresponding author on request.

Acknowledgments: The authors are thankful to Planet Labs and Google Earth for providing satellite data and UNITAR/UNOSAT for reproducing the images. This work is supported by the Office for Developing Future Research Leaders (L-Station), SOUSEI Support Program for Young Researcher Hokkaido University, and Kurata Hitachi Foundation, Japan. The authors would like to thank Global Land Program-Japan Nodal Office and Young Sustainability Symposium (YSS-2020) for their support and collaboration. We are thankful to Rear Admiral Sanatan Kulshrestha (Former Director General Naval Armament Inspection-DGNIA, India) and Avinash Pandey (Asian Human Rights Commission-AHRC, Hong Kong) for their comments and suggestions. The authors extend sincere gratitude to the editor and anonymous reviewers for their constructive comments and valuable suggestions.

Conflicts of Interest: The authors declare no conflict of interest.

References

1. Lessnoff, M. *Social Contract*; Macmillan Education: London, UK, 1986; ISBN 978-0-333-36791-9.
2. Stokes, D.E. *Pasteur's Quadrant: Basic Science and Technological Innovation*; Brookings Institution Press: Washington, DC, USA, 2011; ISBN 978-0-8157-1907-6.
3. Barbier, E. *Natural Resources and Economic Development*; Cambridge University Press: Cambridge, UK, 2007; ISBN 978-0-521-70651-3.
4. The World Bank. *World Development Report 1993: Investing in Health*; The World Bank Group: Washington, DC, USA, 1993; Volume 1.
5. Kääb, A. Remote sensing of mountain environment. In Proceedings of the Projecting Global Change Impacts and Sustainable Land Use and Natural Resources Management in Mountain Biosphere Reserves, Pinos Genil, Spain, 14 March 2005; UNESCO: Pinos Genil, Spain; p. 92.
6. Campbell, J.B. Origins of Aerial Photographic Interpretation, U.S. Army, 1916 to 1918. *Photogramm. Eng. Remote. Sens.* **2008**, *74*, 77–93. [CrossRef]

7. Stichelbaut, B. The application of First World War aerial photography to archaeology: The Belgian images. *Antiquity* **2006**, *80*, 161–172. [CrossRef]
8. Thomas, R.; Joseph, S.S. Emendation of undesirable attack on multiparty data sharing with anonymous Id assignment using AIDA algorithm. *Innov. Syst. Des. Eng.* **2015**, *6*, 6.
9. Lee, R.J.; Steele, S.L. Military use of satellite communications, remote sensing, and global positioning systems in the war on terror. *J. Air L. Com.* **2014**, *79*, 69.
10. Jasani, B.; Pesaresi, M.; Schneiderbauer, S.; Zeug, G. *Remote Sensing from Space: Supporting International Peace and Security*; Springer Science & Business Media: Berlin/Heidelberg, Germany, 2009; ISBN 1-4020-8484-6.
11. Dashora, A.; Lohani, B.; Malik, J.N. A repository of earth resource information—CORONA satellite programme. *Curr. Sci.* **2007**, *92*, 926–932.
12. Roy, S. The Palestinian-Israeli Conflict and Palestinian socioeconomic decline: A place denied. *Int. J. Politics Cult. Soc.* **2003**, *17*, 365–403. [CrossRef]
13. Singhal, A. In Search of Military GIS. *Geospat. World* **2009**. Available online: https://www.geospatialworld.net/article/in-search-of-military-gis/ (accessed on 5 May 2020).
14. United Nations (Ed.) *Handbook on Geospatial Infrastructure in Support of Census Activities*; Studies in Methods, Series F; United Nations: New York, NY, USA, 2009; ISBN 978-92-1-161527-2.
15. Rizi, S.M.M.; Geller, A. *Merging Remote Sensing Data and Population Surveys in Large, Empirical Multiagent Models: The Case of the Afghan Drug Industry*; George Mason University: Fairfax, VA, USA, 2010; p. 8.
16. Brown, I. Assessing eco-scarcity as a cause of the outbreak of conflict in Darfur: A remote sensing approach. *Int. J. Remote. Sens.* **2010**, *31*, 2513–2520. [CrossRef]
17. Soytong, P.; Perera, R. Use of GIS tools for environmental conflict resolution at map ta phut industrial zone in Thailand. *Sustainability* **2014**, *6*, 2435–2458. [CrossRef]
18. Hay, S.I. Remote sensing and disease control: Past, present and future. *Trans. R. Soc. Trop. Med. Hyg.* **1997**, *91*, 105–106. [CrossRef]
19. Sulik, J.J.; Edwards, S. Feature extraction for Darfur: Geospatial applications in the documentation of human rights abuses. *Int. J. Remote Sens.* **2010**, *31*, 2521–2533. [CrossRef]
20. *Annual Report of the United Nations High Commissioner for Human Rights and Reports of the Office of the High Commissioner and the Secretary-General*; Report on the Prevention of Genocide; Human Rights Council: Geneva, Switzerland, 2019.
21. UNEP (United Nations Environment Programme). *The Environmental Food Crisis: The Environment's Role in Averting Future Food Crises: A UNEP Rapid Response Assessment*; UNEP/Earthprint: Nairobi, Kenya, 2009; ISBN 978-82-7701-054-0.
22. Hettling, J.K. The use of remote sensing satellites for verification in international law. *Space Policy* **2003**, *19*, 33–39. [CrossRef]
23. Avtar, R.; Sahu, N.; Aggarwal, A.K.; Chakraborty, S.; Kharrazi, A.; Yunus, A.P.; Dou, J.; Kurniawan, T.A. Exploring renewable energy resources using remote sensing and GIS—A review. *Resources* **2019**, *8*, 149. [CrossRef]
24. Estoque, R.C. A review of the sustainability concept and the state of SDG monitoring using remote sensing. *Remote Sens.* **2020**, *12*, 1770. [CrossRef]
25. Avtar, R.; Aggarwal, R.; Kharrazi, A.; Kumar, P.; Kurniawan, T.A. Utilizing geospatial information to implement SDGs and monitor their Progress. *Environ. Monit. Assess.* **2020**, *192*, 35. [CrossRef]
26. Jasani, B.; Larsson, C. Security implications of remote sensing. *Space Policy* **1988**, *4*, 46–59. [CrossRef]
27. Abramson, S.F.; Carter, D.B. The historical origins of territorial disputes. *Am. Political Sci. Rev.* **2016**, *110*, 675–698. [CrossRef]
28. Pech, L.; Lakes, T. The impact of armed conflict and forced migration on urban expansion in Goma: Introduction to a simple method of satellite-imagery analysis as a complement to field research. *Appl. Geogr.* **2017**, *88*, 161–173. [CrossRef]
29. Sawalhah, M.N.; Al-Kofahi, S.D.; Othman, Y.A.; Cibils, A.F. Assessing rangeland cover conversion in Jordan after the Arab spring using a remote sensing approach. *J. Arid. Environ.* **2018**, *157*, 97–102. [CrossRef]
30. Shatnawi, N.; Weidner, U.; Hinz, S. Monitoring urban expansion as a result of refugee fluxes in north jordan using remote sensing techniques. *J. Urban. Plan. Dev.* **2020**, *146*, 04020026. [CrossRef]
31. Levin, N.; Ali, S.H.; Crandall, D. Utilizing remote sensing and big data to quantify conflict intensity: The Arab Spring as a case study. *Appl. Geogr.* **2018**, *94*, 1–17. [CrossRef]
32. Weidmann, N.B.; Rød, J.K.; Cederman, L.-E. Representing ethnic groups in space: A new dataset. *J. Peace Res.* **2010**, *47*, 491–499. [CrossRef]
33. Avtar, R.; Komolafe, A.A.; Kouser, A.; Singh, D.; Yunus, A.P.; Dou, J.; Kumar, P.; Das Gupta, R.; Johnson, B.A.; Minh, H.V.T.; et al. Assessing sustainable development prospects through remote sensing: A review. *Remote. Sens. Appl. Soc. Environ.* **2020**, *20*, 100402. [CrossRef]
34. Starr, H. Opportunity, willingness and geographic information systems (GIS): Reconceptualizing borders in international relations. *Politi. Geogr.* **2002**, *21*, 243–261. [CrossRef]
35. Branch, J. Geographic information systems (GIS) in international relations. *Int. Organ.* **2016**, *70*, 845–869. [CrossRef]
36. Baldwin, D.A. The concept of security. *Rev. Int. Stud.* **1997**, *23*, 5–26. [CrossRef]
37. Anderson, R. A Definition of peace. *Peace Confl.* **2004**, *10*, 101. [CrossRef]
38. White, N.D. *Keeping the Peace: The United Nations and the Maintenance of International Peace and Security*; Manchester University Press: Manchester, UK, 1997.

39. Galtung, J.; Fischer, D. Positive and negative peace. In *Johan Galtung; SpringerBriefs on Pioneers in Science and Practice*; Springer: Berlin/Heidelberg, Germany, 2013; Volume 5, pp. 173–178. ISBN 978-3-642-32480-2.

40. Thomas, C.; Wilkin, P. *Globalization, Human Security, and the African Experience*; Lynne Rienner Publishers: Boulder, CO, USA, 1999; ISBN 1-55587-699-4.

41. Cederman, L.-E.; Gleditsch, K.S.; Buhaug, H. *Inequality, Grievances, and Civil War*; Cambridge Studies in Contentious Politics; Cambridge University Press: New York, NY, USA, 2013; ISBN 978-1-107-01742-9.

42. Dalen, Ø.; Johannessen, O.; Bjørgo, E.; Babiker, M.; Andersen, G. *Use of ERS SAR Imagery in Refugee Relief*; ESTEC: Nordwijk, The Netherlands, 2000.

43. Bjorgo, E. Using very high spatial resolution multispectral satellite sensor imagery to monitor refugee camps. *Int. J. Remote Sens.* **2000**, *21*, 611–616. [CrossRef]

44. Dysart, M.D. *Remote Sensing and Mass Migration Policy Development*; Air War College, Air University: Montgomery, AL, USA, 2012.

45. Giada, S.; De Groeve, T.; Ehrlich, D.; Soille, P. Information extraction from very high resolution satellite imagery over Lukole refugee camp, Tanzania. *Int. J. Remote Sens.* **2003**, *24*, 4251–4266. [CrossRef]

46. Quinn, J.A.; Nyhan, M.M.; Navarro, C.; Coluccia, D.; Bromley, L.; Luengo-Oroz, M. Humanitarian applications of machine learning with remote-sensing data: Review and case study in refugee settlement mapping. *Philos. Trans. R. Soc. A Math. Phys. Eng. Sci.* **2018**, *376*, 20170363. [CrossRef]

47. Hassan, M.M.; Smith, A.C.; Walker, K.; Rahman, M.K.; Southworth, J. Rohingya refugee crisis and forest cover change in Teknaf, Bangladesh. *Remote Sens.* **2018**, *10*, 689. [CrossRef]

48. Gorsevski, V.; Kasischke, E.S.; Dempewolf, J.; Loboda, T.; Grossmann, F. Analysis of the Impacts of armed conflict on the Eastern Afromontane forest region on the South Sudan—Uganda border using multitemporal Landsat imagery. *Remote Sens. Environ.* **2012**, *118*, 10–20. [CrossRef]

49. Jiang, W.; He, G.; Long, T.; Liu, H. Ongoing conflict makes yemen dark: From the perspective of nighttime light. *Remote Sens.* **2017**, *9*, 798. [CrossRef]

50. Prem, M.; Saavedra, S.; Vargas, J.F. End-of-conflict deforestation: Evidence from Colombia's peace agreement. *World Dev.* **2020**, *129*, 104852. [CrossRef]

51. Bromley, D.W. *Environment and Economy: Property Rights and Public Policy*; Basil Blackwell Ltd., University of Wisconsin-Madison: Madison, WI, USA, 1991.

52. Marx, A.; Loboda, T. Landsat-based early warning system to detect the destruction of villages in Darfur, Sudan. *Remote Sens. Environ.* **2013**, *136*, 126–134. [CrossRef]

53. Prins, E. Use of low cost Landsat ETM+ to spot burnt villages in Darfur, Sudan. *Int. J. Remote Sens.* **2007**, *29*, 1207–1214. [CrossRef]

54. Anderson, D.M.; Lochery, E. Violence and exodus in Kenya's rift valley, 2008: Predictable and preventable? *J. East. Afr. Stud.* **2008**, *2*, 328–343. [CrossRef]

55. Lee, C.; De Vries, W.T. Bridging the semantic gap between land tenure and EO data: Conceptual and methodological underpinnings for a geospatially informed analysis. *Remote Sens.* **2020**, *12*, 255. [CrossRef]

56. Avtar, R.; Singh, D.; Umarhadi, D.A.; Yunus, A.P.; Misra, P.; Desai, P.N.; Kouser, A.; Kurniawan, T.A.; Phanindra, K. Impact of COVID-19 lockdown on the fisheries sector: A case study from three harbors in western India. *Remote Sens.* **2021**, *13*, 183. [CrossRef]

57. Levin, N.; Ali, S.H.; Crandall, D.; Kark, S. World Heritage in danger: Big data and remote sensing can help protect sites in conflict zones. *Glob. Environ. Chang.* **2019**, *55*, 97–104. [CrossRef]

58. Knoth, C. Combining automatic and manual image analysis in a web-mapping application for collaborative conflict damage assessment. *Appl. Geogr.* **2018**, *97*, 25–34. [CrossRef]

59. Hagenlocher, M.; Lang, S.; Tiede, D. Integrated assessment of the environmental impact of an IDP camp in Sudan based on very high resolution multi-temporal satellite imagery. *Remote Sens. Environ.* **2012**, *126*, 27–38. [CrossRef]

60. Casana, J.; Laugier, E.J. Satellite imagery-based monitoring of archaeological site damage in the Syrian civil war. *PLoS ONE* **2017**, *12*, e0188589. [CrossRef]

61. Xu, Y.; Knudby, A.; Côté-Lussier, C. Mapping ambient light at night using field observations and high-resolution remote sensing imagery for studies of urban environments. *Build. Environ.* **2018**, *145*, 104–114. [CrossRef]

62. Do, Q.-T.; Shapiro, J.N.; Elvidge, C.D.; Abdel-Jelil, M.; Ahn, D.P.; Baugh, K.E.; Hansen-Lewis, J.; Zhizhin, M.; Bazilian, M.D. Terrorism, geopolitics, and oil security: Using remote sensing to estimate oil production of the Islamic State. *Energy Res. Soc. Sci.* **2018**, *44*, 411–418. [CrossRef]

63. Benítez, P.C.; McCallum, I.; Obersteiner, M.; Yamagata, Y. Global potential for carbon sequestration: Geographical distribution, country risk and policy implications. *Ecol. Econ.* **2007**, *60*, 572–583. [CrossRef]

64. Hu, Z.D.; Ge, Y.J. Geopolitical Energy Security Evaluation Method and Its Application Based on Politics of Scale. *Int. Arch. Photogramm. Remote. Sens. Spat. Inf. Sci.* **2013**, *XL-4/W3*, 79–81. [CrossRef]

65. Moisio, S.; Harle, V. The limits of geopolitical remote sensing. *Eurasian Geogr. Econ.* **2006**, *47*, 204–210. [CrossRef]

66. Colomina, I.; Molina, P. Unmanned aerial systems for photogrammetry and remote sensing: A review. *ISPRS J. Photogramm. Remote Sens.* **2014**, *92*, 79–97. [CrossRef]

67. Antonsich, M. In Defense of "Geopolitical Remote Sensing": Reply to Moisio and Harle. *Eurasian Geogr. Econ.* **2006**, *47*, 211–215. [CrossRef]
68. Keeley, J.F. The use of commercial satellite imagery and Canadian security needs. In *Commercial Satellite Imagery and United Nations Peacekeeping: A View from Above*; Keeley, J.F., Huebert, R., Eds.; Ashgate: Aldershot, UK, 2004; pp. 87–98.
69. Melillos, G.; Themistocleous, K.; Papadavid, G.; Agapiou, A.; Prodromou, M.; Michaelides, S.; Hadjimitsis, D.G. *Integrated Use of Field Spectroscopy and Satellite Remote Sensing for Defence and Security Applications in Cyprus*; Themistocleous, K., Hadjimitsis, D.G., Michaelides, S., Papadavid, G., Eds.; SPIE: Paphos, Cyprus, 2016; p. 96880F.
70. Al-Rodhan, N.R.F. *Meta-Geopolitics of Outer Space: An Analysis of Space Power, Security and Governance*; Palgrave Macmillan: London, UK, 2012; ISBN 978-1-349-33967-9.
71. Witmer, F.D. Remote sensing of violent conflict: Eyes from above. *Int. J. Remote Sens.* **2015**, *36*, 2326–2352. [CrossRef]
72. Zhang, X.; Xie, J.; Li, C.; Xu, R.; Zhang, Y.; Liu, S.; Wang, J. MEMS-based super-resolution remote sensing system using compressive sensing. *Opt. Commun.* **2018**, *426*, 410–417. [CrossRef]
73. Cantelli, L.; Mangiameli, M.; Melita, C.D.; Muscato, G. UAV/UGV Cooperation for Surveying Operations in Humanitarian Demining. In Proceedings of the 2013 IEEE International Symposium on Safety, Security, and Rescue Robotics (SSRR), Linkoping, Sweden, 21–26 October 2013; IEEE: Linkoping, Sweden; pp. 1–6.
74. Ibrahim, A.A.; Haruna, A. The united nations and the challenges of global peace, security and development. *Res. Humanit. Soc. Sci.* **2014**, *4*.
75. Satellite Sentinel Project (SSP). *Troops in the Demilitarized Zone Confirmation of Violations by Sudan and South Sudan*; Satellite Sentinel Project (SSP); Harvard Humanitarian Initiative: Cambridge, MA, USA, 2013; p. 23.
76. Satellite Sentinel Project (SSP). *Radius of Operations: Sudan Increases Air Attack Capacity*; Satellite Sentinel Project (SSP); Harvard Humanitarian Initiative: Cambridge, MA, USA, 2011; p. 8.
77. Som, V. Chinese troops withdraw 2 Km in Galwan Valley, show new satellite images. *Indian Express* **2020**.
78. Adu-Amanfoh, F. *The Roles of Peace and Security, Political Leadership, and Entrepreneurship in the Socio-Economic Development of Emerging Countries: A Compendium of Lessons Learnt from Sub-Saharan Africa*; AuthorHouse: Bloomington, IN, USA, 2014; ISBN 978-1-4918-9160-5.
79. Russell, S. *Violence by Fire in East Timor, September 8, 1999; Crimes against Humanity*; Yale University: New Haven, CT, USA, 2006; p. 17.
80. Schimmer, R. *Tracking the Genocide in Darfur: Population Displacement as Recorded by Remote Sensing*; Yale Center for International and Area Studies New Haven: New Haven, CT, USA, 2008; p. 52.
81. Marx, A.; Goward, S. Remote sensing in human rights and international humanitarian law monitoring: Concepts and methods. *Geogr. Rev.* **2013**, *103*, 100–111. [CrossRef]
82. Marx, A.; Windisch, R.; Kim, J.S. Detecting village burnings with high-cadence smallsats: A case-study in the Rakhine State of Myanmar. *Remote. Sens. Appl. Soc. Environ.* **2019**, *14*, 119–125. [CrossRef]
83. Madden, M.; Ross, A. Genocide and GIScience: Integrating personal narratives and geographic information science to study human rights. *Prof. Geogr.* **2009**, *61*, 508–526. [CrossRef]
84. Schoepfer, E.; Kranz, O.; Addink, E.; Coillie, F. Monitoring natural resources in conflict using an object-based multi-scale image analysis approach. *Proc. GEOBIA* **2010**.
85. Elvidge, C.D.; Zhizhin, M.; Baugh, K.E.; Hsu, F.-C. Automatic boat identification system for VIIRS low light imaging data. *Remote. Sens.* **2015**, *7*, 3020–3036. [CrossRef]
86. Koch, M.; El-Baz, F. Identifying the effects of the gulf war on the geomorphic features of kuwait by remote sensing and GIS. *Photogramm. Eng. Remote Sens.* **1998**, *64*, 739–746.
87. Yasmi, Y.; Schanz, H. Managing conflict escalation in forestry: Logging versus local community interests in Baru Pelepat village, Sumatra, Indonesia. *Int. J. Biodivers. Sci. Ecosyst. Serv. Manag.* **2010**, *6*, 43–51. [CrossRef]
88. Luo, L.; Wang, X.; Guo, H.; Lasaponara, R.; Shi, P.; Bachagha, N.; Li, L.; Yao, Y.; Masini, N.; Chen, F.; et al. Google Earth as a powerful tool for archaeological and cultural heritage applications: A review. *Remote. Sens.* **2018**, *10*, 1558. [CrossRef]
89. Amnesty International Nigeria. Gruesome Footage Implicates Military in War Crimes. Available online: https://www.amnesty.org/en/latest/news/2014/08/nigeria-gruesome-footage-implicates-military-war-crimes/ (accessed on 19 August 2019).
90. Nunez, A.C.N. *Admissibility of Remote Sensing Evidence before International and Regional Tribunals; Innovations in Human Rights Monitoring*; Amnesty International: New York, NY, USA, 2012.
91. American Association for the Advancement for Science (AAAS). Geospatial Technologies and Human Rights. Available online: https://www.aaas.org/programs/geospatial-technologies (accessed on 15 August 2019).
92. American Association for the Advancement for Science (AAAS). *Satellite Imagery and Possible Mass Graves in Sheberghan, Afghanistan*; Scientific Responsibility, Human Rights & Law Program; American Association for the Advancement of Science: Washington, DC, USA, 2020.
93. American Association for the Advancement for Science (AAAS). *High-Resolution Satellite Imagery and the Conflict in Eastern Burma*; Scientific Responsibility, Human Rights & Law Program; American Association for the Advancement of Science: Washington, DC, USA, 2020.

94. American Association for the Advancement for Science (AAAS). *Geospatial Technologies and Human Rights—North Korea's Prison Camps Case Study Summary*; Scientific Responsibility, Human Rights & Law Program; American Association for the Advancement of Science: Washington, DC, USA, 2020.

95. American Association for the Advancement for Science (AAAS). *Geovisualization for Mapping Human Rights Incidents in Northwestern Pakistan*; Scientific Responsibility, Human Rights & Law Program; American Association for the Advancement of Science: Washington, DC, USA, 2020.

96. Wang, B.Y.; Raymond, N.; Gould, G.; Baker, I. Problems from hell, solution in the heavens?: Identifying obstacles and opportunities for employing geospatial technologies to document and mitigate mass atrocities. *Stability: Int. J. Secur. Dev.* **2013**, *2*, 18. [CrossRef]

97. Satellite Sentinel Project (SSP). *Cover-Up: New Evidence of Three Mass Graves in South Kordofan*; Harvard Humanitarian Initiative: Cambridge, MA, USA, 2011; p. 10.

98. Sandalinas, J. Satellite imagery and its use as evidence in the proceedings of the international criminal court. *ZLW* **2015**, *64*, 666.

99. Pavone, I.R. Infectious diseases as a new threat to international peace and security: The security council and the securitization of health. *Fachinf. Int. Interdiszip. Rechtsforschung Völkerrechtsblog* **2016**. [CrossRef]

100. Dister, S.W.; Beck, L.R.; Wood, B.L.; Falco, R.; Fish, D. The use of GIS and remote sensing technologies in a landscape approach to the study of Lyme disease transmission risk. *Proc. GIS* **1993**, *93*, 15–18.

101. Bhunia, G.S.; Shit, P.K. Introduction to geoinformatics in public health. In *Geospatial Analysis of Public Health*; Bhunia, G.S., Shit, P.K., Eds.; Springer International Publishing: Cham, Switzerland, 2019; pp. 1–27. ISBN 978-3-030-01680-7.

102. Beck, L.R.; Rodriguez, M.H.; Dister, S.W.; Rodriguez, A.D.; Rejmankova, E.; Ulloa, A.; Meza, R.A.; Roberts, D.R.; Paris, J.F.; Spanner, M.A.; et al. Remote sensing as a landscape epidemiologic tool to identify villages at high risk for malaria transmission. *Am. J. Trop. Med. Hyg.* **1994**, *51*, 271–280. [CrossRef] [PubMed]

103. Dlamini, S.N.; Beloconi, A.; Mabaso, S.; Vounatsou, P.; Impouma, B.; Fall, I.S. Review of remotely sensed data products for disease mapping and epidemiology. *Remote. Sens. Appl. Soc. Environ.* **2019**, *14*, 108–118. [CrossRef]

104. Kotchi, S.O.; Bouchard, C.; Ludwig, A.; Rees, E.E.; Brazeau, S. Using Earth observation images to inform risk assessment and mapping of climate change-related infectious diseases. *Can. Commun. Dis. Rep.* **2019**, *45*, 133–142. [CrossRef]

105. Avtar, R.; Kumar, P.; Supe, H.; Dou, J.; Sahu, N.; Mishra, B.K.; Yunus, A.P. Did the COVID-19 lockdown-induced hydrological residence time intensify the primary productivity in lakes? Observational results based on satellite remote sensing. *Water* **2020**, *12*, 2573. [CrossRef]

106. Avtar, R.; Herath, S.; Saito, O.; Gera, W.; Singh, G.; Mishra, B.; Takeuchi, K. Application of remote sensing techniques toward the role of traditional water bodies with respect to vegetation conditions. *Environ. Dev. Sustain.* **2014**, *16*, 995–1011. [CrossRef]

107. Caballero-Leiva, I.; Marrero Betancort, N.; Rodriguez Betancor, J.J.; Rodríguez Esparragón, D.; Marcello Ruiz, F.J. An Approximation to the Relationship Between Climatic Variables Obtained Through Remote Satellite Sensors and Hospital Admissions: A Case Study on Gran Canaria Island. *Sens. Transducers* **2019**, *238*, 80–86.

108. Bhunia, G.S.; Shit, P.K. Exploring ecology and associated disease pattern. In *Geospatial Analysis of Public Health*; Bhunia, G.S., Shit, P.K., Eds.; Springer International Publishing: Cham, Switzerland, 2019; pp. 139–198. ISBN 978-3-030-01680-7.

109. Racault, M.F.; Abdulaziz, A.; George, G.; Menon, N.; Punathil, M.; McConville, K.; Loveday, B.; Platt, T.; Sathyendranath, S.; Vijayan, V. Environmental reservoirs of vibrio cholerae: Challenges and opportunities for ocean-color remote sensing. *Remote Sens.* **2019**, *11*, 2763. [CrossRef]

110. Beck, L. Remote sensing and human health: New sensors and new opportunities. *Emerg. Infect. Dis.* **2000**, *6*, 217–227. [CrossRef]

111. Igarashi, T.; Kuze, A.; Sobue, S.; Yamamoto, A.; Yamamoto, K.; Oyoshi, K.; Imaoka, K.; Fukuda, T. Japan's efforts to promote global health using satellite remote sensing data from the Japan Aerospace Exploration Agency for prediction of infectious diseases and air quality. *Geospat. Health* **2014**, *8*, 603–610. [CrossRef]

112. Jamieson, P.D. *Lucrative Targets: The U.S. Air Force in the Kuwaiti Theater of Operations*; The USAF in the Persian Gulf War; Air Force History and Museums Program: Washington, DC, USA, 2001; ISBN 978-0-16-050958-2.

113. Marolda, E.J. *By Sea, Air, and Land: An Illustrated History of the U.S. Navy and the War in Southeast Asia*; Naval Historical Center, Dept. of the Navy: For Sales by Supt. of Docs.; U.S. G.P.O.: Washington, DC, USA, 1994; ISBN 978-0-945274-09-4.

114. American Association for the Advancement for Science (AAAS). *AAAS, Amnesty Use Geo-Visualization to Shed Light on Human Rights in Pakistan*; Scientific Responsibility, Human Rights & Law Program; American Association for the Advancement of Science: Washington, DC, USA, 2010.

115. Weir, D.; McQuillan, D.; Francis, R.A. Civilian science: The potential of participatory environmental monitoring in areas affected by armed conflicts. *Environ. Monit. Assess.* **2019**, *191*, 1–17. [CrossRef]

116. Letouzé, E.; Meier, P.; Vinck, P. *Big Data for Conflict Prevention: New Technology and the Prevention of Violence and Conflict*; International Peace Institute: New York, NY, USA, 2013; pp. 4–27.

117. Wang, Y.; Chen, Q.; Zhu, Q.; Liu, L.; Li, C.; Zheng, D. A survey of mobile laser scanning applications and key techniques over urban areas. *Remote Sens.* **2019**, *11*, 1540. [CrossRef]

118. Rasti, B.; Ghamisi, P. Remote sensing image classification using subspace sensor fusion. *Inf. Fusion* **2020**, *64*, 121–130. [CrossRef]

119. Avtar, R.; Watanabe, T. *Unmanned Aerial Vehicle: Applications in Agriculture and Environment*; Springer: Berlin/Heidelberg, Germany, 2020.

120. Bassoli, R.; Sacchi, C.; Granelli, F.; Ashkenazi, I. A Virtualized Border Control System Based on UAVs: Design and Energy Efficiency Considerations. In Proceedings of the 2019 IEEE Aerospace Conference, Big Sky, MT, USA, 2 March 2019; pp. 1–11.

121. Koslowski, R.; Schulzke, M. Drones Along Borders: Border Security UAVs in the United States and the European Union. *Int. Stud. Perspect.* **2018**, *19*, 305–324. [CrossRef]

122. Abdelzaher, T.; Ayanian, N.; Basar, T.; Diggavi, S.; Diesner, J.; Ganesan, D.; Govindan, R.; Jha, S.; Lepoint, T.; Marlin, B.; et al. Will Distributed Computing Revolutionize Peace? The Emergence of Battlefield IoT. In Proceedings of the 2018 IEEE 38th International Conference on Distributed Computing Systems (ICDCS), Vienna, Austria, 2 July 2018; pp. 1129–1138.

123. Xiao, L.; Wang, J.; Qiu, X.; Rong, Z.; Zou, X. Dynamic-SLAM: Semantic monocular visual localization and mapping based on deep learning in dynamic environment. *Robot. Auton. Syst.* **2019**, *117*, 1–16. [CrossRef]

124. Cen, R.; Zhang, X.; Tao, Y.; Xue, F.; Zhang, Y. Temporal delay estimation of sparse direct visual inertial odometry for mobile robots. *J. Frankl. Inst.* **2020**, *357*, 3893–3906. [CrossRef]

125. Langford, R.E. *Introduction to Weapons of Mass Destruction: Radiological, Chemical, and Biological*; Wiley-Interscience: Hoboken, NJ, USA, 2004; ISBN 978-0-471-46560-7.

126. Rufer, R. *Disarmament, Demobilisation and Reintegration (DDR): Conceptual Approaches, Specific Settings, Practical Experiences*; Geneva Centre for the Democratic Control of Armed Forces (DCAF): Geneva, Switzerland, 2005; p. 116.

Article

Remotely Sensed Urban Surface Ecological Index (RSUSEI): An Analytical Framework for Assessing the Surface Ecological Status in Urban Environments

Mohammad Karimi Firozjaei [1], **Solmaz Fathololoumi** [2], **Qihao Weng** [3,*], **Majid Kiavarz** [1] and **Seyed Kazem Alavipanah** [1]

[1] Department of Remote Sensing and GIS, Faculty of Geography, University of Tehran, Tehran 1417853933, Iran; mohammad.karimi.f@ut.ac.ir (M.K.F.); kiavarzmajid@ut.ac.ir (M.K.); salavipa@ut.ac.ir (S.K.A.)

[2] Faculty of Agriculture and Natural Resources, University of Mohaghegh Ardabili, Ardabil 5619911367, Iran; fathis@uma.ac.ir

[3] Center for Urban and Environmental Change, Department of Earth and Environmental Systems, Indiana State University, Terre Haute, IN 47809, USA

[*] Correspondence: qweng@indstate.edu; Tel.: +1-81-2237-2255; Fax: +1-81-2237-8029

Received: 30 April 2020; Accepted: 22 June 2020; Published: 24 June 2020

Abstract: Urban Surface Ecological Status (USES) reflects the structure and function of an urban ecosystem. USES is influenced by the surface biophysical, biochemical, and biological properties. The assessment and modeling of USES is crucial for sustainability assessment in support of achieving sustainable development goals such as sustainable cities and communities. The objective of this study is to present a new analytical framework for assessing the USES. This analytical framework is centered on a new index, Remotely Sensed Urban Surface Ecological index (RSUSEI). In this study, RSUSEI is used to assess the USES of six selected cities in the U.S.A. To this end, Landsat 8 images, water vapor products, and the National Land Cover Database (NLCD) land cover and imperviousness datasets are downloaded for use. Firstly, Land Surface Temperature (LST), Wetness, Normalized Difference Vegetation Index (NDVI), and Normalized Difference Soil Index (NDSI) are derived by remote sensing methods. Then, RSUSEI is developed by the combination of NDVI, NDSI, Wetness, LST, and Impervious Surface Cover (ISC) with Principal Components Analysis (PCA). Next, the spatial variations of USES across the cities are evaluated and compared. Finally, the association degree of each parameter in the USES modeling is investigated. Results show that the spatial variability of LST, ISC, NDVI, NDSI, and Wetness is heterogeneous within and between cities. The mean (standard deviation) value of RSUSEI for Minneapolis, Dallas, Phoenix, Los Angeles, Chicago and Seattle yielded 0.58 (0.16), 0.54 (0.17), 0.47 (0.19), 0.63 (0.21), 0.50 (0.17), and 0.44 (0.19), respectively. For all the cities, PC1 included more than 93% of the surface information, which is contributed by greenness, moisture, dryness, heat, and imperviousness. The highest and lowest mean values of RSUSEI are found in "Developed, High intensity" (0.76) and "Developed, Open Space" (0.35) lands, respectively. The mean correlation coefficient between RSUSEI and LST, ISC, NDVI, NDSI, and Wetness, is 0.47, 0.97, −0.31, 0.17, and −0.27, respectively. The statistical significance of these correlations is confirmed at 95% confidence level. These results suggest that the association degree of ISC in USES modeling is the highest, despite the differences in land cover and biophysical characteristics in the cities. RSUSEI could be very useful in modeling and comparing USES across cities with different geographical, climatic, environmental, and biophysical conditions and can also be used for assessing urban sustainability over space and time.

Keywords: Urban Surface Ecological Status (USES); Remotely Sensed Surface Ecological Index (RSUSEI); sustainability; impervious surfaces; US cities; National Land Cover Database (NLCD)

---

1. Introduction

Surface Ecological Status (SES) reflects the structure and function of an ecosystem. SES is influenced by surface biophysical, biochemical, and biological properties [1,2]. SES has wide applicability e.g., in ecological and environmental assessments, including ecosystem management and life quality evaluations [2,3]. SES and its spatial variations are influenced by natural and anthropogenic factors [4,5] e.g., in urban areas. Increased human activity is one of the most important anthropogenic factors affecting the Urban Surface Ecological Status (USES) and its changes [5–7]. Given the high concentration of human activity in urban environments, assessing and modeling USES is crucial for urban environmental management and planning, informing decision-makers and the public about ecosystem services, and sustainability assessment in support of achieving sustainable development goals such as sustainable cities and communities [8].

In previous studies, spectral indices derived from satellite imagery have been widely used to model SES [1,5–7,9–11]. These indices include Normalized Difference Vegetation Index (NDVI), Leaf Area Index (LAI), Normalized Difference Built-up Index (NDBI), Normalized Difference Soil Index (NDSI), Normalized Difference Water Index (NDWI), and Land Surface Temperature (LST) [4–6,10–13]. The advantages of remote sensing (RS) data, which can provide observations over a large area and a long period of time, have been extended to SES modeling on a local, regional, and global scale [14–17]. However, the complexity in the relationship between SES and biophysical and environmental factors makes it difficult to quantify SES based on a single spectral index [4–6]. Aggregated remote sensing indices have shown more advantages than a single index in modeling SES [4,18]. An integrated Remote Sensing-based Ecological Index (RSEI) was developed for the rapid assessment of SES, using satellite data [6]. The advantages of RSEI can be summarized as (a) scalable, (b) visualizable, (c) comparable at different scales, and (d) customizable to minimize error or variation caused by other properties in the weight definitions [4–6]. Despite these valuable benefits, the RSEI was developed solely by using spectral indices related to land surface components and surface climate. The use of index-based built-up areas in Hu and Xu (2018) [6] and subsequent studies showed that the index cannot address the issue of bare land and sparsely vegetated areas, due to spectral confusion with the built-up areas.

Impervious Surface Cover (ISC) is one of the most important factors in distinguishing the characteristics of different types of land use and land cover in urban environments and is responsible for changing the characteristics of surface greenness, moisture, dryness, and heat [19]. ISC has a clear physical meaning in land surface composition, suitable for comparative urban analysis [20–22]. Hence, the inclusion of ISC can potentially increase the accuracy of modeling the USES. Based on the Vegetation-Impervious surface-Soil (V-I-S) model [23], the percentage of each of the three fractions of impervious, vegetation, and soil covers in a pixel indicates the difference in the surface characteristics of different urban land cover/uses. This model assumes that land cover in urban environments is a linear combination of three components.

The objective of this study is to present a new analytical framework for assessing the Remotely Sensed Urban Surface Ecological index (RSUSEI) by integration of surface greenness, moisture, dryness, heat, and imperviousness using Principal Components Analysis (PCA). Based on the V-I-S model, this study intends to assess USES and compare six cities of the U.S.A including Minneapolis, Dallas, Phoenix, Los Angeles, Chicago, and Seattle which have distinct geographical, geological, climatic, environmental, and surface biophysical conditions.

2. Study Area

The new analytical framework for assessing the SES is tested in urban environments comprising the cities of Minneapolis, Dallas, Phoenix, Los Angeles, Chicago, and Seattle (Figure 1). To select these cities, various criteria such as (1) geographical conditions, (2) climatic conditions, (3) surface characteristics, (4) density of population, and (5) physical size of the cities were considered. These cities possess different geographical, climatic, environmental, and biophysical conditions. Based on the Köppen climate classification, the selected cities have various climate types: humid continental (Dfa—Minneapolis,

Chicago), tropical and subtropical desert (Bwh—Phoenix), dry-summer subtropical (Csa—Los Angeles; Csb—Seattle), or humid subtropical (Cfa—Dallas). Thus, the spatial variability of the surface cover and biophysical characteristics of these cities are different and heterogeneous.

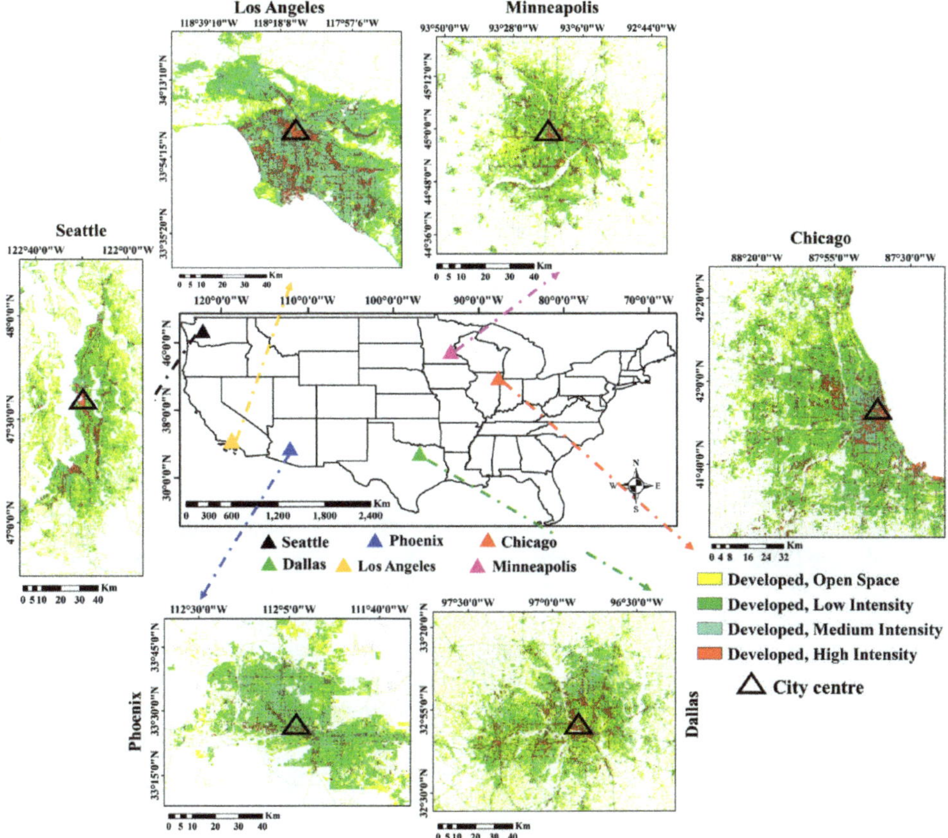

Figure 1. Geographical location of selected cities including Minneapolis, Dallas, Phoenix, Los Angeles, Chicago, and Seattle in the U.S.A and land cover maps of each selected cities.

For the selected cities in the U.S.A, the area and percentage of each land cover are different (Figure 1 and Table 1). The highest area of land cover in Minneapolis, Dallas, Phoenix, Los Angeles, Chicago, and Seattle cities is related to "Developed, Open Space", "Developed, Low Intensity", "Developed, Low Intensity", "Developed, Medium Intensity", "Developed, Low Intensity", and "Developed, Low Intensity", respectively. Among the cities, the highest percentage of "Developed, Open Space", "Developed, Low Intensity", "Developed, Medium Intensity", and "Developed, High Intensity" lands is found in Minneapolis (36.0%), Chicago (45.5%), Los Angeles (44.6%), and Los Angeles (16.3%), respectively.

Table 1. Area (km^2) and percentage (%) of land cover classes of selected cities in the U.S.A.

Land Cover Class	Minneapolis		Dallas		Phoenix		Los Angeles		Chicago		Seattle	
	Area	%	Area	%	Area	%	Area	%	Area	%	Area	%
Developed, Open Space	962.6	36.0	1329.3	27.0	646.5	23.5	616.9	15.2	1042.3	19.1	936.6	31.2
Developed, Low Intensity	859.0	32.1	1682.1	34.1	966.8	35.2	965.9	23.9	2430.7	45.5	1182.1	39.3
Developed, Medium Intensity	599.1	22.4	1305.2	26.5	913.5	33.2	1804.7	44.6	1338.1	24.5	647.5	21.6
Developed, High Intensity	250.9	9.5	606.2	12.4	221.1	8.1	653.1	16.3	650.8	11.9	238.2	7.9
Total area	2671.6	100	4922.8	100	2747.9	100	4040.6	100	5461.9	100	3004.4	100

3. Data and Methods

3.1. Data

A leaf-on season of Landsat 8 image for each city was downloaded for use from the U.S.A Geological Survey (USGS) website (http://www.usgs.gov), including Minneapolis (date: 8 September, 2016), Dallas (8 September, 2016), Phoenix (14 September, 2016), Los Angeles (26 September, 2016), Chicago (12 September, 2016), and Seattle (13 September, 2016). Due to the spatial, temporal, spectral, and radiometric resolution, Landsat images are suitable for modeling and monitoring environmental and ecological conditions [24–26]. These images are georeferenced with the number of rows and paths available. The spatial resolution of Landsat 8 reflective and thermal bands are 30 and 100 m, respectively. Resampled thermal infrared bands based on the Cubic method with a spatial resolution of 30 m are also available on the USGS website. Due to the climatic and seasonal effects on USES, leaf-on clear-sky images were selected for this study. Furthermore, the Moderate Resolution Imaging Spectroradiometer (MODIS) water vapor product (MOD07) with a spatial resolution of 5000 m was used to calculate Land Surface Temperature (LST) from the Landsat images. This product contains the following features: (1) total-ozone burden, (2) atmospheric stability, (3) temperature and moisture profiles, (4) and atmospheric water vapor. In addition, datasets from the National Land Cover Database (NLCD), including land cover and imperviousness for 2016 were used. The imperviousness data were utilized to represent surface biophysical characteristics, while the land cover data to evaluate the impact of land cover on USES. The NLCD land cover and imperviousness datasets with 30 m spatial resolutions for the U.S.A prepared for different years using Landsat time-series images [27,28] are available from USGS at the https://www.mrlc.gov/data website.

3.2. Methods

Firstly, Landsat imagery was preprocessed. In the second step, the Single Channel (SC) algorithm, the Tasseled cap transformation, spectral indices, and the NLCD imperviousness data product were used to derive surface biophysical characteristics, which include NDVI, Wetness, NDSI, LST, and ISC for the selected cities. Next, USES of the selected cities were modeled by the combination of surface biophysical characteristics data using PCA. Then, the spatial variations of USES across different selected cities were evaluated and compared to each other. Finally, the association degree of each surface biophysical characteristic on USES was investigated based on statistical analysis (Figure 2).

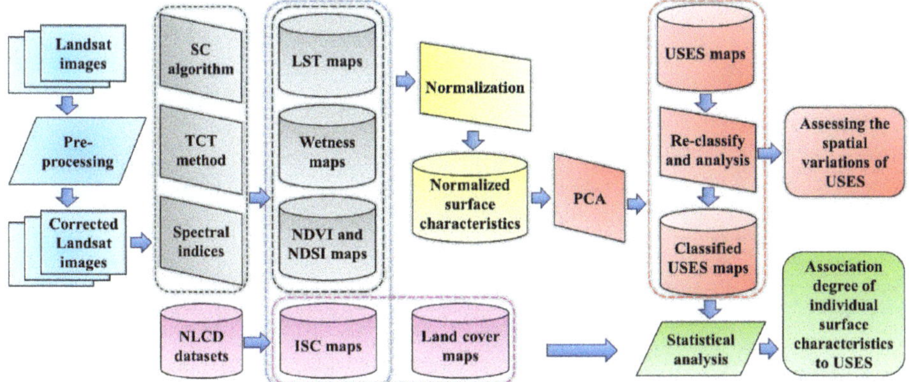

Figure 2. The flowchart of the analytical framework. SC: Single Channel, TCT: Tasseled Cap Transformation, NLCD: National Land Cover Database, LST: Land Surface Temperature, NDVI: Normalized Difference Vegetation Index, NDSI: Normalized Difference Soil Index, ISC: Impervious surface cover, PCA: Principle Components Analysis, USES: Urban Surface Ecological Status.

3.2.1. Landsat Image Preprocessing and Surface Characteristics Modeling

The Fast Line-of-sight Atmospheric Analysis of Spectral Hypercubes (FLAASH) model was applied for the atmospheric correction of Landsat images. This model uses parameters such as satellite overpass time, sensor altitude, geographic location, atmospheric model of the region, and the solar elevation and zenith angles [29,30].

The spectral indices used in this study included NDVI (greenness), Wetness (moisture), NDSI (dryness), and LST (heat), which are shown in Table 2.

Table 2. Spectral indices in this study and their calculation details.

Index	Equation	Reference
NDVI	$\frac{NIR-Red}{NIR+Red}$	[31]
Wetness	Tasseled cap transformation (TCT) component 3	[32,33]
NDSI	$\frac{SWIR1-NIR}{SWIR1+NIR}$	[34]
LST	Single Channel (SC) algorithm	[35]

Land cover and imperviousness maps of the selected cities were obtained from NLCD in 2016. Land cover maps of the urban areas based on this dataset included classes of "Developed, Open Space", "Developed, Low Intensity", "Developed, Medium Intensity", and "Developed, High Intensity". The percentage urban impervious surface was resolved in 1% increments from 0 to 100 for areas identified as urban, in the land cover layer of the database.

3.2.2. Remotely Sensed Urban Surface Ecological Index (RSUSEI)

In this study, RSUSEI was developed to assess the USES. Urban surfaces were assumed to consist of three fractions of impervious, vegetation, and soil covers. Based on the V-I-S model, the percentage of each of these surface covers in a pixel indicated the difference in the surface characteristics of different urban land cover/uses (Figure 3). This model assumes that land cover in urban environments is a linear combination of three components [23]. Therefore, to model accurately the USES and to assess the SES of different land covers in urban environments, it is important to consider the biophysical characteristics related to the fractions of impervious, vegetation, and soil cover.

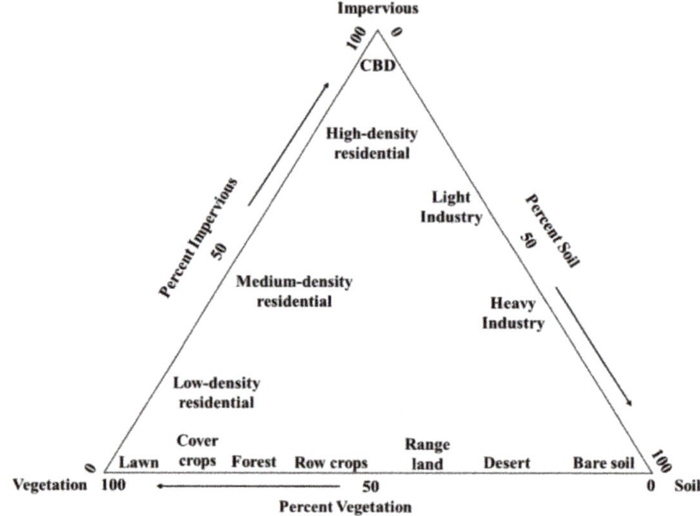

Figure 3. The Vegetation-Impervious surface-Soil model for remote sensing of urban environments [23].

Surface greenness, dryness, moisture, heat, and imperviousness are the biophysical characteristics that were utilized to describe these component surfaces and are integrated for analysis of USES. It can be represented as Equation (1):

$$\text{USES} = f(\text{Greenness, Dryness, Moisture, Heat, Imperviousness}) \tag{1}$$

However, these environmental and surface biophysical parameters may be correlated with each other in a region. The use of PCA can be very useful to solve the collinearity between the predictive variables in the model of USES. To reduce the effect of climatic and meteorological conditions on the results of the RSUSEI, standardized values of LST (heat), NDVI (greenness), NDSI (dryness), Wetness (moisture), and ISC (imperviousness) indices (between 0 and 1) were computed [36]. Then, the PCA method was employed to combine the five indices for assessing the USES. Finally, the PC1 was selected to represent USES in the urban environments. RSUSEI can then be modeled conceptually based on Equation (2).

$$\text{RSUSEI} = \text{PC1}(\text{NDVI, Wetness, LST, NDSI, ISC}) \tag{2}$$

RSUSEI values were normalized between 0 and 1. The maximum value was related to the worst (highest LST, ISC, and NDSI, but lowest NDVI and Wetness) and the minimum value to the best USES (lowest LST, ISC, and NDSI, but highest NDVI and Wetness). To analyze and evaluate the spatial variations of USES, the normalized values of RSUSEI were grouped into five classes: Excellent (0–0.2), Very Good (0.2–0.4), Good (0.4–0.6), Fair (0.6–0.8), and Poor (0.8–1) [4–6]. The mean and Standard Deviation (SD) values of RSUSEI were calculated and the area of USES classes were mapped for each city. Additionally, the Eigenvalues of the five main components of PCA for each city were calculated.

3.2.3. Association Degree of Individual Surface Characteristics to USES

To calculate and assess the association degree of each surface characteristic to USES, the mean value of RSUSEI was calculated by the NLCD land cover class for the six selected cities. The correlation coefficient (r) between RSUSEI and each biophysical variable, i.e., LST, NDVI, NDSI, and Wetness, is calculated for each city. In addition, the mean RSUSEI value for different ISC percentage ranges (1 to 100) was calculated and then the r between the mean RSUSEI values and the percentage of ISC was calculated for each city.

4. Results

4.1. Spatial Distribution of Surface Characteristics

Figure 4 shows maps of surface biophysical characteristics including Normalized LST (nLST), Normalized NDVI (nNDVI), Normalized NDSI (nNDSI), and Normalized Wetness (nWetness) in the selected cities. The spatial pattern was heterogeneous. A large proportion of the central parts of all these selected cities included developed lands (ISC > 50%). The suburbs of Minneapolis, Dallas, Chicago, and Seattle included land covers with a high percentage of vegetation cover, while in Phoenix and Los Angeles there were bare lands with dry surfaces. In all cities, the values of nLST and nNDSI in the central parts of the cities were higher than the other parts, while nNDVI values were lower. In addition, the spatial distribution of ISC values in these cities was heterogeneous. Areas with developed and nature lands had the highest and lowest values of ISC, respectively.

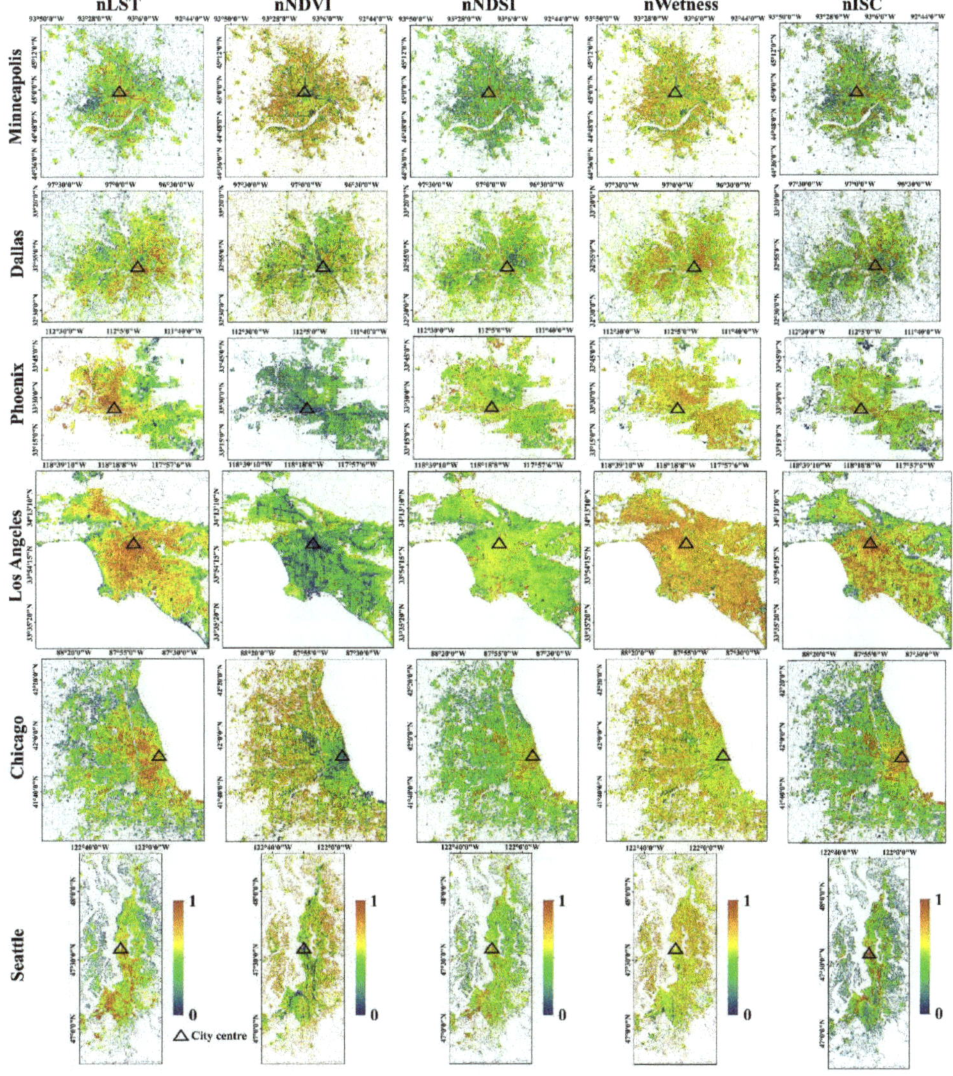

Figure 4. nLST, nNDVI, nNDSI, and nWetness maps of selected cities in the U.S.A.

The mean and SD values of nLST, nNDVI, nNDSI, nWetness, and nISC in the selected cities were different. The difference in the mean values of these spectral indices indicated the difference in the surface biophysical characteristics of these cities. High SD values of these spectral indices indicated a higher degree of spatial variability of the surface biophysical characteristics. The correlation coefficient (r) between the mean values of nLST and nNDVI, nNDSI, nWetness, and nISC for these cities were -0.92, 0.95, -0.22, and 0.88, respectively, which indicated a strong correlation between different surface biophysical characteristics, except Wetness. The statistical significance of these correlations were confirmed at 95% confidence level. Cities with high mean values of nLST and nNDSI tended to have low mean values of nNDVI and vice versa (Table 3). By reducing surface vegetation, the amount of evapotranspiration from the surface decreased, leading to an increase in surface heat and dryness. In addition, a higher value of ISC caused an increase in the value of LST (heat) and NDSI (dryness) and a decrease in the value of NDVI (greenness) and Wetness (moisture).

Table 3. Mean (Standard deviation) values of nLST, nNDVI, nNDSI, nWetness, and nISC in the selected cities in the U.S.A.

Cities	nLST	nNDVI	nNDSI	nWetness	nISC
Minneapolis	0.39 (0.16)	0.66 (0.19)	0.41 (0.16)	0.75 (0.11)	0.36 (0.27)
Dallas	0.54 (0.14)	0.56 (0.19)	0.49 (0.15)	0.55 (0.11)	0.41 (0.28)
Phoenix	0.65 (0.13)	0.25 (0.16)	0.64 (0.09)	0.79 (0.04)	0.53 (0.11)
Los Angeles	0.63 (0.12)	0.29 (0.17)	0.59 (0.13)	0.69 (0.12)	0.52 (0.26)
Chicago	0.42 (0.16)	0.56 (0.23)	0.42 (0.14)	0.65 (0.08)	0.43 (0.25)
Seattle	0.46 (0.14)	0.58 (0.22)	0.50 (0.17)	0.58 (0.12)	0.37 (0.25)

The spatial distribution of surface biophysical characteristic differences can be caused by the differences in the spatial variability of land covers [19,37–40]. Table 4 shows that surface biophysical characteristics including nLST, nNDVI, nNDSI, nWetness, and nISC were different for each land cover. Anthropogenic activities reduce natural surface covers and affect surface characteristics including surface reflection, change in the material's thermal capacity, conductivity, diffusion, albedo, and evapotranspiration [19,37,41–43]. For the selected cities, the highest (lowest) mean values of nLST, nNDSI, and nISC and the lowest (highest) values of nNDVI and nWetness were related to Developed, High Intensity (Developed, Open Space). Therefore, due to the spatial variability of land covers, the spatial variability of surface biophysical characteristics in the selected cities were different.

Table 4. Mean value of nLST, nNDVI, nNDSI, nWetness and nISC by land cover.

Land Cover Class	nLST	nNDVI	nNDBI	nWetness	nISC
Developed, Open Space	0.42	0.62	0.42	0.68	0.06
Developed, Low Intensity	0.49	0.52	0.45	0.64	0.34
Developed, Medium Intensity	0.57	0.40	0.52	0.61	0.62
Developed, High Intensity	0.64	0.22	0.62	0.56	0.89

4.2. Spatial Distribution of USES

The spatial distribution of USES of the selected cities was heterogeneous (Figure 5). A visual survey of the RSUSEI maps shows that Chicago and Los Angeles had higher RSUSEI values than Minneapolis, Dallas, Phoenix, and Seattle. Areas with high values of RSUSEI (red color) had a lower quality of USES, which had high heat (LST), imperviousness (ISC) dryness (NDSI), greenness (NDVI), and moisture (Wetness) values, and vice versa. Figure 6 shows the mean value of RSUSEI for Minneapolis, Dallas, Phoenix, Los Angeles, Chicago, and Seattle to be 0.58, 0.54, 0.47, 0.63, 0.50, and 0.44, respectively.

The difference in the mean value of RSUSEI between the cities indicated a significant difference in their USES. The best and worst USES were Seattle and Los Angeles, respectively. The SD values of RSUSEI for Minneapolis (0.16), Dallas (0.17), Phoenix (0.19), Los Angeles (0.21), Chicago (0.17), and Seattle (0.19) cities were high. These values indicated the high spatial variability of the USES within each selected city. Overall, the SD values of the different cities were very similar. The highest and lowest spatial variations of USES were observed in Minneapolis and Seattle, respectively.

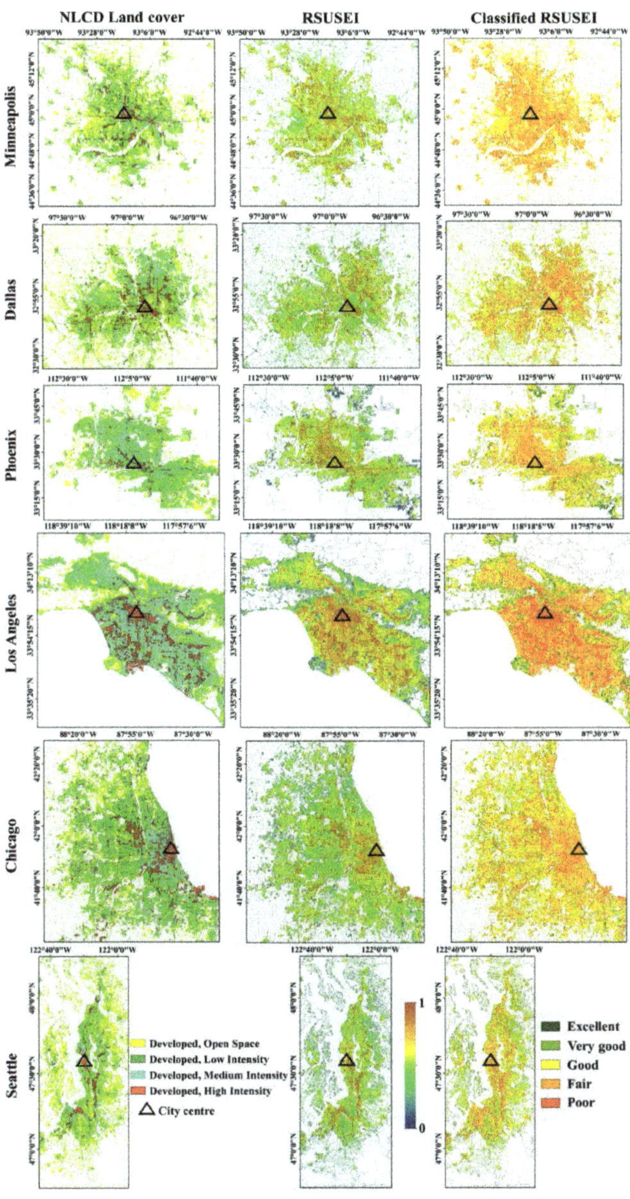

Figure 5. Remotely Sensed Urban Surface Ecological index (RSUSEI), classified RSUSEI and NLCD land cover maps of the selected cities in the U.S.A.

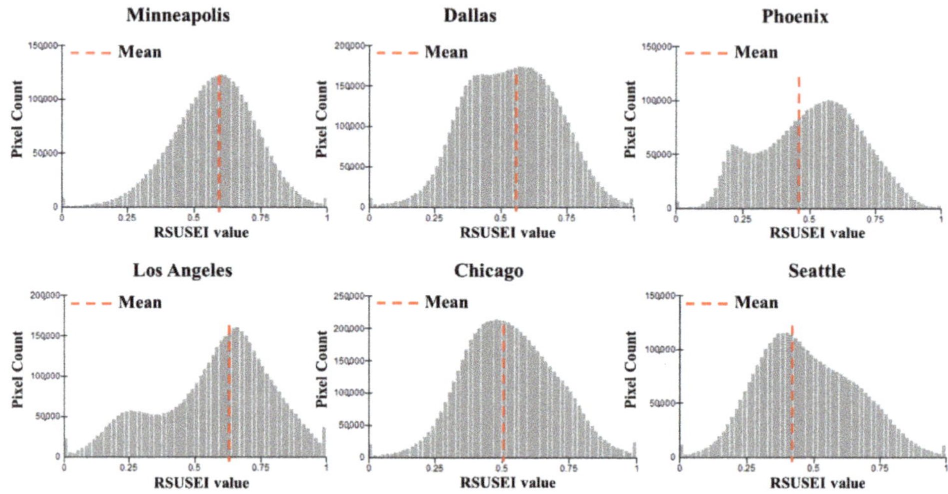

Figure 6. Frequency of RSUSEI values of the selected US cities.

The spatial distribution of the RSUSEI classes further revealed the spatial variability of USES across the selected cities (Figure 7). Overall, the majority of the land in the selected cities possessed the USES class from Very Good to Fair. The highest percentage of USES class for Minneapolis, Dallas, Phoenix, Chicago, and Seattle cities was Good and for Los Angeles city was Fair. The Poor class of USES had better spatial coverage of 5% to 14%, compared to that of the Excellent class from 1% to 6%. In addition, the highest percentage of Excellent, Very Good, Good, Fair, and Poor classes of USES was observed in Los Angeles, Seattle, Chicago, Minneapolis, and Los Angeles, respectively (Figure 7). The spatial heterogeneity of surface biophysical characteristics and anthropogenic activities caused differences in USES among the cities and within each city.

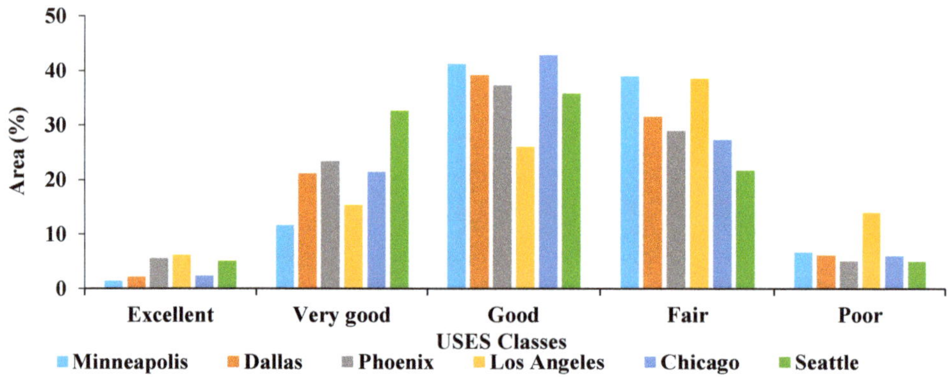

Figure 7. Area of USES classes of selected cities in the U.S.A. (%).

It is worth noting that the Eigenvalues of the PC1 in RSUSEI modeling for the Minneapolis, Dallas, Phoenix, Los Angeles, Chicago, and Seattle cities were 0.33, 0.34, 0.40, 0.39, 0.29, and 0.37, respectively. For all selected cities, PC1 included more than 93% of the main surface information including greenness, moisture, dryness, heat, and imperviousness. Therefore, using PC1 in RSUSEI can well represent the spatial heterogeneity is the USES.

4.3. Association Degree of Surface Biophysical Parameters on the USES Modeling

The mean value of RSUSEI varied across different land covers in the selected cities (Table 5). The mean values of RSUSEI in "Developed, Open Space", "Developed, Low Intensity", "Developed, Medium Intensity", and "Developed, High Intensity" lands were 0.35, 0.49, 0.63, and 0.76, respectively (Table 6). In general, "Developed, high intensity" and "Developed, Open Space" lands detected the highest and lowest RSUSEI values in these cities, respectively. This result suggests the effectiveness of RSUSEI to separate USES by land cover.

Table 5. The mean RSUSEI of land cover classes of selected cities in the U.S.A.

Land Cover Class	Minneapolis	Dallas	Phoenix	Los Angeles	Chicago	Seattle	Mean
Developed, Open Space	0.40	0.35	0.37	0.38	0.34	0.32	0.35
Developed, Low Intensity	0.58	0.51	0.47	0.45	0.45	0.48	0.49
Developed, Medium Intensity	0.63	0.65	0.64	0.59	0.64	0.64	0.63
Developed, High Intensity	0.65	0.76	0.80	0.85	0.76	0.76	0.76

Table 6. Correlation coefficient between RSUSEI and nLST, nNDVI, nNDSI, and nWetness.

City	nLST	nNDVI	nNDSI	nWetness	ISC
Minneapolis	0.41	−0.24	0.3	−0.34	0.90
Dallas	0.4	−0.21	0.06	−0.24	0.99
Phoenix	0.32	−0.27	0.31	−0.56	0.99
Los Angeles	0.59	−0.52	0.01	−0.4	0.99
Chicago	0.53	−0.28	0.21	−0.045	0.98
Seattle	0.57	−0.39	0.14	−0.09	0.98
Mean	0.47	−0.31	0.17	−0.27	0.97

Areas with high surface heat (LST), imperviousness (ISC), dryness (NDSI), low surface vegetation density (NDVI), and moisture (Wetness) exhibited poor USES. The locations of these areas corresponded to "Developed, High Intensity" lands (Figure 5). By contrast, "Developed, Open Space" lands possessed the best USES (Table 4), which discovered low values of LST and NDSI and high values of NDVI and Wetness. Mixed pixels in warm and dry cities tended to include built-up lands and lands with low vegetation cover and low moisture content. Due to the high values of RSUSEI for lands with low vegetation density and low moisture content, these cities discovered poor USES. On the other hand, mixed pixels in humid cities were associated with high vegetation density and surface moisture. Since there were low values of RSUSEI for lands with high vegetation density and high surface moisture, these cities experienced poor USES. These findings suggest that RSUSEI holds an excellent ability to differentiate between USES of different land covers (Figure 4 and Table 5).

The association degree of nLST, nNDVI, nNDSI, and nWetness in RSUSEI in the selected cities varied. The mean r between nLST, nNDVI, nNDSI, and nWetness and RSUSEI was 0.47, −0.31, 0.17, and −0.27, respectively (Table 6). The statistical significance of these correlations were confirmed at 95% confidence level. In modeling the USES, the association degree of nLST was found to be higher than nNDVI, nNDSI, and nWetness. For the RSUSEI modeling in Los Angeles, Chicago, and Seattle cities, the association degree of nNDVI appeared higher than nNDSI and nWetness. In contrast, in Minneapolis and Dallas, the association degree of nWetness was higher than nNDVI and nNDSI.

The correlation coefficient between the mean value of RSUSEI and the mean value of NLCD imperviousness percentage was 0.93 for all cities, but it varied within cities. Minneapolis, Dallas, Phoenix, Los Angeles, Chicago, and Seattle yielded an r value of 0.90, 0.99, 0.99, 0.99, 0.98, and 0.98, respectively (Table 6). The statistical significance of these correlations are confirmed at 95% confidence level. These values indicated a positive strong correlation between ISC and RSUSEI. The spatial variation patterns of RSUSEI and NLCD imperviousness were similar to each other (Figure 5). The RSUSEI value increased with increasing the percentage of impervious surface. In the

USES modeling, the association degree of ISC was highest among all the surface parameters used in this study.

5. Discussion

The SES in urban environments is a function of the surface biophysical, biochemical, and biological properties. Recent studies have used the data of surface greenness, moisture, dryness, and heat for SES modeling [4,6,7]. However, many processes in the urban environments are subject to the impact of surface imperviousness [23,44–46]. ISC has a clear physical meaning in land surface composition, suitable for distinguishing the characteristics of different types of land use and land cover in the urban environments, and is associated with changes in the characteristics of surface greenness, moisture, dryness, and heat [19,20,22,45]. This study shows that the association degree of imperviousness is higher than surface heat, greenness, dryness, and moisture. Therefore, considering surface imperviousness information in USES modeling is very necessary. Other studies have also shown that surface imperviousness affected the SES [4,47,48]. For many cities around the world, surface imperviousness data are available with functional spatial resolution for urban modeling. In this study, five components including surface greenness, moisture, dryness, heat, and imperviousness are considered for RSUSEI development. Results showed that RSUSEI is highly capable in the modeling of the USES spatial heterogeneity in cities with different geographical, climatic, environmental, and biophysical conditions. This index has a high capacity to differentiate between USES of different land covers. Assessment and modeling of USES are crucial in sustainability assessment in support of achieving sustainable development goals such as sustainable cities and communities [8]. Hence, RSUSEI can be used for assessing urban sustainability over space and time.

6. Conclusions

In this study, an analytical framework is proposed for assessing the SES in urban environments and tested in six selected cities in the U.S.A, i.e., Minneapolis, Dallas, Phoenix, Los Angeles, Chicago, and Seattle. This analytical framework is centered on a new index, Remotely Sensed Urban Surface Ecological index (RSUSEI), which integrated satellite derived information on the greenness, moisture, dryness, heat, and imperviousness in a city. The results showed that the spatial distribution of USES varied with the cities and land cover types. In general, land covers with low vegetation density and moisture, and high heat, imperviousness, and dryness exhibit high RSUSEI values and poor USES, and vice versa. The USES in arid regions, such as Los Angeles, are found to be worse than the USES in humid regions, such as Seattle. The association degree of ISC is higher than nLST, nNDVI, nNDSI, and nWetness in the RSUSEI modeling. An increase in surface imperviousness reduces surface vegetation density and moisture while increasing surface dryness and heat degree, thereby worsening USES. Our results show that RSUSEI has a high capability in revealing the differences in USES within and between cities with different geographical, climatic, environmental, and surface conditions. Due to the functional spatial resolution and continuity of Landsat imagery, the results of this study can be very useful in USES modeling in urban environments with different biophysical, geographical, and climatic conditions. In addition, the availability of NLCD data products in the U.S.A is highly beneficial for USES assessment, monitoring, and modeling. RSUSEI can be used for assessing urban sustainability over space and time. It is suggested that in future studies, the efficiency of disaggregation models in improving the spatial resolution of USES maps should be considered. It is also useful to compare the performance of different spectral indices in surface imperviousness modeling to assess USES. In addition, RSUSEI can be used as a time series to monitor and model the long-term changes in a region and to quantify the impact of anthropogenic activities on USES.

Author Contributions: M.K.F. and S.F. conceived and designed the research of, and wrote, the first draft; Q.W. re-designed the research, and revised and edited the paper; M.K. and S.K.A. provided comments. All authors contributed to, and approved, the final manuscript. All authors have read and agreed to the published version of the manuscript.

Funding: This research received no external funding.

Conflicts of Interest: The authors declare no conflict of interest.

References

1. Reza, M.I.H.; Abdullah, S.A. Regional Index of Ecological Integrity: A need for sustainable management of natural resources. *Ecol. Indic.* **2011**, *11*, 220–229. [CrossRef]
2. Frohn, R.C.; Lopez, R.D. *Remote Sensing for Landscape Ecology: New Metric Indicators: Monitoring, Modeling, and Assessment of Ecosystems*; CRC Press: Boca Raton, FL, USA, 2017.
3. Wu, Z.; Zhang, Y. Water Bodies' Cooling Effects on Urban Land Daytime Surface Temperature: Ecosystem Service Reducing Heat Island Effect. *Sustainability* **2019**, *11*, 787. [CrossRef]
4. Xu, H.; Wang, M.; Shi, T.; Guan, H.; Fang, C.; Lin, Z. Prediction of ecological effects of potential population and impervious surface increases using a remote sensing based ecological index (RSEI). *Ecol. Indic.* **2018**, *93*, 730–740. [CrossRef]
5. Xu, H.; Wang, Y.; Guan, H.; Shi, T.; Hu, X. Detecting Ecological Changes with a Remote Sensing Based Ecological Index (RSEI) Produced Time Series and Change Vector Analysis. *Remote Sens.* **2019**, *11*, 2345. [CrossRef]
6. Hu, X.; Xu, H. A new remote sensing index for assessing the spatial heterogeneity in urban ecological quality: A case from Fuzhou City, China. *Ecol. Indic.* **2018**, *89*, 11–21. [CrossRef]
7. Zhu, X.; Wang, X.; Yan, D.; Liu, Z.; Zhou, Y. Analysis of remotely-sensed ecological indexes' influence on urban thermal environment dynamic using an integrated ecological index: A case study of Xi'an, China. *Int. J. Remote Sens.* **2019**, *40*, 3421–3447. [CrossRef]
8. Estoque, R.C. A Review of the Sustainability Concept and the State of SDG Monitoring Using Remote Sensing. *Remote Sens.* **2020**, *12*, 1770. [CrossRef]
9. Lin, T.; Ge, R.; Huang, J.; Zhao, Q.; Lin, J.; Huang, N.; Zhang, G.; Li, X.; Ye, H.; Yin, K. A quantitative method to assess the ecological indicator system's effectiveness: A case study of the Ecological Province Construction Indicators of China. *Ecol. Indic.* **2016**, *62*, 95–100. [CrossRef]
10. Pettorelli, N.; Vik, J.O.; Mysterud, A.; Gaillard, J.-M.; Tucker, C.J.; Stenseth, N.C. Using the satellite-derived NDVI to assess ecological responses to environmental change. *Trends Ecol. Evol.* **2005**, *20*, 503–510. [CrossRef] [PubMed]
11. Willis, K.S. Remote sensing change detection for ecological monitoring in United States protected areas. *Biol. Conserv.* **2015**, *182*, 233–242. [CrossRef]
12. Cohen, W.B.; Goward, S.N. Landsat's role in ecological applications of remote sensing. *Bioscience* **2004**, *54*, 535–545. [CrossRef]
13. Sun, Z.; Chang, N.-B.; Opp, C. Using SPOT-VGT NDVI as a successive ecological indicator for understanding the environmental implications in the Tarim River Basin, China. *J. Appl. Remote Sens.* **2010**, *4*, 043554.
14. Estoque, R.C.; Murayama, Y. Monitoring surface urban heat island formation in a tropical mountain city using Landsat data (1987–2015). *ISPRS J. Photogramm.* **2017**, *133*, 18–29. [CrossRef]
15. He, C.; Gao, B.; Huang, Q.; Ma, Q.; Dou, Y. Environmental degradation in the urban areas of China: Evidence from multi-source remote sensing data. *Remote Sens. Environ.* **2017**, *193*, 65–75. [CrossRef]
16. Fu, P.; Xie, Y.; Weng, Q.; Myint, S.; Meacham-Hensold, K.; Bernacchi, C. A physical model-based method for retrieving urban land surface temperatures under cloudy conditions. *Remote Sens. Environ.* **2019**, *230*, 111191. [CrossRef]
17. Fu, Y.; Li, J.; Weng, Q.; Zheng, Q.; Li, L.; Dai, S.; Guo, B. Characterizing the spatial pattern of annual urban growth by using time series Landsat imagery. *Sci. Total Environ.* **2019**, *666*, 274–284. [CrossRef]
18. King, R.S.; Baker, M.E. Considerations for analyzing ecological community thresholds in response to anthropogenic environmental gradients. *J. N. Am. Benthol. Soc.* **2010**, *29*, 998–1008. [CrossRef]
19. Firozjaei, M.K.; Weng, Q.; Zhao, C.; Kiavarz, M.; Lu, L.; Alavipanah, S.K. Surface anthropogenic heat islands in six megacities: An assessment based on a triple-source surface energy balance model. *Remote Sens. Environ.* **2020**, *242*, 111751. [CrossRef]
20. Weng, Q.; Lu, D.; Liang, B. Urban surface biophysical descriptors and land surface temperature variations. *Photogramm. Eng. Remote Sens.* **2006**, *72*, 1275–1286. [CrossRef]

21. Weng, Q.; Liu, H.; Liang, B.; Lu, D. The spatial variations of urban land surface temperatures: Pertinent factors, zoning effect, and seasonal variability. *IEEE J. Sel. Top. Appl. Earth Obs. Remote Sens.* **2008**, *1*, 154–166. [CrossRef]

22. Lu, D.; Weng, Q. Extraction of urban impervious surfaces from an IKONOS image. *Int. J. Remote Sens.* **2009**, *30*, 1297–1311. [CrossRef]

23. Ridd, M.K. Exploring a VIS (vegetation-impervious surface-soil) model for urban ecosystem analysis through remote sensing: Comparative anatomy for cities. *Int. J. Remote Sens.* **1995**, *16*, 2165–2185. [CrossRef]

24. Firozjaei, M.K.; Sedighi, A.; Kiavarz, M.; Qureshi, S.; Haase, D.; Alavipanah, S.K. Automated Built-Up Extraction Index: A New Technique for Mapping Surface Built-Up Areas Using LANDSAT 8 OLI Imagery. *Remote Sens.* **2019**, *11*, 1966. [CrossRef]

25. Amiri, R.; Weng, Q.; Alimohammadi, A.; Alavipanah, S.K. Spatial–temporal dynamics of land surface temperature in relation to fractional vegetation cover and land use/cover in the Tabriz urban area, Iran. *Remote Sens. Environ.* **2009**, *113*, 2606–2617. [CrossRef]

26. Haashemi, S.; Weng, Q.; Darvishi, A.; Alavipanah, S.K. Seasonal variations of the surface urban heat island in a semi-arid city. *Remote Sens.* **2016**, *8*, 352. [CrossRef]

27. Xian, G.; Homer, C.; Demitz, J.; Fry, J.; Hossain, N. Change of impervious surface area between 2001 and 2006 in the conterminous United States. *Photogramm. Eng. Remote Sens.* **2011**, *77*, 758–762.

28. Yang, L.; Huang, C.; Homer, C.G.; Wylie, B.K.; Coan, M.J. An approach for mapping large-area impervious surfaces: Synergistic use of Landsat-7 ETM+ and high spatial resolution imagery. *Can. J. Remote Sens.* **2003**, *29*, 230–240. [CrossRef]

29. Berk, A.; Conforti, P.; Kennett, R.; Perkins, T.; Hawes, F.; van den Bosch, J. MODTRAN® 6: A major upgrade of the MODTRAN® radiative transfer code. In Proceedings of the 2014 6th Workshop on Hyperspectral Image and Signal Processing: Evolution in Remote Sensing (WHISPERS), Lausanne, Switzerland, 24–27 June 2014; pp. 1–4.

30. Firozjaei, M.K.; Kiavarz, M.; Nematollahi, O.; Karimpour Reihan, M.; Alavipanah, S.K. An evaluation of energy balance parameters, and the relations between topographical and biophysical characteristics using the mountainous surface energy balance algorithm for land (SEBAL). *Int. J. Remote Sens.* **2019**, *40*, 1–31. [CrossRef]

31. Tucker, C.J. Red and photographic infrared linear combinations for monitoring vegetation. *Remote Sens. Environ.* **1979**, *8*, 127–150. [CrossRef]

32. Baig, M.H.A.; Zhang, L.; Shuai, T.; Tong, Q. Derivation of a tasselled cap transformation based on Landsat 8 at-satellite reflectance. *Remote Sens. Lett.* **2014**, *5*, 423–431. [CrossRef]

33. Mijani, N.; Alavipanah, S.K.; Hamzeh, S.; Firozjaei, M.K.; Arsanjani, J.J. Modeling thermal comfort in different condition of mind using satellite images: An Ordered Weighted Averaging approach and a case study. *Ecol. Indic.* **2019**, *104*, 1–12. [CrossRef]

34. Zhao, H.; Chen, X. Use of normalized difference bareness index in quickly mapping bare areas from TM/ETM+. In Proceedings of the International geoscience and remote sensing symposium, Seoul, Korea, 25–29 July 2005; p. 1666.

35. Jiménez-Muñoz, J.C.; Sobrino, J.A.; Skoković, D.; Mattar, C.; Cristóbal, J. Land surface temperature retrieval methods from Landsat-8 thermal infrared sensor data. *IEEE Geosci. Remote Sens. Lett.* **2014**, *11*, 1840–1843. [CrossRef]

36. Firozjaei, M.K.; Alavipanah, S.K.; Liu, H.; Sedighi, A.; Mijani, N.; Kiavarz, M.; Weng, Q. A PCA–OLS Model for Assessing the Impact of Surface Biophysical Parameters on Land Surface Temperature Variations. *Remote Sens.* **2019**, *11*, 2094. [CrossRef]

37. Fu, P.; Weng, Q. A time series analysis of urbanization induced land use and land cover change and its impact on land surface temperature with Landsat imagery. *Remote Sens. Environ.* **2016**, *175*, 205–214. [CrossRef]

38. Firozjaei, M.K.; Fathololoumi, S.; Kiavarz, M.; Arsanjani, J.J.; Alavipanah, S.K. Modelling surface heat island intensity according to differences of biophysical characteristics: A case study of Amol city, Iran. *Ecol. Indic.* **2020**, *109*, 105816. [CrossRef]

39. Firozjaei, M.K.; Kiavarz, M.; Alavipanah, S.K.; Lakes, T.; Qureshi, S. Monitoring and forecasting heat island intensity through multi-temporal image analysis and cellular automata-Markov chain modelling: A case of Babol city, Iran. *Ecol. Indic.* **2018**, *91*, 155–170. [CrossRef]

40. Karimi, F.M.; Kiavarz, M.M.; Alavi, P.S.K. Monitoring and predicting spatial-temporal changes heat island in babol city due to urban sprawl and land use changes. *Eng. J. Geospat. Inf. Technol.* **2017**, *5*, 123–151.

41. Liu, H.; Weng, Q. Seasonal variations in the relationship between landscape pattern and land surface temperature in Indianapolis, USA. *Environ. Monit. Assess.* **2008**, *144*, 199–219. [CrossRef]

42. Lu, D.; Weng, Q. Spectral mixture analysis of ASTER images for examining the relationship between urban thermal features and biophysical descriptors in Indianapolis, Indiana, USA. *Remote Sens. Environ.* **2006**, *104*, 157–167. [CrossRef]

43. Weng, Q.; Hu, X.; Quattrochi, D.A.; Liu, H. Assessing intra-urban surface energy fluxes using remotely sensed ASTER imagery and routine meteorological data: A case study in Indianapolis, USA. *IEEE J. Sel. Top. Appl. Earth Obs. Remote Sens.* **2014**, *7*, 4046–4057. [CrossRef]

44. Weng, Q. Remote sensing of impervious surfaces in the urban areas: Requirements, methods, and trends. *Remote Sens. Environ.* **2012**, *117*, 34–49. [CrossRef]

45. Weng, Q.; Lu, D. A sub-pixel analysis of urbanization effect on land surface temperature and its interplay with impervious surface and vegetation coverage in Indianapolis, United States. *Int. J. Appl. Earth Obs.* **2008**, *10*, 68–83. [CrossRef]

46. Weng, Q.; Rajasekar, U.; Hu, X. Modeling urban heat islands and their relationship with impervious surface and vegetation abundance by using ASTER images. *IEEE Trans. Geosci. Remote Sens.* **2011**, *49*, 4080–4089. [CrossRef]

47. Wickham, J.; Wade, T.; Norton, D. Spatial patterns of watershed impervious cover relative to stream location. *Ecol. Indic.* **2014**, *40*, 109–116. [CrossRef]

48. Beck, S.M.; McHale, M.R.; Hess, G.R. Beyond impervious: Urban land-cover pattern variation and implications for watershed management. *Environ. Manag.* **2016**, *58*, 15–30. [CrossRef] [PubMed]

 **remote sensing**

Article

Assessment of Urban Ecological Quality and Spatial Heterogeneity Based on Remote Sensing: A Case Study of the Rapid Urbanization of Wuhan City

Jingye Li [1], Jian Gong [1,*], Jean-Michel Guldmann [2] and Jianxin Yang [1]

[1] Department of Land Resource Management, School of Public Administration, China University of Geosciences, Wuhan 430074, China; jingye.li@cug.edu.cn (J.L.); yangjianxin@cug.edu.cn (J.Y.)
[2] Department of City and Regional Planning, The Ohio State University, Columbus, OH 43210, USA; guldmann.1@osu.edu
* Correspondence: gongjian@cug.edu.cn

Abstract: Rapid urbanization significantly affects the productivity of the terrestrial ecosystem and the foundation of regional ecosystem services, thereby detrimentally influencing the ecological environment and urban ecological security. The United Nations' Sustainable Development Goals (SDGs) also require accurate and timely assessments of where people live in order to develop, implement and monitor sustainable development policies. Sustainable development also emphasizes the process of protecting the ecological environment for future generations while maintaining the current needs of mankind. We propose a comprehensive evaluation method for urban ecological quality (UEQ) using Landsat TM/ETM+/OLI/TIRS images to extract remote sensing information representing four ecological elements, namely humidity, greenness, heat and dryness. An improved comprehensive remote sensing ecological index (IRSEI) evaluation model is constructed by combining the entropy weight method and principal component analysis. This modeling is applied to the city of Wuhan, China, from 1995 to 2020. Spatial autocorrelation analysis was conducted on the geographic clusters of the IRSEI. The results show that (1) from 1995 to 2015, the mean IRSEI of Wuhan city decreased from 0.60 to 0.47, indicating that environmental deterioration overwhelmed improvements; (2) the global Moran's I for IRSEI ranged from 0.535 to 0.592 from 1995 to 2020, indicating significant heterogeneity in its spatial distribution, highlighting that high and low clusters gradually developed at the edge of the city and at the city center, respectively; (3) the high clusters are mainly distributed in the Huangpi and Jiangxia districts, and the low clusters at the city center, which exhibits a dense population and intense human activity. This paper uses remote sensing index methods to evaluate UEQ as a scientific theoretical basis for the improvement of UEQ, the control of UEQ and the formulation of urban sustainable development strategies in the future. Our results show that the UEQ method is a low-cost, feasible and simple technique that can be used for territorial spatial control and spatiotemporal urban sustainable development.

Keywords: remote sensing ecological index; ecological protection; principal component analysis; entropy value method; spatial autocorrelation; sustainable development; Wuhan city

Citation: Li, J.; Gong, J.; Guldmann, J.-M.; Yang, J. Assessment of Urban Ecological Quality and Spatial Heterogeneity Based on Remote Sensing: A Case Study of the Rapid Urbanization of Wuhan City. *Remote Sens.* **2021**, *13*, 4440. https://doi.org/10.3390/rs13214440

Academic Editor: Ronald C. Estoque

Received: 25 August 2021
Accepted: 2 November 2021
Published: 4 November 2021

Publisher's Note: MDPI stays neutral with regard to jurisdictional claims in published maps and institutional affiliations.

1. Introduction

Urban ecological quality (UEQ) evaluation is an important field of urban ecology research and the basis of urban planning and ecological management. With the continuous expansion of urbanization, China's cities have achieved medium-high quality development. However, social problems, such as resource exhaustion, an imbalance of economic structure and environmental pollution, do appear frequently. It is urgent to improve the capacity to implement urban sustainable development. In 2015, United Nations (UN) member states unanimously committed to achieving the Sustainable Development Goals (SDGs) by

2030 [1]. Although the urbanization process has improved people's living standards, promoted the sustainable development of productive forces and provided economic benefits, it has also broken the balance between human society and the natural environment, and has brought great challenges to UEQ [2,3]. According to the 2018 Revision of World Urbanization Prospects [4,5], the urban population will account for 68% of the global population by 2050, which is an increase of 13% from 2018, and China's urban population will increase by 255 million people. Cities cover less than 2% of the Earth's surface, but consume 78% of the energy generated and produce 60% of greenhouse gas emissions [5]. Additionally, urban land consumption outpaces population growth by approximately 50% [6].

Such changes affect human survival and the sustainable development of the social economy [7–10]. Using UEQ measures to determine the status of the ecological environment could promote the sustainable development of regional economies [8,11–14]. Therefore, the quantitative description and assessment of the spatiotemporal dynamics of urban ecological environments are emerging as leading research topics [11,15].

Numerous studies have been conducted on such an assessment from different perspectives, and several evaluation methods have been suggested. The pressure–state–response model and fuzzy evaluation methods are commonly used in ecological quality assessment. In recent years, geographic information system (GIS) and remote sensing (RS) technologies have provided efficient monitoring and analysis methods for ecological quality research and sustainable development. Progress in satellite-based Earth observation systems facilitates assessing the state of an ecosystem from local to global scales. The scale and scope of this research are expanding constantly. Index systems have been constructed using GIS to conduct strategic environmental assessment for regional and land-use planning [16,17]. In China, research on the ecological environment is based on the Technical Specifications for Ecological Environmental Assessment, promulgated by the National Environmental Protection Agency in 2006 [18]. According to these specifications, the ecological environment index (EI) should encompass biological richness, air pollution, water network density, vegetation cover, land degradation and related factors. The EI is the main tool used to evaluate the quality of the ecological environment [19]. However, as climatic and geological conditions differ across regions, the weight of each index must be adjusted accordingly. Currently, researchers mostly use manual processing, as weight allocation is not strictly required and evaluation criteria vary, making it extremely difficult to accurately compare urban ecological conditions. Therefore, a scientific and logical ecological quality assessment method is required.

The acceleration of urbanization has led to a series of ecological and environmental effects, such as reduced surface water transpiration and water quality. It is generally difficult to monitor these natural processes with on-site instruments. However, remote sensing technologies can provide quantitative physical data with high spatial and temporal resolutions to facilitate the quantitative monitoring and analysis of environmental effects. Among all of the environmental effects of urbanization, the thermal environment has received more attention. The urban thermal environment is an important representative indicator of the urban environment. It is influenced by the physical properties of the urban surface and human social and economic activities, and is a comprehensive summary and embodiment of urban ecosystems. Vegetation is another important component of urban ecosystems. Urban vegetation can selectively absorb and reflect solar radiation energy, adjust the latent and sensible heat exchange, regulate urban air, reduce pollution and other processes that affect the city's natural environment and is another highly comprehensive index of urban ecological evaluation. The spatial distribution and richness of vegetation in cities have always been considered to have important effects on the evolution of the urban ecological environment.

The remote sensing ecological index (RSEI) combines humidity, greenness, heat and dryness indices obtained from RS, and facilitates the monitoring and evaluation of the UEQ. The RSEI, which was first proposed by Hu and Xu [18], could aid in visualizing spatial and temporal analyses and predictions of change in the regional environment, thereby compensating for the deficiencies of the EI. This paper uses existing research from a new perspective to more accurately study urban socio-economic activity intensity and its relationship with the regional ecological environment. Using the RSEI will help in studying the interactions between human activities and natural ecology, and the resulting knowledge of theory, concepts and methods is expected to benefit local governments [20]. In recent years, the RSEI has been applied in ecological quality monitoring in 35 cities of China [19,21,22], Eurasia [23] and America [21,23]. The RSEI and the results of principal component analysis (PCA) have been combined to develop an ecological index [19,24]. However, using the PCA results in insufficient information utilization, as the adaptive nature of PCA algorithms inevitably limits the full use of the available information. For example, the RSEI obtained in two studies using only the first component for normalization ranged from 60% to 90%, which cannot guarantee adequate contribution rates.

Accordingly, the aim of the current study is to improve the RSEI calculation method by proposing an improved-comprehensive remote sensing ecological index (IRSEI) constructed by employing PCA and equal weights (EW). Our study overcomes the shortcomings of previous studies, which only considered the application of PCA in ecological quality assessment, and the resolved knowledge gaps are reflected in the comprehensive consideration of EW and the PCA method to determine the UEQ. The contribution rates of the eigenvalues of PCA and EW are taken as the weights. This method enables the full use of the available data and ensures that the value of the calculated IRSEI is ecologically optimal. In addition, more indicators could be integrated and the IRSEI reduces noise interference and makes optimal use of practical image information. These factors facilitate the reliable and quantitative monitoring of the regional ecological environment.

A comparison and evaluation of the differences in quality in large cities can improve the cognitive ability of the internal mechanism of the reciprocal feed-back relationship between the construction of megacities and regional ecological balance, and can provide a scientific reference for controlling the scale of urban sustainable development and ecological planning and regulation. Wuhan is one of the fastest growing cities in central China, but few studies have been conducted on quantitative UEQ monitoring based on remote sensing data. Therefore, we used a series of parameters obtained from remote sensing imagery to construct the IRSEI for the evaluation of the UEQ of Wuhan city from 1995 to 2020. In addition to the UEQ, we determined the temporal and spatial changes in the city. We present a discussion of the ecological changes caused by economic and social developments and natural conditions. Finally, we provide theoretical guidance and a scientific basis for ecological construction in Wuhan city.

The objectives of this study are to:

(1) Use GIS and RS technology to construct the IRSEI efficiently by integrating multiple sensors, including the Landsat Thematic Mapper (TM), Operational Land Imager (OLI) and Thermal Infrared Sensor (TIRS);
(2) Monitor spatial and temporal changes in UEQ in Wuhan from 1995 to 2020;
(3) Explore the spatial differentiation characteristics of the IRSEI in Wuhan.

2. Materials and Methods

2.1. Study Area and Data Preprocessing

We select the rapidly urbanizing city of Wuhan as study area for ecological monitoring and assessment. Wuhan is the capital city of Hubei Province. Its geographical location is $29°58'$–$31°22'$N and $113°41'$–$115°05'$E (Figure 1). From the perspective of Wuhan's geographical location and the location of its basin, the development of Wuhan has had great impact on the environment of the whole Yangtze River basin, and even the whole country. Therefore, ecological assessment and policy-based restoration and protection in Wuhan are vital for the ecological restoration of the Yangtze River basin. The city has jurisdiction over six central urban areas and seven distant urban areas. The land area comprises 8494.41 km^2. The permanent population was 10.91 million in 2018. The Yangtze and Han rivers meet there, forming a geographical pattern referred to as "two rivers and three towns". Wuhan has a subtropical humid monsoon climate, with abundant rainfall and sufficient heat throughout the year. The average annual temperature is 15.8 to 17.5 °C. The area is rich in ecological resources, with nearly 40% green coverage and more than 10 m^2 of green space per capita. These ecological resources are crucial for Wuhan to build an ecological civilization city and, therefore, are critical factors in the protection of the ecological environment.

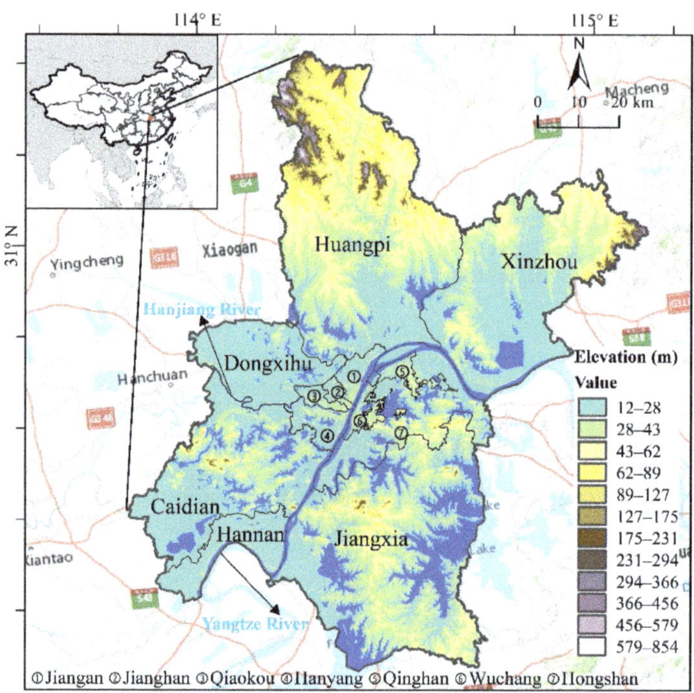

Figure 1. Location of Wuhan city.

In order to consider the quality of the remote sensing data, such as cloud cover and vegetation condition, we use data from Landsat 5 TM in 1995, Landsat 5 ETM in 2005 and Landsat 8 OLI in 2015 and 2020 as the main remote sensing data. RS data are particularly useful because they can be used for temporal and spatial monitoring [25]. Details on the satellite images used in this study are provided in Table 1. The source dates of the images are relatively close; therefore, differences caused by different seasons and vegetation growth states can be ignored. Owing to topographic differences in images at different times and the influence of illumination and atmospheric factors on surface reflectance, the selected images required preprocessing with radiometric calibration and atmospheric and geometric correction prior to the calculation of the IRSEI. The corrections were performed using the Environment for Visualizing Images (ENVI) software. The Fast Line-of-sight Atmospheric Analysis of Hypercubes (FLAASH) model was used for atmospheric correction to eliminate the radiation error caused by atmospheric absorption and scattering. The accuracy of radiation calibration was more than 95%, and that of atmospheric correction exceeded 85%. Further, the error in geometric correction was controlled to less than 1 pixel. The quadratic polynomial and the nearest neighbor methods were used to correct the geometry of the images and the preprocessed images of the study area were clipped using the vector data of the administrative districts of Wuhan. Other data sources included the administrative zoning map of Wuhan, digital elevation model data of Wuhan from the geospatial data cloud (http://www.gscloud.cn/sources/accessdata/310?pid=302 (accessed on 15 May 2021)) and the cloud platform of geographical national condition monitoring of China (http://www.dsac.cn/DataProduct/Search?&cateID=2010&areaID=18 (accessed on 15 May 2021)). Nighttime light data were obtained from the national geophysical data center (https://www.ngdc.noaa.gov/eog/dmsp/downloadV4com-posites.html (accessed on 15 May 2021)).

Table 1. Data used and their source.

Data Used	Data Acquisition Data	Spatial Resolution	Source
LANDSAT TM	24 October 1995	30 × 30	http://earthexplorer.usgs.gov/ (accessed on 15 May 2021).
LANDSAT ETM	11 September 2005	30 × 30	
LANDSAT OLI	28 September 2015	30 × 30	
	29 October 2020	30 × 30	

2.2. Methodology

2.2.1. Modeling Framework

We combine principal component analysis (PCA) and the entropy value method to design synthetic indicators that facilitate quick and quantitative assessment of UEQ, based on humidity, greenness, dryness and the heat index. Using this method enables prioritizing the natural factors of the ecological evaluation system. The overall framework of IRSEI modeling, as shown in Figure 2, includes four main steps. First, we obtain Landsat Enhanced Thematic Mapper Plus (ETM+)/OLI/TIRS images and perform preprocessing, including atmospheric correction, radiometric calibration and image mosaic (see Section 2.1). Second, we derive four remote sensing indicators: humidity, greenness, dryness and heat. Third, we calculate the PCA components, obtain PC1 and use the entropy method to calculate the results that are used for the construction of the IRSEI. Finally, the characteristics of the spatial and temporal changes in the ecological quality of Wuhan over the past 25 years are determined, and the spatial heterogeneity of the city is analyzed.

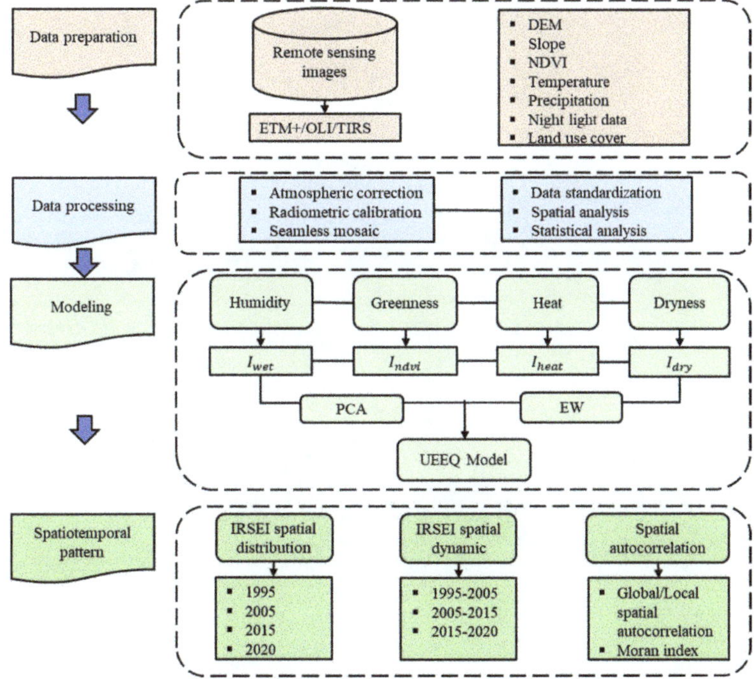

Figure 2. Overall methodological framework.

2.2.2. Calculation of Component Indices

- Humidity index (I_{wet})

The Kauth–Thomas transform (also called the tasseled hat transform) is a linear transformation method based on multispectral imaging [26,27]. This method is widely used in ecological monitoring for data compression and removal of redundancy. The moisture component obtained by this transform reflects moisture information in the soil and vegetation. A low humidity value indicates severe land degradation, low vegetation cover and a poor ecological environment. A high humidity value indicates sufficient soil moisture, rich surface vegetation cover and a good ecological environment.

In this study, I_{wet} was chosen as the humidity index [28], which is expressed as land surface moisture and is generated from Landsat TM, ETM+ and OLI image reflectance using Equations (1)–(3) [10,23,29]:

$$I_{wetTM} = 0.0315\rho1 + 0.2021\rho2 + 0.3102\rho3 + 0.1594\rho4 - 0.6806\rho5 - 0.6109\rho7 \tag{1}$$

$$I_{wetETM+} = 0.2626\rho1 + 0.2141\rho2 + 0.0926\rho3 + 0.0656\rho4 - 0.7629\rho5 - 0.5388\rho7 \tag{2}$$

$$I_{wetOLI} = 0.1511\rho1 + 0.1973\rho2 + 0.3283\rho3 + 0.3407\rho4 - 0.7117\rho5 - 0.4559\rho7 \tag{3}$$

where $\rho1$, $\rho2$, $\rho3$, $\rho4$, $\rho5$ and $\rho7$ represent reflectance in bands 1, 2, 3, 4, 5 and 7 of Landsat TM/ETM+ images and reflectance in bands 2, 3, 4, 5, 6 and 7 of Landsat OLI data, respectively.

- Greenness index (I_{ndvi})

The normalized difference vegetation index (NDVI) is often used to monitor vegetation growth [30] and directly reflects the quality of the regional ecological environment. This index is used in the classification of regional land cover, environmental change and vegetation. The NDVI greenness index is computed as follows [31]:

$$I_{ndvi} = (\rho4 - \rho3)/(\rho4 + \rho3) \tag{4}$$

where $\rho4$ represents the reflectance of the near-infrared band and $\rho3$ represents the reflectance of the red band.

- Heat index (I_{heat})

Land surface temperature (LST) refers to heat, which is related closely to vegetation growth, crop yield, surface water circulation, urbanization, other natural phenomena and processes and human activities [32]. LST can be used as a heat index to reflect the surface ecological environment. Several algorithms use thermal infrared technology to retrieve LST, including the atmospheric correction, single-window and single-channel algorithms. Comparison between LST retrieval results obtained using the atmospheric correction method and the actual measurement of LST indicates that the error is within 1 °C, thereby meeting research accuracy requirements. LST is generated using Equations (5)–(9) [33,34]:

$$L = \text{gain} \times DN + \text{bias} \tag{5}$$

$$Tb = K2/ln(K1/L + 1) \tag{6}$$

$$LST = Tb/\{1 + [(\lambda Tb)/\rho]ln\varepsilon\} - 273.15 \tag{7}$$

where DN is the pixel gray value, gain and bias are thermal infrared band excursions and L is the radiation brightness value.

Equation (7) is a simplified form of the inverse function of Planck's formula, with $K1$ and $K2$ being the calibration parameters. All of the parameter values are available from the metadata file (MTL) of the satellite data. ε is the specific infrared emissivity and is calculated with the method proposed by Min [35]. λ is the central wavelength of the thermal infrared band and ρ = s 1.438 10^{-2} mK.

$$\begin{cases} \epsilon_{water} = 0.995 \; (NDVI \leq 0) \\ \epsilon_{building} = 0.9589 + 0.086 \times F_{veg} - 0.0671 \times F_{veg}^2 \; (0 < NDVI < 0.7) \\ \epsilon_{natural} = 0.9625 + 0.0614 \times F_{veg} - 0.0461 \times F_{veg}^2 \; (NDVI \geq 0.7) \end{cases} \tag{8}$$

Vegetation coverage (F_{veg}) refers to the ratio (%) of the vertical projection area of vegetation on the ground to the total statistical area. P_{veg} is based on Landsat NDVI and adopts the dichotomy model of mixed pixels [36]. The calculation formula is as follows [37]:

$$P_{veg} = \frac{NDVI - NDVI_{soil}}{NDVI_{veg} - NDVI_{soil}} \tag{9}$$

where NDVI is the normalized vegetation index, $NDVI_{soil}$ is the normalized vegetation index value of bare land and $NDVI_{veg}$ is the normalized vegetation index value of complete vegetation coverage. $NDVI_{soil}$ and $NDVI_{veg}$ were selected as $NDVI_{max}$ and $NDVI_{min}$ with a confidence level of more than 95%.

- Dryness index (I_{dry})

The dryness index refers to the quantification of soil desiccation, which is a condition detrimental to the ecological environment. As most urban construction land is located in our study area, the dryness index can be represented by combining the bare soil index (SI) and the built-up index (IBI) into a normalized building–bare-soil index (NDBSI) [29]. We proposed extracting the bare soil and building area by setting an appropriate threshold and, subsequently, calculating the NDBSI as a weighted average and employing the area ratio as the weight.

$$NDBSI = (SI + IBI)/2 \tag{10}$$

$$SI = [(\rho 5 + \rho 3) - (\rho 4 + \rho 1)]/[(\rho 5 + \rho 3) + (\rho 4 + \rho 1)] \tag{11}$$

$$IBI = \frac{\left[\frac{2\rho 5}{\rho 5 + \rho 4} - \left(\frac{\rho 4}{\rho 4 + \rho 3} + \frac{\rho 2}{\rho 2 + \rho 5}\right)\right]}{\left[\frac{2\rho 5}{\rho 5 + \rho 4} + \left(\frac{\rho 4}{\rho 4 + \rho 3} + \frac{\rho 2}{\rho 2 + \rho 5}\right)\right]} \tag{12}$$

where $\rho 1$, $\rho 2$, $\rho 3$, $\rho 4$ and $\rho 5$ have been defined earlier in the context of the humidity index.

2.2.3. Water Mask and Standardization

The humidity index reflects the moisture of the vegetation and soil. The area covered by water in the study area occupies a large proportion of the I_{wet}, which reduces the advantage of vegetation and soil in I_{wet}. Therefore, the calculated I_{wet} is not a true reflection of the vegetation and soil moisture, and it is necessary to mask the water bodies present in the study area. We use a modified normalized difference water index (MNDWI) to mask these water bodies. The formula is:

$$MNDWI = (\rho_{Green} - \rho_{MIR})/(\rho_{Green} + \rho_{MIR}) \tag{13}$$

where ρ_{Green} represents the reflectance of the near-infrared band and ρ_{MIR} represents the reflectance of the red band.

2.2.4. Construction of the Improved Remote Sensing Ecological Index (IRSEI) Evaluation Model

First, we obtain the primary remote sensing ecological index based on PCA. The four indices are standardized to the range [0–1] and PCA is used to combine these indices. PCA1 is obtained from the four RSEIs to build a preliminary assessment model. Generally, the first PCA collects most of the information on the four indicators, and PC1 can be used to represent the characteristics of the regional ecological environment. Therefore, we use only one PC in further analyses. To facilitate index measurement and comparison, the initial RSEI is standardized, as follows:

$$RSEI_{PCA} = 1 - f\left(I_{\text{wet}}, I_{\text{ndvi}}, I_{\text{heat}}, I_{\text{dry}}\right) \tag{14}$$

$$f = \sum_{i=1}^{4}(e_i \times PC1) \tag{15}$$

where I_{ndvi} represents the green component; I_{wet} represents the humidity component; I_{heat} represents heat; I_{dry} represents dryness; and PC1 is the first principal component. The obtained RSEI value is within the [0–1] range. e_i is the characteristic value contribution rate of the index corresponding to PC1. The closer RSEI is to 1, the better the UEQ of the region. The first principal component analysis index values are listed in Table 2. A detailed description of the calculation steps is available in the relevant literature [11,22,24,29].

Table 2. Principal component analysis index and eigenvalue.

Year	PC1	Eigenvalue	Contribution/%	Accumulation/%
1995	NDVI	0.0441	88.6768	88.6768
	WET	0.0048	9.5508	98.2276
	NDBSI	0.0002	0.4187	98.6463
	LST	0.0006	1.3537	100
2005	NDVI	0.0464	81.3557	81.3557
	WET	0.0071	12.5046	93.8603
	NDBSI	0.0003	0.5021	94.3624
	LST	0.0032	5.6376	100
2015	NDVI	0.0476	96.3065	96.3065
	WET	0.0012	2.3826	98.6891
	NDBSI	0.0001	0.131	98.8201
	LST	0.0006	1.1799	100
2020	NDVI	0.04	97.4195	97.4195
	WET	0.0007	1.7021	99.1216
	NDBSI	0.0001	0.035	99.1566
	LST	0.0003	0.8434	100

Second, we introduce the entropy value method, which determines the weight of each index according to the information provided by the observed values of each index [38,39]. The evaluation index system includes N indices (NDVI, WET, NDBSI and LST). This is a problem that consists of m samples (cell) and uses N indicators for comprehensive evaluation. The initial data matrix A of the evaluation system is formed and X_{ij} is the value in i cell of the j remote sensing ecological indicator. The detailed procedures of the entropy method are described as follows [22,38,40]:

$$A = \begin{pmatrix} X_{11} & \cdots & X_{1m} \\ \vdots & \vdots & \vdots \\ X_{n1} & \cdots & X_{nm} \end{pmatrix}_{n \times m}$$

1. Proportion of the value in i cell of the indicator j.

$$P_{ij} = \frac{X_{ij}}{\sum_{i=1}^{n} X_{ij}} \quad (j = 1, 2, \cdots m) \tag{16}$$

2. Entropy value of the j th index.

$$e_j = -\frac{1}{lnm} \times \sum_{i=1}^{n} P_{ij} ln(P_{ij}) \quad k > 0, e_j \geq 0, 0 \leq e_j \leq 1$$

3. Difference coefficient of the first index.

For the j th index, the more significant the difference is in the index value X_{ij}, the greater the effect on the scheme evaluation and the smaller the entropy value.

$$g_j = 1 - e_j$$

The larger the g_j value, the more critical the indicator.

4. Weight.

$$W_j = \frac{g_j}{\sum_{j=1}^{m} g_j} , j = 1,2 \cdots m \tag{17}$$

5. Ecological index score based on the entropy method.

$$RSEI_{EW} = \sum_{j=1}^{m} W_j \times P_{ij} \ (i = 1,2, \cdots n) \tag{18}$$

The PCA effectively removes redundant information between bands and compresses multiband image information into a few independent bands that are more effective than the original band. The entropy method can effectively remove deficiencies caused by a lack of PCA information. The weights for all of the indicators are listed in Table 3.

Table 3. Weights of indicators.

Year	Indicators	Effect Direction	Weight
1995	Humidity index	+	0.7463
	Greenness index	+	0.0144
	Heat index	-	0.1315
	Dryness index	-	0.1078
2005	Humidity index	+	0.7918
	Greenness index	+	0.0538
	Heat index	-	0.1201
	Dryness index	-	0.0343
2015	Humidity index	+	0.734
	Greenness index	+	0.0048
	Heat index	-	0.1465
	Dryness index	-	0.1147
2020	Humidity index	+	0.9401
	Greenness index	+	0.003
	Heat index	-	0.0004
	Dryness index	-	0.0565

Finally, the IRSEI integrates humidity, greenness, heat and dryness through PCA and EW, which is calculated according to Equation (19):

$$IRSEI = (RSEI_{PCA} + RSEI_{EW})/2 \tag{19}$$

In this formula, $RSEI_{PCA}$ is the main component, $RSEI_{EW}$ is the weighted result of the entropy method and the final IRSEI is calculated as their arithmetic average. The IRSEI for each year has to be standardized to accurately compare the remote sensing images of different time frames. The closer IRSEI is to 1, the better the UEQ (and vice versa). The IRSEI for the four years is classified into five groups employing the ArcGIS software (Esri, USA). Referring to previous studies [22–24,29,41], these groups are labeled "Excellent, Good, Moderate, Fair, and Poor" and they facilitate comparisons across the study area (Table 4).

Table 4. Grades of ecological indicators.

Grades	I	II	III	IV	V
	Excellent	Good	Moderate	Fair	Poor
IRSEI indicator	[0.8–1.0]	[0.6–0.8]	[0.4–0.6]	[0.2–0.4]	[0–0.2]

2.2.5. Spatial Autocorrelation Analysis of IRSEI

Global spatial autocorrelation (SA) measures the average correlation, spatial distribution pattern and significance of all of the objects in the entire study area. SA visualizes spatial aggregations and exceptions to the IRSEI. The Moran's index is commonly used to calculate SA [42]. The main calculation indices for spatial autocorrelation are the global Moran's index and the local Moran's index. We analyze both the "global" spatial clustering and the "local" spatial clustering of the IRSEI. The formula for calculating the global Moran's index is:

$$I = \frac{\sum_{i=1}^{n} \sum_{j=1}^{n} W_{ij}(x_i - \bar{x})(x_j - \bar{x})}{S^2 \times \sum_{i=1}^{n} \sum_{j=1}^{n} W_{ij}} \text{ (SA)} \tag{20}$$

where n is the total number of grid cells in the study area (500 m $\times$ 500 m); W_{ij} represents the spatial weight of elements i and j; x_i and x_j are the attribute values of cell i and cell j, respectively; $\bar{x}$ represents the average value of the attributes across all cells; and S^2 is the sample variance.

The value of the global Moran index I varies between -1 and 1, where $I > 0$ indicates positive SA, i.e., a high value corresponds to high-value clusters, whereas a low value corresponds to low-value clusters. The closer I is to 1, the smaller the overall spatial difference. When $I < 0$, there is negative SA, i.e., there is significant spatial difference between a cell and its surrounding cells. The closer I is to -1, the greater the overall spatial difference. When $I = 0$, there is no SA.

Due to the fact that the global Moran's index describes the overall aggregation situation, it cannot accurately determine where the place of aggregation is located and is unable to indicate the hot spots and cold spots of the entire region. Accordingly, we use the local indicator of SA to measure local SA and determine hot and cold spots. The formula for the local Moran's I_i for cell i is:

$$I_i = \frac{(x_i - x)}{S^2} \sum_{j=1}^{n} W_{ij}(x_j - x) \tag{21}$$

When the local Moran index $I_i > 0$, the spatial difference between the cell and its surrounding cells is minor. When the local Moran index $I_i < 0$, the spatial difference between cell i and its surrounding cells is significant. When the local Moran index $I_i = 0$, there is no spatial difference between cell i and its surrounding cells. In this study, we use the software GeoDA to calculate and obtain the global and local Moran's indices.

3. Results

3.1. Attributing Factors

A comparison of the spatial distributions of the four ecological factors in the study area (Figure 3) shows high levels of land surface moisture close to and alongside the Yangtze River, which extends in the central part of Wuhan from west to east. The NDVI is high on the northeast side, along the Yangtze and Han rivers, the central part of Wuhan and in patches in the south and east. Comparing the NDVI, LST and moisture maps shows that moderate temperature and moisture are the most favorable conditions for vegetation growth, whereas extreme weather conditions can damage plant vitality. The temperature and moisture conditions are moderate in the study area and the NDVI is remarkably high. A high LST is detected in the southern part of Wuhan, with some patches in the north and east. A moderate LST is detected in the central part of Wuhan. The NDBSI does not display much variation, as most of the study area is covered by agricultural land (Figure 3).

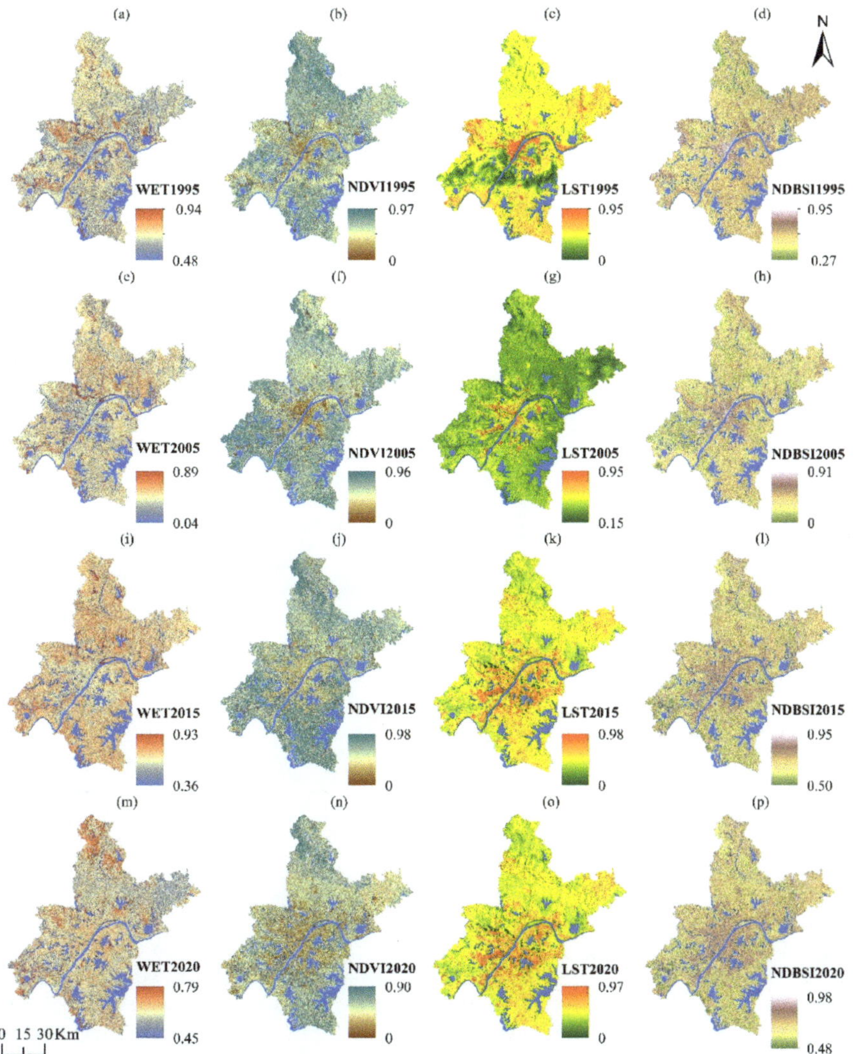

Figure 3. Spatial distribution of ecological indicators, 1995–2020. (**a–d**) indicators 1995, (**e–h**) indicators 2005, (**i–l**) indicators 2015, (**m–p**) indicators 2020.

To test the representativeness of the index IRSEI, we calculate the correlation coefficients among IRSEI, WET, NDVI, NDSI and LST in the same period (Table S1, Supplementary Materials), and test the applicability of the model through average correlations. From 1995 to 2015, the average correlation of IRSEI with the other variables is the highest, ranging from 0.60 to 0.70. The mean correlation of IRSEI over this period was 0.64, which indicates that IRSEI integrates most of the information embodied in all four indicators. It is more representative than any single indicator and can better reflect the ecological situation.

3.2. Spatial and Temporal Distribution of UEQ in Wuhan

Generally, higher IRSEI values are associated with higher levels of greenness and moisture, whereas lower IRSEI values are directly proportional to dryness and temperature. This implies that high IRSEI values represent positive ecological conditions.

As shown in Figure 4, the IRSEI increases from 0.79 to 0.98 from 2010 to 2015, indicating improved ecological conditions. However, from the second half of 2015 up to 2020, its value drops to 0.82, indicating deterioration. Comparing the values from 2010 to 2020 indicates overall improved conditions, as the IRSEI increases from 0.79 to 0.82. However, the maximum values (1.09, 1.03 and 0.96) decline continuously, indicating that high-quality IRSEI conditions are declining continuously. Further, low-quality IRSEI conditions improve in the first half of the study period (1995 to 2005); however, in the second half (2005 to 2020), these conditions decline and reach their previous stage. Our findings also show maximal variation in the median IRSEI values, i.e., indicating the recovery of favorable conditions (moderate to high temperature, moderate to low moisture and higher vegetation) for all factors during the study period.

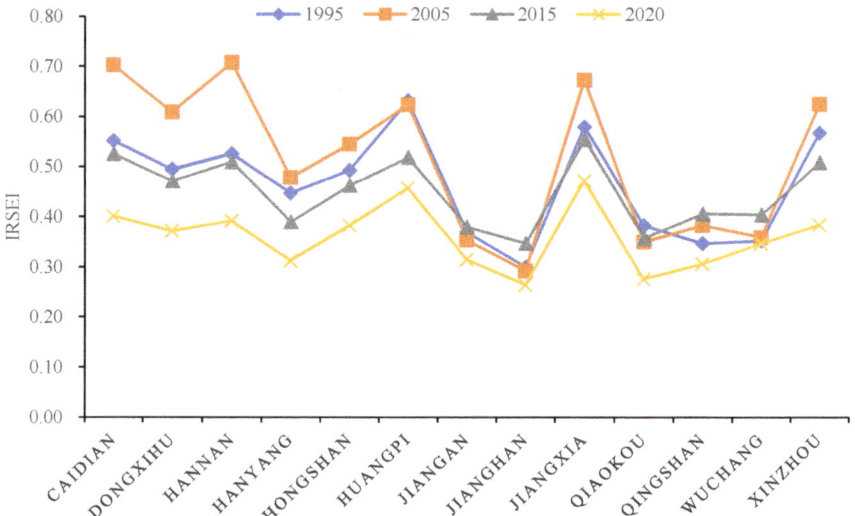

Figure 4. Changing trend of the IRSEI in each district of Wuhan from 1995 to 2020.

The mean IRSEI value and the area and percentage of each evaluation grade in Wuhan from 1995 to 2020 are displayed on Figures 4 and 5. Overall, the proportion of areas with average and good IRSEI ratings is the highest during the study period (>57%). The proportions of average and above average regions are 82.33%, 87.14%, 74.21% and 57.34%, indicating that the ecological environment of Wuhan was unstable from 1995 to 2020, with ecological conditions first improving and subsequently deteriorating. The UEQ of the Xinzhou, Hanyang, Qiaokou, Huangpi and Caidian districts show the most obvious decline, with reduction rates of 32.32%, 30.18%, 27.84%, 27.67% and 27.24%, respectively.

From the perspective of a single year (see Table 5), the area share of good ecological environment in 1995 was the highest, reaching 43.67% of the total area. The area share of poor ecological environment was the lowest in 1995, comprising an area of only 371 km^2, or less than 5% of the total area. The share of poor ecological environment was approximately 12% of the total area. The area with a good ecological environment rating in 2005 was larger than that of 1995 and accounted for the highest proportion (45.57%), comprising an area of 3494 km^2. The percentage of area rated excellent was the smallest (6.23%) after 1995. In 2020, the poor ecological environment generally accounted for the highest proportion (38.92%), comprising an area of 2984 km². The good ecological environment rating accounted for only 18.32%.

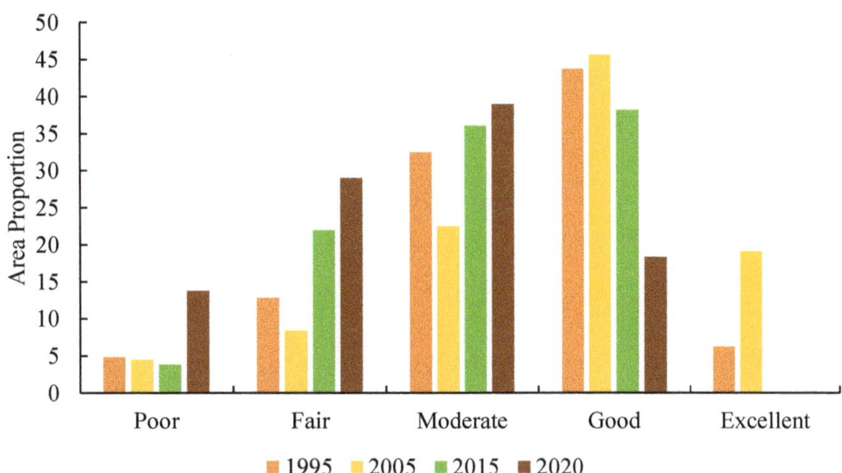

Figure 5. Changing area proportion of the IRSEI in Wuhan from 1995 to 2020.

Table 5. Area and proportion of the IRSEI over 1995 to 2020 in Wuhan city (unit: km², %).

IRSEI	1995		2005		2015		2020	
	Area	Proportion	Area	Proportion	Area	Proportion	Area	Proportion
0–0.2	371	4.84%	344	4.49%	294	3.84%	1051	13.71%
0.2–0.4	981	12.80%	643	8.38%	1684	21.96%	2221	28.97%
0.4–0.6	2486	32.43%	1725	22.49%	2764	36.05%	2984	38.92%
0.6–0.8	3348	43.67%	3494	45.57%	2926	38.16%	1404	18.32%
0.8–1.0	478	6.23%	1463	19.08%	0	0.00%	8	0.10%

The changing trend during the research period shows that the mean IRSEI values in 1995, 2005, 2015 and 2020 decreased year by year (0.60, 0.67, 0.58 and 0.47, respectively). The declining values indicate that the ecological environment of Wuhan has deteriorated continuously, probably owing to the rapid economic development of the city. According to the Wuhan Municipal Bureau of Statistics, the gross domestic product (GDP) increased from CNY 3.991 billion in 1978 to CNY 134.10 billion in 2017. The permanent resident population increased from 8.58 million people in 2004 to 10.33 million people in 2014. Ecological problems ascribed to human activities, such as vegetation damage and soil pollution, have become increasingly prominent.

As governments and social organizations have become increasingly aware of environmental protection, Wuhan has strengthened its enforcement of ecologically relevant laws and regulations, effectively halting the trend of environmental deterioration. This is reflected in the varying ecological evaluation grades. The differences in rating reflect an increase in area from 643 km² in 2005 to 1684 km² in 2015 (area expansion of 14%) to 2221 km² in 2020 (area expansion of 7%).

The spatial distribution (Table 6 and Figure 6) shows that areas with a good ecological environment are distributed mainly in the surrounding urban areas of Wuhan. These areas have a relatively weak economy and the land-use types are mainly cultivated land and woodland, with rich vegetation and high biodiversity levels. The areas with poor ecological environments are concentrated in Hongshan, Hanyang, Wuchang and Qingshan. According to the different functions of each administrative region of Wuhan, Hongshan is based mainly on the education industry. Several colleges and universities are located in the area, and it is densely populated. Qingshan, Hanyang and Wuchang are primarily industrial areas. Heavy industrial companies, such as Wuhan Iron & Steel Co., Ltd., Wushi Chemical Co., Ltd. and Dongfeng Motor Co., Ltd., are located in these areas. Industrial production and human economic activities have a direct detrimental effect on the environment of these areas.

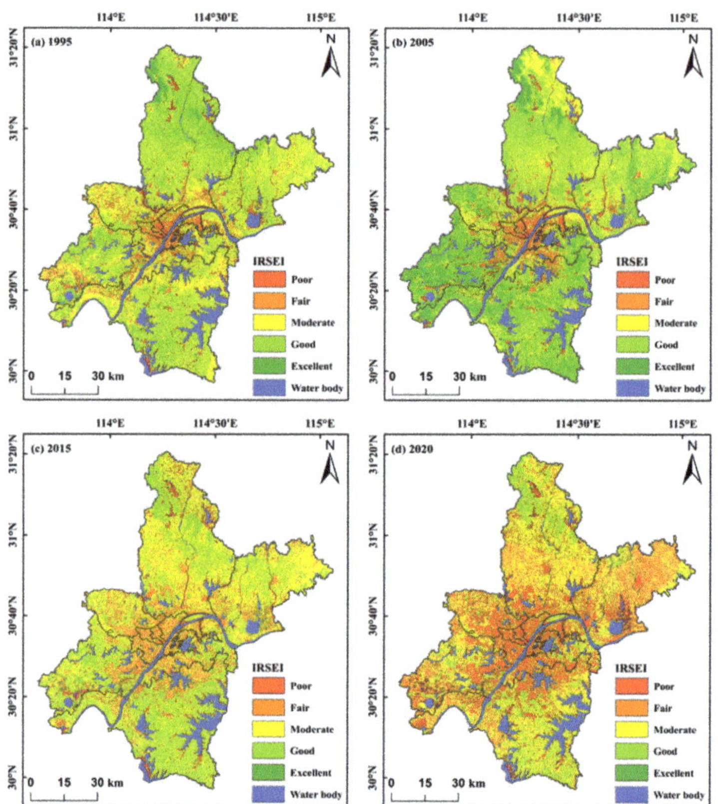

Figure 6. Grading map of UEQ from 1995 to 2020 in Wuhan city. (**a**) IRSEI 1995, (**b**) IRSEI 2005, (**c**) IRSEI 2015 and (**d**) IRSEI 2020.

Table 6. Area statistics of UEQ evaluation grade from 1995 to 2020 in Wuhan city (unit: km^2, %).

	IRSEI	1995		2005		2015		2020	
		Area	Proportion	Area	Proportion	Area	Proportion	Area	Proportion
	0–0.2	33.34	3.41	44.30	4.53	43.34	4.43	159.00	16.25
	0.2–0.4	182.83	18.70	73.39	7.50	204.01	20.85	302.66	30.94
Caidian	0.4–0.6	318.68	32.59	115.98	11.86	301.04	30.77	356.95	36.49
	0.6–0.8	394.58	40.36	315.74	32.28	429.84	43.94	159.47	16.30
	0.8–1.0	48.29	4.94	428.85	43.84	0.03	0.00	0.26	0.03
	0–0.2	26.76	5.63	26.11	5.49	6.62	1.39	93.09	19.55
	0.2–0.4	107.11	22.52	58.36	12.26	161.61	33.96	160.44	33.70
Dongxihu	0.4–0.6	194.95	40.98	114.47	24.05	191.74	40.29	172.55	36.25
	0.6–0.8	144.37	30.35	176.46	37.08	115.93	24.36	49.77	10.45
	0.8–1.0	2.48	0.52	100.51	21.12	0.01	0.00	0.22	0.05

Table 6. *Cont.*

	IRSEI	1995		2005		2015		2020	
		Area	Proportion	Area	Proportion	Area	Proportion	Area	Proportion
Hannan	0–0.2	12.66	4.91	9.69	3.76	13.55	5.26	42.96	16.68
	0.2–0.4	42.45	16.45	15.87	6.16	55.44	21.51	83.13	32.28
	0.4–0.6	111.99	43.40	31.36	12.17	88.47	34.33	97.70	37.93
	0.6–0.8	81.25	31.49	94.55	36.69	100.24	38.90	33.61	13.05
	0.8–1.0	9.70	3.76	106.23	41.22	0.00	0.00	0.17	0.07
Hanyang	0–0.2	9.14	9.17	11.42	11.46	5.12	5.14	31.27	31.40
	0.2–0.4	31.18	31.30	27.49	27.59	52.91	53.10	36.47	36.63
	0.4–0.6	35.36	35.48	27.14	27.24	28.77	28.87	24.28	24.39
	0.6–0.8	23.74	23.83	27.28	27.38	12.84	12.89	7.53	7.57
	0.8–1.0	0.22	0.22	6.30	6.32	0.00	0.00	0.01	0.01
Hongshan	0–0.2	42.60	8.93	41.29	8.66	24.48	5.13	104.50	21.91
	0.2–0.4	93.16	19.53	81.07	16.99	158.38	33.20	143.26	30.04
	0.4–0.6	184.41	38.66	129.99	27.25	159.77	33.49	147.52	30.93
	0.6–0.8	151.94	31.85	181.63	38.07	134.42	28.18	81.66	17.12
	0.8–1.0	4.88	1.02	43.07	9.03	0.01	0.00	0.02	0.01
Huangpi	0–0.2	77.87	3.67	63.18	2.97	86.94	4.09	216.05	10.17
	0.2–0.4	122.09	5.75	105.77	4.98	368.63	17.35	483.26	22.74
	0.4–0.6	488.82	23.02	595.10	28.01	865.43	40.74	960.08	45.19
	0.6–0.8	1187.44	55.92	1173.85	55.25	803.40	37.82	462.81	21.78
	0.8–1.0	247.18	11.64	186.56	8.78	0.07	0.00	2.57	0.12
Jiangan	0–0.2	17.08	23.72	16.66	23.21	2.93	4.08	22.01	30.71
	0.2–0.4	23.97	33.28	28.60	39.84	40.42	56.30	27.62	38.53
	0.4–0.6	20.21	28.06	18.12	25.23	22.22	30.95	16.14	22.51
	0.6–0.8	10.70	14.85	7.76	10.81	6.22	8.66	5.91	8.24
	0.8–1.0	0.06	0.09	0.65	0.91	0.00	0.00	0.00	0.00
Jianghan	0–0.2	9.42	34.28	9.08	33.07	1.13	4.10	10.69	38.95
	0.2–0.4	11.41	41.50	12.48	45.46	18.63	67.85	11.69	42.58
	0.4–0.6	4.84	17.60	4.16	15.16	6.54	23.83	4.17	15.20
	0.6–0.8	1.82	6.61	1.61	5.86	1.16	4.22	0.90	3.26
	0.8–1.0	0.00	0.01	0.12	0.45	0.00	0.00	0.00	0.00
Jiangxia	0–0.2	77.13	4.60	58.66	3.50	68.87	4.11	173.97	10.37
	0.2–0.4	153.61	9.16	92.52	5.52	285.30	17.01	368.37	21.96
	0.4–0.6	581.14	34.66	283.28	16.89	429.99	25.64	643.15	38.35
	0.6–0.8	757.98	45.21	806.03	48.06	892.77	53.23	487.53	29.07
	0.8–1.0	106.69	6.36	436.59	26.03	0.16	0.01	4.21	0.25
Qiaokou	0–0.2	6.98	18.06	8.96	23.25	1.63	4.23	14.54	37.70
	0.2–0.4	16.31	42.17	16.03	41.61	24.51	63.61	15.28	39.61
	0.4–0.6	8.19	21.19	8.91	23.13	10.09	26.20	7.36	19.08
	0.6–0.8	6.93	17.92	3.99	10.36	2.30	5.97	1.39	3.61
	0.8–1.0	0.25	0.66	0.64	1.66	0.00	0.00	0.00	0.00
Qingshan	0–0.2	10.40	20.39	8.53	16.66	1.58	3.09	15.73	30.69
	0.2–0.4	23.10	45.31	20.76	40.55	25.24	49.30	20.53	40.06
	0.4–0.6	12.99	25.47	15.01	29.32	18.78	36.68	11.98	23.37
	0.6–0.8	4.38	8.59	6.19	12.09	5.60	10.93	3.02	5.89
	0.8–1.0	0.12	0.24	0.71	1.39	0.00	0.00	0.00	0.00
Wuchang	0–0.2	10.64	20.73	11.20	21.70	3.01	5.83	13.72	26.48
	0.2–0.4	23.56	45.91	22.57	43.72	24.67	47.78	18.84	36.37
	0.4–0.6	10.70	20.85	10.82	20.97	16.45	31.86	12.13	23.41
	0.6–0.8	6.25	12.17	5.90	11.44	7.50	14.53	7.11	13.72
	0.8–1.0	0.18	0.35	1.12	2.18	0.00	0.00	0.00	0.01
Xinzhou	0–0.2	37.12	2.78	34.87	2.61	35.06	2.62	153.92	11.51
	0.2–0.4	150.44	11.26	87.89	6.57	264.28	19.76	549.54	41.09
	0.4–0.6	514.20	38.48	370.17	27.68	624.33	46.69	530.22	39.65
	0.6–0.8	577.11	43.18	692.87	51.82	413.46	30.92	103.51	7.74
	0.8–1.0	57.51	4.30	151.36	11.32	0.03	0.00	0.18	0.01

3.3. Dynamic Monitoring of UEQ in Wuhan

Based on the IRSEI grade classification, the detected changes were divided further into nine levels and seven classes. The range for the levels of detected changes was −4 to +4, with a positive value indicating that the UEQ had improved, 0 indicating no change and a negative value indicating deterioration. For the classes with no detected changes, level 0 was classified as unchanged, level −4 as significantly worse and levels −2 and −3 as worse; level −1 as slightly worse; level 1 as slightly better; levels 2 and 3 as better; and level 4 as significantly better (Table 7).

Table 7. Change in the ecological index grade.

Change Grade	Level Change
Significantly worse	−4 (Excellent to Poor)
Obviously worse	−3 (Excellent to Fair/Good to Poor)
	−2 (Excellent to Moderate/Good to Fair/Moderate to Poor)
Slightly worse	−1 (Excellent to Good/Good to Moderate/Moderate to Fair/Fair to Poor)
No change	0 (no level change, eg. Excellent to Excellent)
Slightly better	1 (Above, and vice versa)
	2 (Above, and vice versa)
Obviously better	3 (Above, and vice versa)
Significantly better	4 (Above, and vice versa)

Table 8 presents the ecological changes in Wuhan from 1995 to 2020. The size of the area representing both UEQ and ecological deterioration (obviously worse and slightly worse) is 3636 km^2, accounting for the highest proportion (39.44%) over 2015–2020. The size of the area with the same UEQ (no change) is 2984 km^2, accounting for 35.56% of the total area. Among the areas with deteriorating UEQ, most (69.51%) deteriorated by one grade. Deterioration in UEQ accounted for 25.60%. Most of the areas showing improved environmental conditions improved by one grade, accounting for 79.41% of the entire improved area. The areas improving by two grades account for 18%. The areas representing levels 3 or 4 are relatively small, indicating gradual changes. The areas with significant changes are related to direct economic activities, such as the transformation of cultivated land and woodland into construction and industrial land. The spatial distribution of UEQ (Figure 7) shows that the deteriorating areas are located mainly around cities and most water bodies. The deterioration of the ecological environment around water bodies is related to a leakage of urban domestic sewage and enterprise wastewater and a rise in aquaculture in recent years. Moreover, the areas with a deteriorating ecological environment are expanding along both sides of the Yangtze and Han rivers. Except for the water area, the UEQ in the central metropolitan area remains mainly unchanged and several areas show signs of improvement. This result indicates that environmental governance in the main urban area of Wuhan has played a positive role in recent years.

Table 8. Change in the ecological index grade from 1995 to 2020.

Change Grade	1995–2005		2005–2015		2015–2020	
	Area	Percentage	Area	Percentage	Area	Percentage
Significantly worse	3	0.04%	20	0.26%	0	0.00%
Obviously worse	373	4.86%	1299	17.11%	981	12.85%
Slightly worse	1157	15.07%	2644	34.85%	2655	34.76%
No change	2952	38.45%	2634	34.70%	2984	39.08%
Slightly better	2165	28.20%	815	10.74%	816	10.69%
Obviously better	1008	13.13%	177	2.33%	199	2.61%
Significantly better	20	0.26%	0	0.00%	0	0.00%

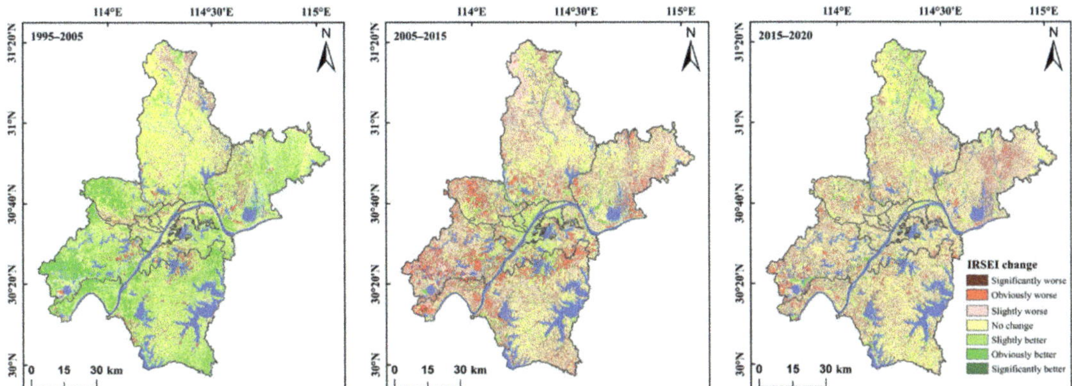

Figure 7. Spatial transfer distributions of the ecological levels of the IRSEI in Wuhan from 1995 to 2020.

3.4. Spatial Autocorrelation Analysis

We explore the spatial autocorrelation (SA) of the IRSEI at a grid cell scale of 500 m × 500 m and our results indicate the existence of SA. The Moran's I was 0.568 in 1995, and 0.535 in 2020. All four IRSEI maps (1995, 2005, 2015 and 2020) display an extremely low probability (*p*-value < 0.01) of completely random spatial distribution. Therefore, the statistical significance test shows that SA exists for all of the ecological factors. The IRSEI increased in places where spatial distribution was favorable to the UEQ. In 1995, high-value clustering of the IRSEI in Wuhan was distributed mainly in the south and north of the study area, whereas low-value clustering was concentrated in the middle of the study area. In 2005, high IRSEI values started gathering gradually in the southern region, and low IRSEI values became more concentrated in the clustering distribution. By 2015, the high/high clustering and low/low clustering of the IRSEI in the study area became more dispersed and tended to spread in every direction. In 2020, low/low clusters had spread from the middle to the east and west, whereas high/high clusters were concentrated mainly in the south and north of Wuhan City.

The Moran's I scatter graph is divided into four quadrants, corresponding to four different spatial distribution types (Figure 8). The first quadrant represents high/high clustering, the second quadrant low value and high-value aggregation, the third quadrant low/low aggregation and the fourth quadrant high-value and low-value aggregation. The IRSEI of Wuhan is concentrated mainly in the first and third quadrants. This result indicates that the IRSEI spatial distribution in Wuhan represents positive spatial autocorrelation, and high IRSEI agglomeration zones are mainly distributed in outer suburban areas, mainly in the north and southeast Wuhan.

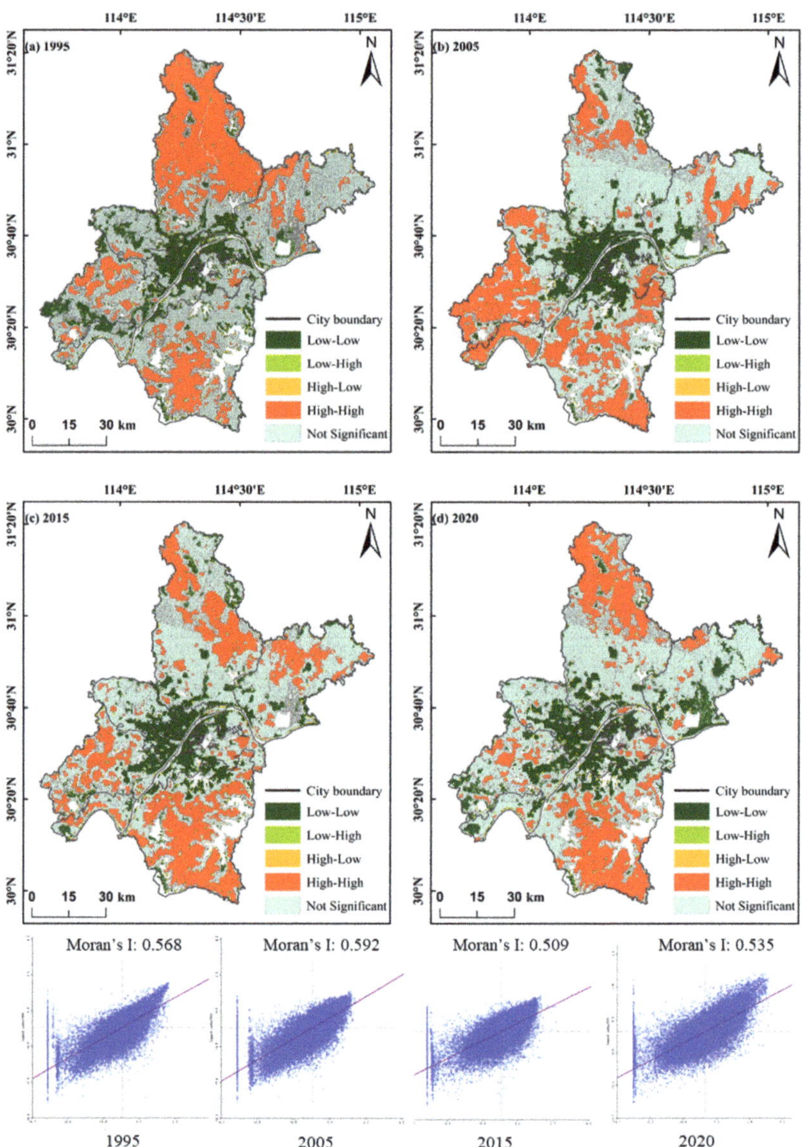

Figure 8. Spatial correlation and Moran index scatterplot of IRSEI from 1995 to 2020 in Wuhan city. (**a**) SA with 1995, (**b**) SA with 2005, (**c**) SA with 2015 and (**d**) SA with 2020. (Note: SA—Spatial autocorrelation).

4. Discussion

4.1. Literature, Policy and Practice

We have reviewed previous studies and demonstrated that it is feasible to evaluate the quality of the urban ecological environment through remote sensing. This research proposes a feasible method. Other remote sensing images could also have been used as data in this research, such as Tiangong-2 WIS images [11]. In terms of method improvement, we mainly improved the integration of quantitative factors. A related similar index, RSUSEI, has primarily increased remote sensing ecological factors by adding the impervious surface cover (ISC) [15]. ISC is also one of the most important factors that distinguish different types of land use/land cover characteristics in urban environments, and has a strong impact on UEQ. However, in our study we also consider the dryness index (NDBSI), the bare soil index (SI), the building index (IBI) and the normalized buildings–bare-soil index. However, there are strong correlations between the impervious surface, bare soil and building indices. Previous studies have found that the relationship between ISC and LST has the form of an exponential function, rather than a simple linear function, as commonly believed [43]. This exponential relationship has been confirmed by many subsequent studies [44,45]. Our IRSEI index takes into account the bare soil, building index and surface temperature. We suggest that the correlations of remote sensing ecological indicators affecting the regional ecological environment should be introduced into comprehensive indicators, or different indicators should be set according to the characteristics of the study region.

There are few high-quality ecological environment patches in Wuhan (IRSEI > 0.8), with close to zero over the past five years, and most of the patches are in the center of the ecological environment. Therefore, we propose a policy whereby Wuhan would focus on protecting forest land and gardens, build high-quality ecological corridors and coordinate the management of rivers in the future, so as to guide sustainable urban development and achieve sustainable development goals (such as SDG 11, sustainable cities and communities). Lake and wetland protection and ecological restoration and management will optimize the pattern of ecological security. Further analyses of the results indicate that there was a negative correlation between LSI, NDBSI and urban ecological quality. The ecological environment in areas with a high surface temperature, such as the Wuhan downtown area and coastal area around the Yangtze River, has tended to deteriorate; however, the humidity indices in these areas were also relatively high, which is conducive to ecological protection. Low vegetation index values in the central urban area also affect the quality of the ecological environment of Wuhan to a certain extent. The IRSEI can macro-evaluate the quality of the regional ecological environment, which is more convenient and efficient. In the future, higher precision can be introduced at the block level. Data, such as Google Street View data, could be used with machine learning algorithms to further identify the proportion of regional urban green space, trees, etc., and improve the accuracy of ecological environment assessment. The index has a high ability to distinguish between different land cover uses. The framework can also be easily extended to a global scale or to map other gridded socio-economic variables (such as GDP and population) to monitor and assess progress towards the SDGs [25]. The assessment and modelling of uses is critical to supporting sustainability assessment in achieving Sustainable Development Goals (SDGs), such as sustainable cities and communities. Therefore, IRSEI can be used to assess the spatial and temporal sustainability of cities.

4.2. Analysis of the Factors Affecting the UEQ

The regression least squares method (OLS) can be used to quantitatively describe the relationship between the ecological index and natural, economic and social factors in Wuhan. The data include temperature, precipitation, elevation, slope and DMSP as explanatory variables. The night light variable reflects the human footprint and fundamentally affects the urban ecological environment. Impervious surfaces and roads and a high population density are not conducive to UEQ. The regression coefficients represent the contribution of six independent variables to the dependent variable. The regression coefficients of precipitation and elevation are equal to 0.522 and 0.441, respectively, indicating that precipitation and elevation positively contribute to the IRSEI.

In contrast, the regression coefficients of night light and slope are negative, indicating that these variables contribute negatively to the IRSEI. The night light variable has a regression coefficient of −0.619, indicating a negative effect. The R^2 is 0.901 and $p < 0.05$, indicating that climate, precipitation, elevation, slope and night light data account for 90% of the variations of the IRSEI.

The regression equation between the IRSEI and the independent variables is as follows:

$$IRSEI = 0.926 + 0.148 \times Temperature + 0.522 \times Precipitation + 0.441 \times Slope - 0.001 \times Elevation - 0.619 \times DMSP\ (R = 0.901).$$

4.3. Method Framework and Validation Analysis

Weighting is an important process in the development of aggregated ecological indices that help promote sustainability. Different weighting methods have different characteristics, and the method employed could reflect the subjectivity of the decision makers. However, such methods combined with remote sensing index data can facilitate decisions and reduce the calculations required.

PCA is widely used in the evaluation of the RSEI. In several studies, the three principal components obtained after dimensionality reduction did not show any obvious effects (contribution was below 15%). However, including all of the pixels in extensive data calculations is a time-consuming process. The RSEI employs a covariance-based (unstandardized) PCA to determine the importance of each indicator involved. The weight of each indicator can be assigned objectively and automatically based on the load (contribution) of each indicator to PC1. In this study, we used PCA and EW to comprehensively calculate the IRSEI. After improvement, the combined method was able to reflect the degree of change in the index, and the calculation was quick and uncomplicated. The spatial distribution of the UEQ over the study period (1995–2020) is consistent with the information in the bulletin on the eco-environmental situation in China in that year. The current, more popular assessment method is based on habitat quality (HQ) [46–50]. In further research, we intend to include HQ in this quantitative assessment.

4.4. Limitations and Future Prospects

The proposed UEQ evaluation model is feasible and straightforward, providing a new idea for ecological protection and comprehensively reflecting the changes in UEQ in Wuhan. From 1995 to 2020, the UEQ of Wuhan declined overall, probably owing to a combination of natural factors and human activities. However, the ecological level in the eastern and southeastern mountainous areas has increased because of the influence of forest resource protection, desertification land management and the warm and humid climate. In contrast, the regional ecological level has declined, owing to the overexploitation and overgrazing of lake resources in the northwest and southwest of Wuhan. The constantly rising levels of urbanization and construction over nearly 20 years have resulted in a downward trend in the UEQ. Overall, the ecology of Wuhan is in a fragile state. In 2020, the proportion of areas with poor ecological environment grades remained high, accounting for 42.68% of the total area.

In future social and economic development, we should follow the laws of nature, prioritize protection and rationally develop and utilize natural resources. The IRSEI effectively revealed the spatial distribution of and change in the UEQ in Wuhan, based on remote sensing images. Although four types of ecological factors closely related to the ecological environment were selected in the calculation process, the ecological environment is a complex and comprehensive variable. Areas with a deteriorating ecological environment tend to be spread along the Yangtze and Han rivers and around the central urban area. Urban expansion has damaged the ecological environment, and urban planning should integrate more ecological concepts to promote a harmonious coexistence and sustainable development for humans, nature and society.

Comprehensive quantitative evaluation requires selecting several impact factors that reflect the actual situation in the study area. We aimed to conduct the UEQ evaluation by employing a scientific, objective and feasible method. Nevertheless, choosing the UEQ evaluation index remains exploratory work. Determining the index weight affects the accuracy of the evaluation results. Accordingly, expanding research to a more scientific multifunction performance index system and determining the index weights require further work. Furthermore, the limited availability of data and a lack of longitudinal comparison of urban data have affected the scientific nature of our research results. In addition, our next step will be exploring how the IRSEI changes at different spatial scales. The rapid development of cities will inevitably lead to a series of ecological and environmental problems, and the deterioration of the ecological environment may further affect the surrounding environment, forming a cycle and harming urban sustainability. This study also demonstrates that IRSEI is characterized by spatial heterogeneity; that is, the poor UEQ patches will focus on areas where the ecological environment is poor and the urbanization is also highest.

5. Conclusions

In this study, the IRSEI model was used to evaluate and monitor the ecological environment in Wuhan from 1995 to 2020. The IRSEI is an ecological environmental quality assessment method based on remote sensing technology. The method has many advantages, such as the ease of obtaining parameters, a large time sequence span and a wide evaluation range. The UEQ method employing remote sensing technology is feasible and simple, and provides a new tool for territorial spatial control and spatiotemporal urban sustainable development. Our proposed UEQ assessment framework can also help to develop potentially relevant additional sub-indicators, which could help to address one of the current challenges in SDG monitoring, namely how to implement SDG indicators. We have implemented the proposed workflow in this study based on an open-source platform and free satellite data, making it an appealing option that is applicable in almost all countries.

The main conclusions from the results of this study are:

- The mean IRSEI value in Wuhan decreased annually from 2005 to 2015. The UEQ continued to decline, mainly because of the rapid economic development of Wuhan, reduction in vegetation coverage caused by human activities, gradual decrease in lake area and transformation of the land-use structure caused by urban expansion;
- From the perspective of spatial patterns, the UEQ of the central urban areas, such as Qingshan, Hanyang, Hongshan and Wuchang, was lower than that of the surrounding metropolitan areas, such as Huangpi, Jiangxia and Caidian. In terms of time series, the UEQ in the central city of Wuhan has been mainly unchanged or improved, indicating that the management of the ecological environment in the central city had achieved specific results;
- The global Moran's I value range from 0.535 to 0.592 from 1995 to 2020, respectively, indicating that the IRSEI spatial distribution displays significant spatial heterogeneity. This finding indicates that high clustering gradually developed to the edge of the city, whereas low clustering gradually developed to the center of the city. The spatial correlation and local index cluster diagram of the IRSEI show that the high points are located mainly in the Huangpi and Jiangxia districts;
- The UEQ evaluation model constructed in this study is feasible and simple and could be implemented at no or negligible cost, making it applicable to most regional areas. Our model, therefore, could be considered a new tool for ecological management and protection, and for assessing progress toward urban sustainable development.

Supplementary Materials: The following are available online at https://www.mdpi.com/article/10.3390/rs13214440/s1, Table S1: Correlation matrix of IRSEI and four factors.

Author Contributions: Conceptualization, J.L. and J.G.; methodology, J.L. and J.G.; validation, J.L., J.Y. and J.-M.G.; data curation, J.L. and J.Y.; writing—original draft preparation, J.L.; writing—review and editing, J.L. and J.-M.G.; funding acquisition, J.G. All authors have read and agreed to the published version of the manuscript.

Funding: This research was funded by the National Natural Science Foundation of China, grant number 41871172 and 41701228 and supported by the Fundamental Research Funds for National Universities, China University of Geosciences (Wuhan).

Institutional Review Board Statement: Not applicable.

Informed Consent Statement: Not applicable.

Data Availability Statement: The satellite images used in this study are obtained from http://earthexplorer.usgs.gov, accessed on 15 May 2021.

Conflicts of Interest: The authors declare no conflict of interest.

References

1. Van Tulder, R.; Rodrigues, S.B.; Mirza, H.; Sexsmith, K. The UN's sustainable development goals: Can multinational enterprises lead the decade of action? *J. Int. Bus. Policy* **2021**, *4*, 1–21. [CrossRef]
2. Keeler, B.L.; Hamel, P.; McPhearson, T.; Hamann, M.H.; Donahue, M.L.; Meza Prado, K.A.; Arkema, K.K.; Bratman, G.N.; Brauman, K.A.; Finlay, J.C.; et al. Social-ecological and technological factors moderate the value of urban nature. *Nat. Sustain.* **2019**, *2*, 29–38. [CrossRef]
3. Elmqvist, T.; Andersson, E.; Frantzeskaki, N.; McPhearson, T.; Olsson, P.; Gaffney, O.; Takeuchi, K.; Folke, C. Sustainability and resilience for transformation in the urban century. *Nat. Sustain.* **2019**, *2*, 267–273. [CrossRef]
4. United Nations. 2018 Revision of World Urbanization Prospects. Available online: https://www.un.org/development/desa/publications/2018-revision-of-world-urbanization-prospects.html (accessed on 16 May 2018).
5. United Nations. Cities and Pollution. Available online: https://www.un.org/en/climatechange/climate-solutions/cities-pollution (accessed on 14 July 2020).
6. Worldbank. Urban Development. Available online: https://www.worldbank.org/en/topic/urbandevelopment/overview#1 (accessed on 20 April 2020).
7. Dale, V.H. The relationship between land-use change and climate change. *Ecol. Applic.* **1997**, *7*, 753–769. [CrossRef]
8. McDaniel, J.; Alley, K.D. Connecting local environmental knowledge and land use practices: A human ecosystem approach to urbanization in West Georgia. *Urban Ecosys.* **2005**, *8*, 23–38. [CrossRef]
9. de Andrés, M.; Barragán, J.M.; García Sanabria, J. Relationships between coastal urbanization and ecosystems in Spain. *Cities* **2017**, *68*, 8–17. [CrossRef]
10. Yang, C.; Zhang, C.; Li, Q.; Liu, H.; Gao, W.; Shi, T.; Liu, X.; Wu, G. Rapid urbanization and policy variation greatly drive ecological quality evolution in Guangdong-Hong Kong-Macau Greater Bay Area of China: A remote sensing perspective. *Ecol. Indic.* **2020**, *115*, 106373. [CrossRef]
11. Ariken, M.; Zhang, F.; Liu, K.; Fang, C.; Kung, H.-T. Coupling coordination analysis of urbanization and eco-environment in Yanqi Basin based on multi-source remote sensing data. *Ecol. Indic.* **2020**, *114*, 106331. [CrossRef]
12. Bonilla-Bedoya, S.; Mora, A.; Vaca, A.; Estrella, A.; Herrera, M.Á. Modelling the relationship between urban expansion processes and urban forest characteristics: An application to the Metropolitan District of Quito. *Comput. Environ. Urban Syst.* **2020**, *79*, 101420. [CrossRef]
13. Kerr, J.T.; Ostrovsky, M. From space to species: Ecological applications for remote sensing. *Trends Ecol. Evol.* **2003**, *18*, 299–305. [CrossRef]
14. Weber, T.; Sloan, A.; Wolf, J. Maryland's Green Infrastructure Assessment: Development of a comprehensive approach to land conservation. *Landsc. Urban Plan.* **2006**, *77*, 94–110. [CrossRef]
15. Firozjaei, M.K.; Fathololoumi, S.; Weng, Q.; Kiavarz, M.; Alavipanah, S.K. Remotely sensed urban surface ecological index (RSUSEI): An analytical framework for assessing the surface ecological status in urban environments. *Remote Sens.* **2020**, *12*, 2029. [CrossRef]
16. Rapport, D.J.; Costanza, R.; McMichael, A.J. Assessing ecosystem health. *Trends Ecol. Evol.* **1998**, *13*, 397–402. [CrossRef]
17. Sfriso, A.; Facca, C.; Ghetti, P.F. Validation of the macrophyte quality index (MaQI) set up to assess the ecological status of italian marine transitional environments. *Hydrobiologia* **2009**, *617*, 117–141. [CrossRef]
18. Hu, X.; Xu, H. A new remote sensing index for assessing the spatial heterogeneity in urban ecological quality: A case from Fuzhou City, China. *Ecol. Indic.* **2018**, *89*, 11–21. [CrossRef]
19. Shan, W.; Jin, X.; Ren, J.; Wang, Y.; Xu, Z.; Fan, Y.; Gu, Z.; Hong, C.; Lin, J.; Zhou, Y. Ecological environment quality assessment based on remote sensing data for land consolidation. *J. Clean. Prod.* **2019**, *239*, 118126. [CrossRef]
20. Cai, B.; Shao, Z.; Fang, S.; Huang, X.; Huq, M.E.; Tang, Y.; Li, Y.; Zhuang, Q. Finer-scale spatiotemporal coupling coordination model between socioeconomic activity and eco-environment: A case study of Beijing, China. *Ecol. Indic.* **2021**, *131*, 108165. [CrossRef]
21. Yue, H.; Liu, Y.; Li, Y.; Lu, Y. Eco-environmental quality assessment in china's 35 major cities based on remote sensing ecological index. *IEEE Access* **2019**, *7*, 51295–51311. [CrossRef]
22. Liao, W.; Jiang, W. Evaluation of the Spatiotemporal Variations in the Eco-environmental Quality in China Based on the Remote Sensing Ecological Index. *Remote Sens.* **2020**, *12*, 2462. [CrossRef]
23. Wang, J.; Liu, D.; Ma, J.; Cheng, Y.; Wang, L. Development of a large-scale remote sensing ecological index in arid areas and its application in the Aral Sea Basin. *J. Arid Land* **2021**, *13*, 40–55. [CrossRef]
24. Xu, H.; Wang, Y.; Guan, H.; Shi, T.; Hu, X. Detecting ecological changes with a remote sensing based ecological index (RSEI) produced time series and change vector analysis. *Remote Sens.* **2019**, *11*, 2345. [CrossRef]

25. Estoque, R.C. A Review of the Sustainability Concept and the State of SDG Monitoring Using Remote Sensing. *Remote Sens.* **2020**, *12*, 1770. [CrossRef]
26. Collins, J.B.; Woodcock, C.E. An assessment of several linear change detection techniques for mapping forest mortality using multitemporal landsat TM data. *Remote Sens. Environ.* **1996**, *56*, 66–77. [CrossRef]
27. Yarbrough, L.D.; Easson, G.; Kuszmaul, J.S. Proposed workflow for improved Kauth–Thomas transform derivations. *Remote Sens. Environ.* **2012**, *124*, 810–818. [CrossRef]
28. Huang, C.; Wylie, B.; Yang, L.; Homer, C.; Zylstra, G. Derivation of a tasselled cap transformation based on Landsat 7 at-satellite reflectance. *Int. J. Remote Sens.* **2002**, *23*, 1741–1748. [CrossRef]
29. Xu, H.; Wang, M.; Shi, T.; Guan, H.; Fang, C.; Lin, Z. Prediction of ecological effects of potential population and impervious surface increases using a remote sensing based ecological index (RSEI). *Ecol. Indic.* **2018**, *93*, 730–740. [CrossRef]
30. Robinson, N.P.; Allred, B.W.; Jones, M.O.; Moreno, A.; Kimball, J.S.; Naugle, D.E.; Erickson, T.A.; Richardson, A.D. A dynamic landsat derived normalized difference vegetation index (NDVI) product for the conterminous united states. *Remote Sens.* **2017**, *9*, 863. [CrossRef]
31. Townshend, J.R.G.; Justice, C.O. Analysis of the dynamics of African vegetation using the normalized difference vegetation index. *Int. J. Remote Sens.* **1986**, *7*, 1435–1445. [CrossRef]
32. Sobrino, J.A.; Jiménez-Muñoz, J.C.; Paolini, L. Land surface temperature retrieval from LANDSAT TM 5. *Remote Sens. Environ.* **2004**, *90*, 434–440. [CrossRef]
33. Estoque, R.C.; Murayama, Y.; Myint, S.W. Effects of landscape composition and pattern on land surface temperature: An urban heat island study in the megacities of Southeast Asia. *Sci. Total Environ.* **2017**, *577*, 349–359. [CrossRef] [PubMed]
34. Artis, D.A.; Carnahan, W.H. Survey of emissivity variability in thermography of urban areas. *Remote Sens. Environ.* **1982**, *12*, 313–329. [CrossRef]
35. Min, M.; Lin, C.; Duan, X.; Jin, Z.; Zhang, L. Spatial distribution and driving force analysis of urban heat island effect based on raster data: A case study of the Nanjing metropolitan area, China. *Sustain. Cities Soc.* **2019**, *50*, 101637. [CrossRef]
36. Carlson, T.N.; Ripley, D.A. On the relation between NDVI, fractional vegetation cover, and leaf area index. *Remote Sens. Environ.* **1997**, *62*, 241–252. [CrossRef]
37. Kerr, Y.H.; Lagouarde, J.P.; Imbernon, J. Accurate land surface temperature retrieval from AVHRR data with use of an improved split window algorithm. *Remote Sens. Environ.* **1992**, *41*, 197–209. [CrossRef]
38. Amiri, V.; Rezaei, M.; Sohrabi, N. Groundwater quality assessment using entropy weighted water quality index (EWQI) in Lenjanat, Iran. *Environ. Earth Sci.* **2014**, *72*, 3479–3490. [CrossRef]
39. Liu, G.-X.; Wu, M.; Jia, F.-R.; Yue, Q.; Wang, H.-M. Entropy-weighted comprehensive evaluation of petroleum flow in China during 1980–2015. *J. Cleaner Prod.* **2018**, *195*, 593–604. [CrossRef]
40. Islam, A.R.M.T.; Al Mamun, A.; Rahman, M.M.; Zahid, A. Simultaneous comparison of modified-integrated water quality and entropy weighted indices: Implication for safe drinking water in the coastal region of Bangladesh. *Ecol. Indic.* **2020**, *113*, 106229. [CrossRef]
41. Yang, J.Y.; Wu, T.; Pan, X.Y.; Du, H.T.; Li, J.L.; Zhang, L.; Men, M.X.; Chen, Y. Ecological quality assessment of Xiongan New Area based on remote sensing ecological index. *J. Appl. Ecol.* **2019**, *30*, 277–284. [CrossRef]
42. Overmars, K.d.; De Koning, G.; Veldkamp, A. Spatial autocorrelation in multi-scale land use models. *Ecol. Modell.* **2003**, *164*, 257–270. [CrossRef]
43. Xu, H. Analysis of impervious surface and its impact on urban heat environment using the normalized difference impervious surface index (NDISI). *Photogramm. Eng. Remote Sens.* **2010**, *76*, 557–565. [CrossRef]
44. Meng, F.; Shan, B.; Liu, M. Remote-sensing evaluation of the relationship between urban heat islands and urban biophysical descriptors in Jinan, China. *J. Appl. Remote Sens.* **2014**, *8*, 083693. [CrossRef]
45. Balçik, F.B. Determining the impact of urban components on land surface temperature of Istanbul by using remote sensing indices. *Environ. Monit. Assess.* **2014**, *186*, 859–872. [CrossRef]
46. Chen, T.; Hui, E.C.M.; Wu, J.; Lang, W.; Li, X. Identifying urban spatial structure and urban vibrancy in highly dense cities using georeferenced social media data. *Habitat Int.* **2019**, *89*, 102005. [CrossRef]
47. Li, F.; Wang, L.; Chen, Z.; Clarke, K.C.; Li, M.; Jiang, P. Extending the SLEUTH model to integrate habitat quality into urban growth simulation. *J. Environ. Manag.* **2018**, *217*, 486–498. [CrossRef]
48. Liu, J.; Zhang, G.; Zhuang, Z.; Cheng, Q.; Gao, Y.; Chen, T.; Huang, Q.; Xu, L.; Chen, D. A new perspective for urban development boundary delineation based on SLEUTH-InVEST model. *Habitat Int.* **2017**, *70*, 13–23. [CrossRef]
49. Romero-Calcerrada, R.; Luque, S. Habitat quality assessment using Weights-of-Evidence based GIS modelling: The case of Picoides tridactylus as species indicator of the biodiversity value of the Finnish forest. *Ecol. Modell.* **2006**, *196*, 62–76. [CrossRef]
50. Terrado, M.; Sabater, S.; Chaplin-Kramer, B.; Mandle, L.; Ziv, G.; Acuña, V. Model development for the assessment of terrestrial and aquatic habitat quality in conservation planning. *Sci. Total Environ.* **2016**, *540*, 63–70. [CrossRef] [PubMed]

Article

Sustainable Development of Life Service Resources: A New Framework Based on GIScience and Spatial Justice

Ze Xu [1], Lu Niu [1], Zhengfeng Zhang [1,*], Jing Huang [1], Zhaodi Lu [1], Yufan Huang [1], Yangyang Wen [1], Chu Li [1] and Xiaokun Gu [2,3]

[1] School of Public Administration and Policy, Renmin University of China, Beijing 100872, China; xuze_bj@126.com (Z.X.); niulu@ruc.edu.cn (L.N.); huangjing_ruc@163.com (J.H.); luzhaodi1996@sina.com (Z.L.); 2016200510@ruc.edu.cn (Y.H.); wenyang2016@ruc.edu.cn (Y.W.); lichu@ruc.edu.cn (C.L.)
[2] China Institute for Urban Governance, Shanghai Jiao Tong University, Shanghai 200030, China; guxk1980@sjtu.edu.cn
[3] School of International and Public Affairs, Shanghai Jiao Tong University, Shanghai 200030, China
* Correspondence: zhangzhengfeng@ruc.edu.cn or zhengfengzh@sina.com

Abstract: The sustainable development goals (SDGs) reflect the pursuit of achieving spatial justice. Both SDG 1.4 and SDG 11.1 reflect a concern for urban services. Life service resources, which are the new concept proposed by the Chinese government, also call for sustainable development path. However, few studies have focused on the realization of spatial justice in life service resources. This paper proposes a two-level, four-step analysis framework composed of quantity, structure, pattern, and coupling coordination to perceive the spatial justice of life service resources. Based on remote sensing technology and geographic information science, this paper acquires and analyses multi-source data including population density, building outlines, point of interests, subway lines, etc. Furthermore, the case study in downtown Beijing found the following: (1) The total life service resources are extensive and varying in type; (2) regional differences are evident and low-level equilibrium and high-level priority development coexist; (3) life service resources are concentrated in contiguous and multi-centre clusters with a greater north–south than east–west difference; (4) the overall level of life service resources is low, specifically for "high in the centre and low in the periphery" and "high in the east and low in the west". Future management should consider narrowing the development gap and formulating industry development plans to improve spatial justice. Finally, the comparison between Beijing and London and more cities in the future needs to consider the urban development stage, population density, and other aspects.

Keywords: sustainable development path; life service resources; spatial justice; SDGs; GIScience; Beijing

Citation: Xu, Z.; Niu, L.; Zhang, Z.; Huang, J.; Lu, Z.; Huang, Y.; Wen, Y.; Li, C.; Gu, X. Sustainable Development of Life Service Resources: A New Framework Based on GIScience and Spatial Justice. *Remote Sens.* **2022**, *14*, 2031. https://doi.org/10.3390/rs14092031

Academic Editor: Ronald C. Estoque

Received: 8 March 2022
Accepted: 21 April 2022
Published: 23 April 2022

Publisher's Note: MDPI stays neutral with regard to jurisdictional claims in published maps and institutional affiliations.

1. Introduction

The *2030 Agenda for Sustainable Development* proposes 169 sustainable development goals (SDGs) and a new model of "people, planet, prosperity, peace, and partnership" placing humanism at the core of sustainable development [1]. Specifically, SDG 1.4 states, "By 2030, ensure that all men and women, in particular the poor and the vulnerable, have equal rights to economic resources, as well as access to basic services". SDG 11.1 presents, "By 2030, ensure access for all to adequate, safe and affordable housing and basic services and upgrade slums". These SDGs clearly point to a common issue, namely, that of urban services. In 2018, the World Urbanization Prospects indicated that, by 2050, the global urban population would increase by another 2.5 billion, and the urbanisation rate would increase to 68% [2]. This means that the disparity between the demands of the urban population and the supply of service facilities will continue to increase. This disparity is reflected in the supply shortage and the upgrading of service facilities' types. In other words, with the development of urbanisation, the needs of residents are not limited to traditional public services (e.g., national defence, fire protection, medical treatment, transportation, education) but have grown to the life service resources (LSRs). This is particularly evident in China. In 2010, the output value of China's life service industry was about CNY 3491.05 billion, accounting for 9% of the gross domestic product (GDP) [3]. This shows that the life service industry has long become a pillar industry in China, and the considerable demand for life service facilities cannot be ignored.

To ensure that everyone can share the benefits of urbanisation, the Chinese government developed the life service industry to establish a basic service system with "extensive coverage, rich business forms, and a reasonable layout". In 2015, China's State Council stated that "the improvement of the national income level has expanded the new demand for life service consumption, the continuous breakthrough of information network technology has expanded the new channels of life service consumption, and the implementation of major national strategies such as new urbanisation has expanded the new space for life service consumption. People's demand for life services is increasing, the demand for service quality is beefing up, and the life service consumption contains huge potential great potential" [4]. In 2016, China's Ministry of Commerce, in the Thirteenth Five-Year Plan for the Development of Residents' Life Service Industry, define the different types of LSRs for the first time [5]. In 2019, China's Ministry of Finance and State Administration of Taxation jointly issued a document that aimed to reduce life service enterprises' financial burden [6]. In 2021, China's National Development and Reform Commission presented several opinions on the problems of "insufficient effective supply, insufficient convenience sharing, and inadequate implementation of policies" for LSRs [7]. These policies demonstrate China's increasing attentiveness to the fair allocation of LSRs.

Existing literature shows that LSRs can impact people's daily lives in three dimensions: quantity, structure, and pattern. First, in terms of quantity, insufficient LSRs decrease residents' satisfaction and social welfare [8,9]. From the viewpoint that everyone should meet specific living standards [10], it is unfair that some people do not have access to the resources necessary to achieve this goal [11]. Second, from the structure dimension, low-quality LSRs (e.g., fast-food restaurants, tobacco, alcohol, and gambling outlets) tend to be more prevalent in communities with a low socio-economic status [12–14], resulting in higher obesity, disease, mortality, and crime rates [15,16]. However, gentrified communities have more influence and resources for keeping low-quality LSRs out of their areas than socio-economically disadvantaged communities [12], which may exacerbate social spatial isolation [17]. Third, from the perspective of patterns, the uneven spatial distribution of LSRs is reflected in a concentration of areas with large passenger flow, strong road centrality, and traffic accessibility [18–20] and has become a common problem faced by many cities [9,21]. This not only hinders residents from enjoying life services conveniently [22], intensifying the spatial deprivation of poor communities and marginal groups [17], but it also inhibits the healthy development of the urban economy [23]. Fortunately, some studies have found several important factors affecting the spatial distribution of LSRs, such as urban planning [12], racism [24], consumption level [25], etc.

Although past studies have focused on characteristics of LSRs in different dimensions [25,26], few have integrated these scattered aspects. More importantly, there is a lack of a "people-oriented" theoretical basis in the existing studies. Therefore, the optimized configuration of LSRs failed to return to the goal of meeting human needs in time. Spatial justice (SJ) theory emphasizes the balanced distribution and equal access to public goods or services [27,28] and advocates that human beings should have equal access to various social resources in space [29,30]. We note that this theory seems to be no longer limited to managing basic public service facilities but has gradually extended to some fields initially configured only by the market mechanism, such as energy utilization [31,32]. The Chinese government has only recently proposed LSRs, which have the characteristics of high daily use frequency and low use price. This indicates that LSRs may be generated with the goal of making profits; however, in the long run, due to its social resource nature, government intervention (especially for socialist countries such as China) and the improvement in people's demand level (after meeting basic public services, people hope to improve the quality of life), the realistic situation and realisation path of LSRs' SJ should also be studied in depth.

It is worth noting that the research on SJ is showing a quantitative trend. Usually, scholars link urban facilities, population, income, and other socioeconomic factors for analysis. There are three representative approaches. The first is to design new indices. For example, residents living in the suburbs are forced to incur more commuting costs than those living in the downtown area. This further leads to differences in affordability for residents, thereby exacerbating spatial injustice. Thus, some scholars have designed a new index named H+T, which is the ratio of living costs (housing costs, transportation costs) to income [33]. The second is to identify key factors. Some scholars have used lots of methods (literature analysis, group discussions, expert interviews, and questionnaires) to determine the relevant factors that affect the SJ of public open space. Then, they determine the importance levels of SJ influencing factors by Exploratory Factor Analysis (EFA) and Fuzzy Synthetic Evaluation (FSE) methods [34]. The third is to construct complex models to analyse the drivers of SJ. For instance, one study has assessed the differential impact of various factors on SJ through Random Forest and SHAP Tree Explainer. It shows that Mean Commute Time can enhance SJ, while Medical Facilities Count and Food Desert Count will reduce SJ [35]. In another study, it is revealed by using structural equation modelling (SEM) that the upward mobility of compact regions is significantly higher than that of sprawling areas [36]. Consistent with the above-mentioned research, this paper aims to quantify SJ. However, due to the particularity of the research object (LSR is a new concept recently proposed by the Chinese government, and it is difficult to find the alternative data), the paper explores the SJ of LSR from the "distribution perspective"; that is, residents in different regions can share the same or similar LSRs. As an exploratory study, we look forward to obtaining more data for in-depth analysis in the future.

To sum up, this paper addresses the following three questions: (1) Why should SJ be considered when studying LSRs? (2) How should SJ be applied and interpreted for LSRs? Finally, (3) what strategies should the government use to optimise the allocation of LSRs? This paper will include a discussion of the specific characteristics of LSRs, explain why SJ should be considered, and describe the current status of SJ for LSRs. Furthermore, an SJ evaluation framework will be developed for LSRs that fully integrates the advantages of big data and geographic information systems (GIS) [37–39]. Using data from downtown Beijing, China, this paper provides a feasible framework for optimising the allocation of LSRs that can be used as a reference for other regions.

This paper responds to SDG 1.4 and SDG 11.1 proposed by the United Nations and uses a variety of geographic information science methods to analyse the optimal spatial allocation of LSRs in the downtown Beijing. The multi-source data (population density, building outline, points of interest, subway lines, etc.) and geographic information science methods (nearest neighbour, kernel density, standard deviation ellipse) used in this paper reflect the important value of GIScience and remote sensing for the sustainable management of urban services.

2. Theoretical Analysis Framework

2.1. Why Should SJ Be Considered When Studying LSRs?

LSRs refer to various service activities and services provided to meet the daily needs of residents and are closely related to living consumption. In 2016, China's Ministry of Commerce divided LSRs into eight categories: (1) catering, (2) accommodation, (3) housekeeping, (4) dyeing, (5) bathing, (6) beauty salons, (7) appliance repairs, and (8) portrait photography [5]. SJ theory was first used to describe the inequity caused by insufficient public services in the 1960s [40]. Since then, many scholars have improved this theory. Henri Lefebvre criticises the opinion that "space is container and field" while advocating that social space is a product of society [41]. David Harvey proposes the concept of "territory redistribution justice"—the fair, just geographical distribution of social resources [42]. To condemn space deprivation and exclusion, Edward Soja encourages marginalised groups to fight for urban rights [43]. At present, SJ is generally regarded as social equity and justice regarding space rights and the interests of citizens in the urban field. Specifically, urban residents have the right to equal participation in all urban space production processes and all kinds of social life; all residents enjoy the benefits provided by urban life (especially high-quality urban centres); various forms of spatial isolation and restriction should be prevented; spatial marginalisation of vulnerable groups should be avoided; and the cultural discrimination and repression of space should be eliminated.

From the perspective of demand, LSRs should be configured based on SJ because the balanced allocation of LSRs is in line with the goal of SJ. LSRs are closely related to people's daily lives and are frequently used service facilities. Abraham Maslow's Hierarchy of Needs divides human needs into five levels from bottom to top: physiological, safety/security, love/belonging, esteem, and self-actualization [44]. Examining SJ and LSRs from the level of physiological needs, if a person is hungry and needs to eat, they will often turn to what is readily available (e.g., fast food restaurants, dessert shops, pastry shops). Furthermore, from the level of safety/security, when people are faced with problems, such as pipeline blockage or electrical damage, they need to obtain services, such as professional home appliance maintenance services, which may be determined by the proximity to their location. Therefore, meeting these types of needs are inseparable from the placement of LSRs. This also aligns with SJ's goals, such as enabling them to enjoy the benefits brought by urban life, especially in high-quality urban centres [30]. In addition, SJ should be considered when establishing LSRs targeting the multilevel needs of residents, as the spatial injustice of LSRs will affect the total demand and long-term demand for these services. Compared to services such as the national defence and police and fire control, people use LSRs more frequently. Regardless of which LSRs are lacking, residents' happiness and social stability will decrease. The differential production of LSRs by capital will eventually result in spatial plunder or spatial injustice of LSRs, such as spatial isolation, right occupation, or the social reconstruction of vulnerable groups, which leads to the decline of cities [42].

From the supply perspective, SJ theory can inform the sustainable supply of LSRs. The used price of LSRs is low so that most people can enjoy life services, including catering, accommodation, beauty salons, etc. This is in line with the SJ principle of "universal benefit". However, maintaining this principle is challenging. Based on the current situation in China, the main suppliers of LSRs are individuals and enterprises, who are most often interested in making a profit. Obviously, relying solely on the market mechanism is prone to uneven space supply of LSRs. This supply issue further restricts people's space rights. Hence, in addition to the market, we also need SJ theory to inform the reconstruction of spaces with diversity and diversification as the core [45].

From the perspective of SJ theory, life services can be regarded as a unique spatial production. In the process of practice, life services produce not only natural spaces (e.g., restaurants, hotels, and dry cleaners) but also social spaces generated by activities (e.g., guided use and after-sales service). In short, the space shaped by life services is based on the unity of natural space and social space. The problems exposed in life services, such as unreasonable service radius and lack of service items, involve natural space and social space. The SJ theory provides an "internal basis of legitimacy and rationality" and "effective value norms" for the balanced allocation of LSRs [30].

2.2. How Should SJ Be Applied and Interpreted for LSRs?

The existing SJ analysis framework is mainly measured from three dimensions: (1) quantity, (2) structure, and (3) pattern. These dimensions allow us to understand the SJ of service facilities, but some shortcomings still exist. These include a lack of systematic perspective, which is not conducive to the formation of a complete understanding of service facilities; insufficient research content where most of the existing studies have focused on descriptive analyses of service facilities but failed to further correlate with the needs of residents; and a limited effective evaluation strategies with most of the existing studies using ethnography, investigation, and interview to evaluate SJ, which require high time and cost resources [46].

To bridge the above gaps, this paper proposes a two-level, four-step analysis framework (Figure 1). The framework is used to achieve the distribution justice of LSRs; that is, residents in different regions can share the same or similar LSRs. The applicability of this framework will be described in detail below.

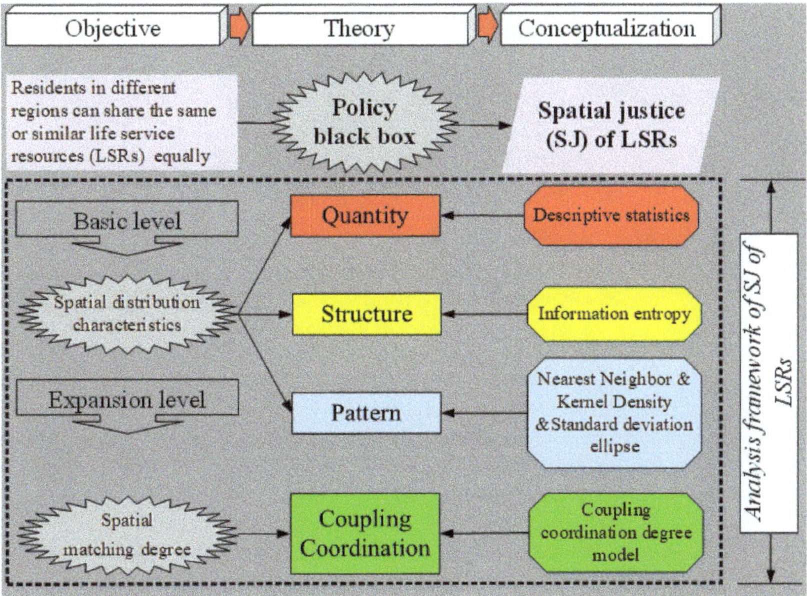

Figure 1. Multidimensional perception of SJ for LSRs.

First, the basic level consists of three dimensions: quantity, structure, and pattern. Quantity refers to the number of LSRs, where the total amount of service facilities in a region determines the overall level of residents' enjoyment of services. If the number of service facilities is insufficient, residents will not be able to make full use of service facilities, further resulting in a sense of injustice. This quantitative difference is a crucial reason for the difference in justice between the rich and the poor [47]. Structure refers to the configuration of different LSRs, where different service facilities have their own characteristics and can meet the diverse needs of residents. A balanced service facility structure can meet the diversified needs of residents. If the structure of LSRs is unreasonable, the multilevel needs of residents cannot be met simultaneously, which will create negative emotions. Hence, emphasising a balanced combination of different service facilities has become an essential trend in modern urban management [48]. Pattern refers to the spatial distribution pattern of LSRs, which affects residents' access to services. If certain types of service facilities are concentrated in one area, it is difficult for residents in other areas to conveniently use such facilities. This violates the just principle that residents in different regions should enjoy services equally [49]. The above three dimensions combine quantitative statistical analysis (descriptive statistics, information entropy) and spatial analysis methods (nearest neighbour, kernel density, standard deviation ellipse).

Second, the expansion level aims to assess the degree of spatial matching between the LSRs and population distribution. Previous studies generally believe that the lack of resources or uneven spatial distribution is an injustice phenomenon [40]. This simple understanding is easy to accept but difficult to verify. The analysis based on "quantity, structure, and pattern" provides credible evidence to evaluate SJ, but it needs further development. In other words, it cannot be taken for granted that every resident is satisfied simply because there are a large number of LSRs. Therefore, LSRs should be associated with population demand and assessed by their coordinating degrees in space. Using this type of "coupling coordination", we can identify the matching degree between each LSRs type and the local population simultaneously. Urban managers can use the above results to guide the spatial allocation of LSRs.

This new analysis framework integrates the steps of "quantity, structure, and pattern" and directly connects LSRs with population demand. There is currently no formal method for evaluating the SJ of LSRs in China; however, the combination of GIS data and spatial analysis can be called a strict and comprehensive method [50].

3. Data and Methods

This section may be divided by subheadings. It should provide a concise and precise description of the experimental results, their interpretation, as well as the experimental conclusions that can be drawn.

3.1. Area Description

Beijing is the centre of China's political, cultural, and international exchanges and technological innovation. It has experienced significant population growth within a limited area in past decades, creating a massive and diverse demand for LSRs. Therefore, exploring the quantity, structure, pattern, and population matching of Beijing's LSRs has important practical significance.

Influenced by the city's transforming functional structure, foreign investment, and migrants, Beijing's social polarisation and new urban poverty have increased [51]. From 2000 to 2010, Beijing's industrial structure adjustment and spatial evolution further reconstructed the urban social space [52]. In 2017, the urban sustainable development goals of an "intensive and efficient production space, moderate living space, and beautiful ecological space" were proposed. Downtown Beijing was chosen as the current study area for three reasons. First, the economy is strong. In 2018, the GDP here accounted for more than 70% of the city's GDP. Second, the concentrated population is large. In 2018, the permanent population here accounted for about 55% of the city. Third, there is variability in the supply and demand of LSRs. The living needs of different groups of people are intertwined. According to the Beijing City Master Plan (2016–2035), the downtown area includes the Dongcheng, Xicheng, Chaoyang, Haidian, Fengtai, and Shijingshan districts (Figure 2).

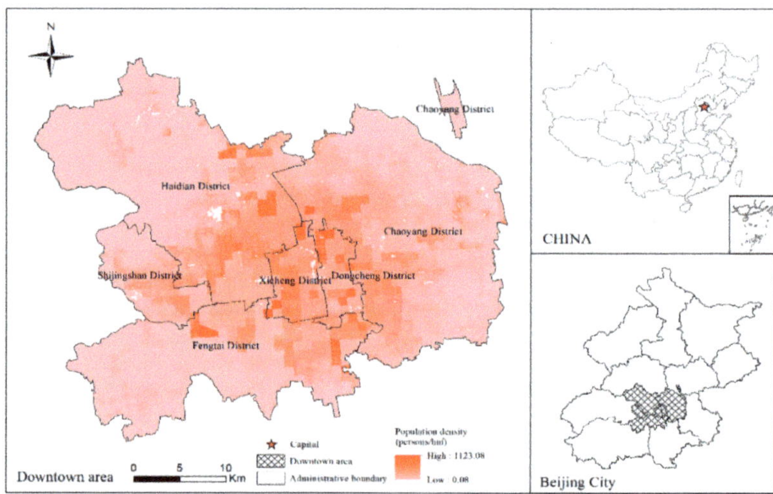

Figure 2. An overview of Beijing.

3.2. Research Methods

3.2.1. Descriptive Statistics

We used descriptive statistics to find the quantity of various LSRs. The LSRs data obtained are counted, and their proportion is calculated.

3.2.2. Information Entropy

This paper calculates balance and dominance based on previous studies [53]. First, we use Shannon's information entropy Equation to define the information entropy of the LSRs as:

$$H = -\sum_i^N P_i ln P_i = -\sum_{i=1}^N \left(A_i / \sum_i^N A_i \right) ln \left(A_i / \sum_i^N A_i \right), \tag{1}$$

In Equation (1), A is the total number of point of interests (POIs) of LSRs in an area. LSRs in this area are divided into N types, and the number of POIs of each type is A_i ($i = 1, 2, \ldots, N$). P_i is the percentage of the number of POIs per service type (equivalent to the probability of occurrence of this type of service), and the value is normalised. H is the information entropy of the residents' life service system, reflecting the balance of the types of LSRs within the region. A higher entropy value means more abundant types of LSRs. Notably, when the number of POIs for each type is entirely equal, the occurrence probability for each type is equivalent to $1/N$, and the maximum information entropy is $H_m = lnN$.

Second, the ratio between the measured value and the maximum value of information entropy can be regarded as the equilibrium degree of the life service structure:

$$J = H/H_m = -\sum_i P_i ln P_i / lnN, \tag{2}$$

In Equation (2), J is the equilibrium degree, and the value location is (0, 1). The remaining parameters are the same as in Equation (1). The closer the J value is to 1, the life service structure of LSRs moves closer to being balanced.

Finally, the concept of dominance is introduced to assess the concentration level of the life service structure:

$$I = 1 - J, \tag{3}$$

In Equation (3), I is the dominance, and the value location is (0, 1). A value closer to 1 indicates that there is one or more dominant type(s) of life service(s) in the area.

3.2.3. Spatial Analysis Method

First, we investigate the spatial concentration of LSRs. The average nearest neighbour index is calculated by ArcGIS 10.1. It compares the average observation distance of a particular type of POI with the expected average distance based on a random distribution to determine the aggregation or dispersion of features. The Equations are as follows [54].

$$R = d_i / d_e = d_i / \left(0.5 / \sqrt{N/A} \right), \tag{4}$$

$$Z = (d_i - d_e) / \left(0.26136 / \sqrt{n^2 / A} \right), \tag{5}$$

In Equations (4) and (5), d_i and d_e are the average observation distance and the expected observation distance, respectively; N is the total number of POIs; A is the area of the research area; R is the nearest neighbour index; Z-score and p-value can determine whether the null hypothesis can be rejected in a statistical sense.

Second, to explore whether LSRs gather, kernel density is used to analyse point density. Plot density and Voronoi diagram density are also commonly used for this purpose; however, they face problems of uniform density in the unit space and abrupt changes in density at the joints of the unit. Kernel density analysis comprehensively considers the difference in focus intensity of different internal points and the continuity of spatial phenomena, so it has more advantages [55]. The Equation is as follows:

$$f(s) = \sum_{i=1}^{n} \frac{1}{h^2} k\left(\frac{s - c_i}{h}\right),$$

(6)

In Equation (6), n is the number of points whose distance at the same position s is not greater than h; h is the distance attenuation threshold; k is the spatial weight function; and $f(s)$ is the kernel density estimate at position s. The above Equations describes the interaction between the kernel density value, and the radiation distance from the centre point—the distance attenuation effect of the centre point outward. Studies have shown that the spatial weight function has a limited impact on the point mode and, more importantly, on determining a suitable search radius (distance attenuation threshold) [56].

This paper uses 1500 m as the search radius for two reasons. One is the research experience of other cities in China. In the case of the central Chongqing city, the search radius is 1500 m. This case confirmed that 1500 m can not only identify small-scale POI aggregation areas, but also reflect the macro-scale polycentric pattern, and has a good smoothing effect [57]. The other is the scope of downtown Beijing. Based on the scope, ArcGIS software can automatically calculate the default search radius. For downtown Beijing, the default radius is roughly 1516 m. To facilitate comparison, 1516 m is simplified to 1510 m in this study.

Finally, we are interested in the expansion trend of LSRs. In 1926, D. Welty Lefever proposed the standard deviation ellipse analysis method to describe the direction distribution of points with parameters such as centre, azimuth, long axis, and short axis [58]. The centre is the relative position of the space occupied by a factor. The azimuth describes the main direction of development, and the direction, length, and length ratio of the major and minor axes, represent the primary and secondary development trends, the degree of dispersion in these trends, and spatial distribution morphology, respectively [59]. This paper uses the standard deviation ellipse tool of ArcGIS 10.1 to analyse spatial statistics.

3.2.4. Coupling Coordination Degree Model

The coupling coordination degree model can be used to measure the mutual influence between two subsystems (here, the matching effect of LSRs and residents' needs) [60]. Usually, the population density in the area can be used to express residents' needs [61]. Therefore, we substitute the kernel density value of the LSRs and the resident population density value into the coupling coordination degree model for calculation (the data are normalised in GIS in advance). Based on previous studies [60], the calculation Equation of coupling coordination degree is as follows:

$$C = \sqrt{f(M) \times g(N)/([f(M) + g(N)]/2)^2},$$

(7)

$$D = \sqrt{C \times T}, T = \alpha f(M) + \beta g(N),$$

(8)

where C represents the coupling degree, $f(M)$ is the kernel density value of the LSRs, and $g(N)$ is the resident population density, where T is the comprehensive coordination index, and α and β are the contribution weight of the kernel density value of the LSRs and resident population density, respectively. In this paper, α and β are each 0.5, which is equally important. D stands for the degree of coupling coordination, and the value range of D is (0, 1). A higher D value represents a higher coupling degree between LSRs and residents' needs. Furthermore, the degree of coupling coordination of LSRs and residents' needs was divided into five levels (Table 1).

Table 1. Classification of the degree of coupling coordination between LSRs and residents' needs.

D	Coupling Coordination State
$0.8 \leq D\ 1$	Highly coupled coordination
$0.6 \leq D\ 0.8$	Moderate coupled coordination
$0.4 \leq D\ 0.6$	Low coupled coordination
$0.2 \leq D\ 0.4$	Moderate uncoupled coordination
$0 \leq D\ 0.2$	Severely uncoupled coordination

3.3. Data Sources and Processing

We introduce POI data from the following aspects: (1) Information. The POI data obtained from Amap include the type, name, administrative district, longitude, and latitude of LSRs. (2) Source. The POI data come from a big data innovation enterprise under Shanghai Economic and Information Commission (https://www.metrodata.cn/) (accessed on 20 April 2022). (3) Time. We investigate downtown Beijing with data from March 2018. Over time, the number of POIs will change, such as the closure or relocation of some hotels. Objectively speaking, the research results of this paper can only reflect the LSRs in the central urban area of Beijing in March 2018. Therefore, this study has certain limitations. In the future, POI data at more time points will be obtained for tracking analysis. (4) Accuracy. We take two methods to test the accuracy of the data. The first is to compare the data provided by the Resource and Environmental Science and Data Center of the Chinese Academy of Sciences (https://www.resdc.cn/) (accessed on 20 April 2022). The second is to conduct manual random comparison on Baidu Maps. (5) Data processing. We cleaned the data, such as removing duplicate content, correcting coordinates, etc. By doing this, 14 categories of point data were initially obtained.

Combined with government documents, the actual situation in Beijing, and data availability, this paper finally determines 6 types of LSR. Specifically, in 2016, China's Ministry of Commerce divided LSRs into 8 categories: catering, accommodation, housekeeping, dyeing, bathing, beauty salons, appliance repairs, and portrait photography. However, due to the following reasons, this paper cannot obtain enough POIs to represent housekeeping. This is because the housekeeping service is characterized by door-to-door service, is less dependent on physical stores than other services, the industry is still in the early stages of development, the supply is relatively small, and the profit of the housekeeping industry in downtown Beijing is limited as it is difficult to afford the high housing and rent prices. This paper also excludes bathing services because they are often difficult to separate from accommodation or beauty salon services, and the industry is small. Therefore, we categorize LSRs into 6 groups: catering, accommodation, dyeing (this refers to the business of laundry, dry cleaning, ironing, dyeing, darning, stain removal, etc.), beauty salons, appliance repairs, and portrait photography.

In addition, subway stations, subway lines, and building outlines data were obtained from Amap. The Amap collects data through vehicle, walking, and aerial photogrammetry. The POI, building outlines, and other data used in this paper are inseparable from the help of remote sensing technology. The population density raster data comes from Worldpop's 2018 prediction of China's population raster data, with a resolution of 100 m $\times$ 100 m (https://www.worldpop.org/) (accessed on 20 April 2022). As the data development team said, "this data is suitable where the accuracy of the satellite-based mapping of settlements is uncertain". This means that it is very suitable for use in the capital of China (most of the core parts of downtown Beijing belong to the central government office area and military management area, where it is difficult to obtain remote sensing high-resolution data). The administrative boundaries come from the Resource and Environment Data Cloud Platform (http://www.resdc.cn) (accessed on 20 April 2022).

4. Results

4.1. Quantity Dimension

The number of catering facilities is the largest, accounting for more than half of the total facilities in the study area (Table 2). The number of beauty salons and accommodation facilities is second, with 17,239 and 15,677 establishments, respectively. The other three types of LSRs are relatively small, less than 10%.

Table 2. Characteristics of POI data of LSRs in downtown Beijing.

Category	Content	Number	Proportion (%)
Catering	Chinese restaurant, foreign restaurant, casual restaurant, cold drink shop, dessert shop, fast food restaurant, coffee shop, pastry shop, tea house, etc.	42,530	51.4362
Accommodation	Residential areas, hotels, tourist hotels, business residences, and commercial and residential buildings, etc.	15,677	18.9599
Dyeing	Laundry	2204	2.6655
Beauty salons	Beauty salon store	17,239	20.8490
Appliance Repair	Appliance repair store	2421	2.9280
Portrait photography	Photo print store	2614	3.1614
Total		82,685	100

4.2. Structure Dimension

Figure 3 shows that: (1) For all districts, facilities such as catering, accommodation, and beauty salons are plentiful, while dyeing establishments, appliance repairs, and portrait photography shops are lacking. (2) As for the information entropy, the order of its value from high to low is Shijingshan, Fengtai, Haidian, Xicheng, Chaoyang, and Dongcheng. The maximum information entropy is 1.37 in Shijingshan. This shows that the quantity difference of various LSRs is the smallest in Shijingshan, and the structure of LSR tends to be balanced numerically here. In other words, although Shijingshan has the least number of LSRs, it is in a low-level equilibrium state. (3) For the equilibrium degree, the ranking result of its value is consistent with that of information entropy. This is because the equilibrium degree is calculated by dividing the information entropy by the constant (see Equation (2)). (4) From the perspective of dominance degree, the ranking result of its value is just opposite to that of equilibrium. This is because it is calculated by subtracting the equilibrium from 1 (see Equation (3)). For example, compared with other districts, Dongcheng has the smallest degree of equilibrium (0.69) and the largest degree of dominance (0.31). This indicates that there is the most obvious dominant LSR here.

4.3. Pattern Dimension

Table 3 shows the results of the average nearest neighbour analysis. The Z scores and p-values indicate that all kinds of LSRs pass the significance test ($p < 0.001$) (i.e., have significant spatial agglomeration). The R value is in descending order for dyeing, accommodation, appliance repair, portrait photography, beauty salon, and catering. This shows that the concentration of dyeing and accommodation services is relatively high, while the concentration of beauty salons and catering services is relatively low.

The kernel density analysis found that the various LSRs in downtown Beijing are expressed as a spatial form of "centralised contiguous and multi-centre clusters" (Figure 4). Specifically, the catering and accommodation service facilities are concentrated in Dongcheng and Xicheng districts, and the concentration of contiguous areas is significantly larger than other types of LSRs. Dongcheng and Xicheng districts are the core functional areas of the capital, with high development intensity and urbanisation levels. Catering and accommodation services occupy an essential position and are used relatively often. Therefore, the services mentioned above are significantly concentrated in the functional core areas of downtown Beijing.

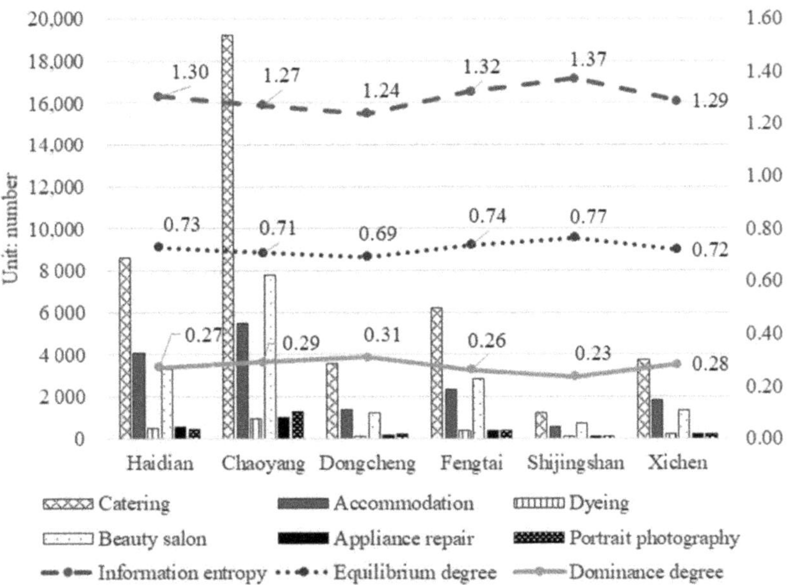

Figure 3. Number of POIs of LSRs and their spatial equilibrium parameters in downtown Beijing.

Table 3. Nearest neighbour analysis of LSRs in downtown Beijing.

Name	Average Observation Distance (d_i)	Expected Average Distance (d_e)	Nearest Neighbour Index (R)	Z-Score
Catering	34.625	108.3	0.32	−268.393
Accommodation	116.455	173.598	0.671	−78.846
Dyeing	295.407	440.337	0.671	−29.561
Beauty salons	61.962	167.085	0.371	−158.032
Appliance Repair	249.173	416.087	0.599	−37.761
Portrait photography	223.872	383.432	0.584	−40.702

Chaoyang, Haidian, Fengtai, and Shijingshan districts are urban function expansion areas, as identified in the 2012 Beijing Major Function Zone Planning. Although these districts belong to this zone, the overall development level of Chaoyang and Haidian districts is significantly higher than that of Fengtai and Shijingshan districts. This objectively limits the concentration of life service functions turning them into multi-centred groups. Specifically, dyeing, beauty salons, appliance repair, and portrait photography services have a higher nuclear density at the junction of Dongcheng, Chaoyang, and Fengtai districts. We found that life services are primarily concentrated in transportation hubs in or near large residential and business districts. The large flow of people usually translates into higher demand for services. For example, Wangjing Station and China World Trade Centre Station are both located in a typical international business centre area. Many universities surrounding Wudaokou Station and Zhongguancun Station gather many high-tech enterprises, which constitute the life services core in Haidian District Circle. Babaoshan and Gucheng stations are close to Haidian District, so their nuclear density is relatively high. The Capital University of Economics and Business District, Muxiyuanqiao South Station, and Fangzhuang Station are adjacent to schools, railway stations, and large residential areas, so there is a large population and flow and strong demand for life services.

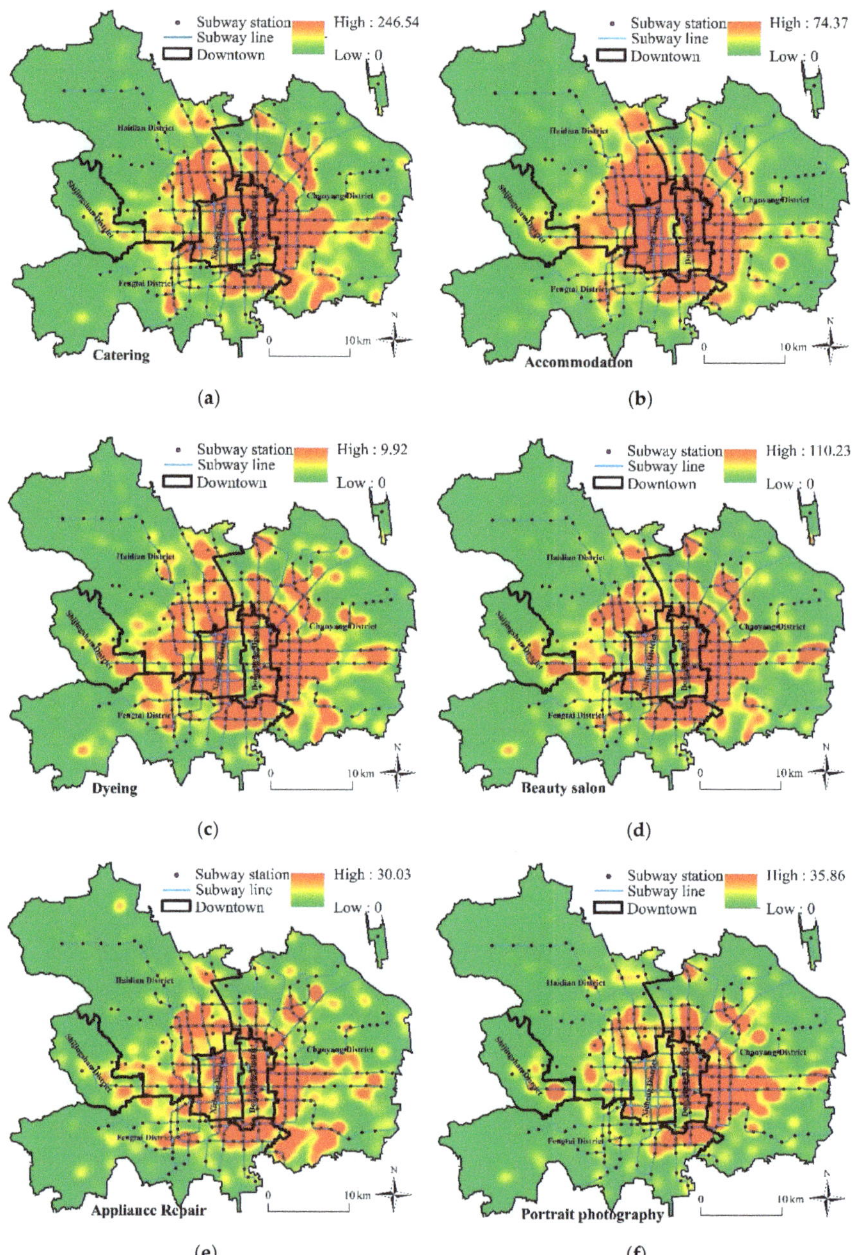

(**a**) (**b**)

(**c**) (**d**)

(**e**) (**f**)

Figure 4. Kernel density analysis of LSRs in downtown Beijing: (**a**) Catering; (**b**) Accommodation; (**c**) Dyeing; (**d**) Beauty salon; (**e**) Appliance repair; (**f**) Portrait photography.

The centrality index shows that the difference in coordinates at the centre of the standard deviation ellipse for various LSRs are slight; the maximum difference in longitude and latitude are only 0.03° and 0.01°, respectively (Table 4). This difference reflects the proximity relationship between the centres of various LSRs in downtown Beijing. The perimeter and area of the spread ability index jointly describe the coverage of the standard deviation ellipse. The perimeter and area of each life service are in the same order, from largest to smallest: dyeing, appliance repair, beauty salon, catering, accommodation, and portrait photography. The long axis and the short axis indicate the direction and scope of the distribution of various life services, respectively. From the length difference between the long and short axis of the ellipse, the maximum difference in beauty salon services is 2.94 km, and the minimum difference in accommodation services is 2.23 km. This means that the flatness of the beauty salon service ellipse is larger, and the directionality of the data is more obvious than other services; the directional difference of the accommodation service ellipse is not statistically significant, indicating that service is more balanced in all directions. Generally, the difference between the long and short axis of the life service ellipse is within 3 km. The azimuth is the angle between the true geographic north and the X-axis of the ellipse. The maximum azimuth of the accommodation service ellipse is 98.93°, the minimum is 88.76° for dyeing services, and the average for all LSRs is 93.04°. Combined with the standard deviation ellipse chart, it is found that the elliptical shape of various LSRs does not show a clear northwest–southeast trend. However, it appears to be an approximately horizontal–vertical standard form. This is mainly because Beijing has gradually formed a typical spatial pattern of axial development over a long time (the power of the City Master Plan).

Table 4. Standard deviational ellipse parameter of LSRs in downtown Beijing.

Category	Centre Longitude (°)	Centre Latitude (°)	Perimeter (km)	Area (km²)	Length of Long Axis (km)	Length of Short Axis (km)	Azimuth (°)
Catering	116.40E	39.93N	68.06	360.23	12.13	9.46	95.16
Accommodation	116.38E	39.93N	65.83	339.03	11.56	9.33	98.93
Dyeing [1]	116.39E	39.92N	70.10	382.61	12.46	9.78	88.76
Beauty salon	116.39E	39.92N	68.50	363.18	12.32	9.38	90.15
Appliance repair	116.39E	39.92N	69.23	371.36	12.43	9.51	93.58
Portrait photography	116.41E	39.92N	64.60	323.26	11.60	8.87	91.66

[1] Note: Except for the dyeing services, the standard deviation ellipse long axis of other LSRs is on the X axis, and the short axis is on the Y axis.

Figure 5 shows the standardised ellipse of LSRs in the study area. Combining the overall picture and the detailed map, each life service ellipse is divided into four parts by the planning development axis. The area difference of each part intuitively shows the north–south difference and the east–west difference. The centres of all services are in the Xicheng and Dongcheng districts, all located north of Chang'an Avenue. This shows that the six types of LSRs are more suited to residents' needs north of Chang'an Avenue, and the supply of life services south of Chang'an Avenue, especially in Fengtai District, is insufficient. The current life services have an overall and more vital "north-south difference". The accommodation service centre is located at the northwest corner of Tiananmen Square. In contrast, the catering and portrait photography centre occupies the northeast, and other life service centres are on the central axis. This shows that accommodation services are developing westward, catering and portrait photography are expanding eastward, while other services have not yet formed a clear development trend. Thus, there are hierarchical and low-level "east-west differences" in life services.

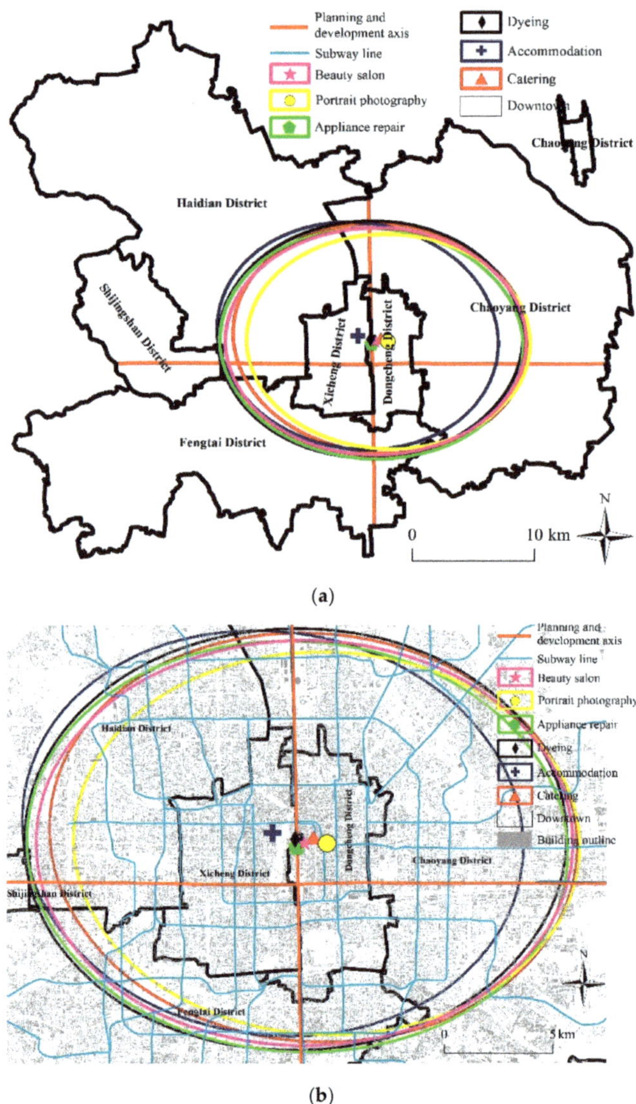

Figure 5. Standard deviational ellipse analysis of LSRs in downtown Beijing; (**a**) Overall picture; (**b**) Detailed picture.

4.4. Coupling Coordination Dimension

Figure 6 shows that the current coupling and coordination of the six types of LSRs and population density has not yet reached the ideal level, and the overall characteristics are "high in the centre and low in the periphery" and "high in the east and low in the west".

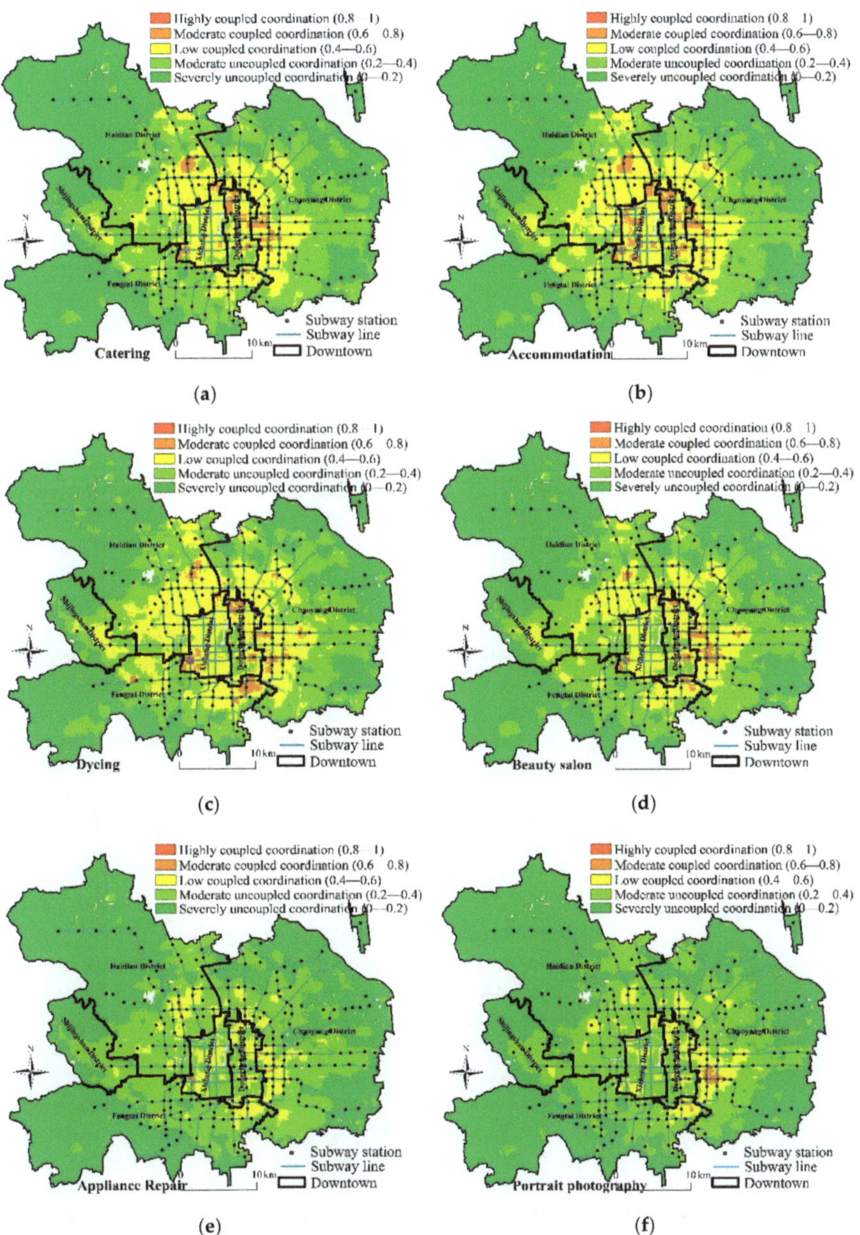

Figure 6. Coupling and Coordination Degree of LSRs in downtown Beijing: (**a**) Catering; (**b**) Accommodation; (**c**) Dyeing; (**d**) Beauty salon; (**e**) Appliance repair; (**f**) Portrait photography.

Specifically, the medium-coupling coordination areas of catering and accommodation facilities are similar, mainly concentrated around subway stations such as Wudaokou, South Luogu Lane, and China World Trade Centre. Wudaokou is close to top universities, such as Peking University and Tsinghua University. South Luogu Lane is a famous tourism alley in Beijing. The China World Trade Centre is in the city's central business district, which contains many Fortune 500 companies. Students, tourists, and corporate employees stimulate the demand for catering and accommodation in the region, resulting in a relatively highly coupled coordination. The moderate coupled coordination areas of dyeing and beauty salons are similar, where "the east is higher than the west". Compared with the other types, the moderate coupled coordination area of appliance repair and portrait photography facilities have the smallest area and tend to be more scattered. Most importantly, this study finds that most peripheral areas show moderate uncoupled coordination and severely coupled coordination.

5. Discussion

5.1. SJ of LSRs Reflected by Coupling Coordination Degree

Currently, there are two approaches to evaluating SJ. One is the qualitative method. For example, a case study approach is used to analyse the factors and mechanisms of SJ in rural Finland [62]. The other is the quantitative method, especially spatial analysis methods. For instance, using kernel density estimation, standard deviation ellipse, and social performance evaluation can explore the SJ of community sports and fitness venues in Shanghai [63]. Quantitative methods enable overlaying the analyses of population and other socioeconomic data to spatially identify areas of SJ [64]. Therefore, this paper uses quantitative methods to discuss SJ. These methods include descriptive statistics, information entropy, nearest neighbour, kernel density, standard deviation ellipse, and coupling coordination degree.

The results of the coupling coordination degree directly reflect the SJ: (1) For catering, accommodation, dyeing, and beauty salons, Dongcheng and Xicheng have the highest degree of SJ. The junction between the above two districts and other regions also has a high level of SJ. Dongcheng and Xicheng are the office areas of the Central Government and the Beijing Municipal Government. Hence, these areas are densely populated and require a wider variety of LSRs. (2) For appliance repair and portrait photography, the southeast of Haidian and the southwest of Chaoyang have the highest degree of SJ. Haidian District is the centre of education and technology, with a large concentration of schools and research institutes. Chaoyang District is an economic centre and an industrial base, with a well-known embassy area and a commercial and trade area. Therefore, some types of LSRs with higher technical requirements and more luxury will be concentrated here.

5.2. The LSRs among Beijing and London: Distributive Injustice

It is difficult to compare the LSRs supply globally for two main reasons directly. First, the Chinese government proposed the concept of LSRs in 2016. It originated from a life service industry with distinct Chinese characteristics. Therefore, it is difficult to find comparable programs in other countries. To our knowledge, even in China, there are few empirical studies on LSRs. The second reason is limited data sources. This paper uses POI data to measure LSRs, different from the previous field investigations. Hence, differences in data sources reduce comparability among various studies. In sum, these reasons pose challenges for effectively comparing the supply of LSRs in Beijing to other regions.

POI data currently have limited attributes (name, type, address, and coordinate information) but lack details, such as floor height, building area, and the number of households. Therefore, for accommodation services, it is inaccurate to directly compare the number of POIs because the number of people that can be accommodated in buildings of different heights varies. For instance, high-rise apartments accommodate more people than villas. However, for catering services, it is feasible to use the POI number of catering service facilities per capita as a benchmark for comparison for two main reasons. The first is that the catering service facilities are usually independent spatial units, such as a street containing many dining establishments (e.g., South Luogu Lane) or a luxuriously decorated hotel (e.g., QUANJUDE Peking Roast Duck). Although some large shopping malls can accommodate several restaurants on different floors simultaneously (e.g., Vanke Plaza, Wudaokou Shopping Centre), the name field of POIs can, generally, effectively distinguish different restaurants in similar locations. The second reason is that eating is a fixed demand for every citizen. Although there are no direct statistics, dining out constitutes a considerable proportion of residents' daily lives for the capital city. The certainty of the demand for dining out and the need for restaurant reception capacity implies that a larger population will need more restaurants. In summary, the number of catering service facilities can roughly reflect the degree of satisfaction of residents' dining out needs.

Inner London is a current representative case with a common benchmark with Beijing. There are 42,530 catering facilities in downtown Beijing (Table 3). According to the Beijing City Master Plan (2016–2035), the total area is about 1378 km^2, and the population density in 2016 was 14,000 persons/km^2. Studies have shown that as of November 2017, the Food & Shop in Inner London has 7355 POIs [65]. It also includes shops, a mall, a marketplace, vending machines, a pharmacy, fast food, a café, and a restaurant. In addition, statistics from the Greater London Authority (GLA) [66] show that the area of Inner London is 319.29 km^2, and the population density in 2017 was 11,070 persons/km^2. Comparing per capita catering service facilities in downtown Beijing to Inner London, the former has about 22.05 restaurants per 10,000 people, and the latter has 20.81 restaurants per 10,000 people. These two places have similar population densities, and the catering service facilities per population are close.

Although the population densities of these two places are similar, one limitation of making this type of comparison is that we do not know the income differences among people. High-income groups have the means to eat out more, while low-income groups may choose to cook at home. Therefore, there is an objective difference in the frequency of catering service facilities due to income. Furthermore, factors such as cultural differences in eating habits need to be considered. For British people, the habit of afternoon tea undoubtedly increases the frequency of visits to coffee shops. Even though the number of POIs for catering facilities per capita in London and Beijing is close, there are still cultural differences in residents' total demand and use frequency of different catering facilities. In addition, there are differences in the stages of urban development between these two cities. In the late 1990s, the British government began the gentrification of Inner London [67], while the wave of urban renewal in China focused on rebuilding urban villages [68]. It was not until 2008 that Beijing began to construct two pilot projects to reconstruct urban villages and proposed to rebuild 50 key villages in the suburbs of Beijing over the next few years [69]. Thirty-two key villages have been rebuilt in downtown Beijing [70]. These differences demonstrate that the LSRs supply is insufficient in areas like downtown Beijing. However, the coupling and coordination degrees for the current study show that there is a low level of coupling and coordination between the catering facilities and the population distribution (Figure 6a). It indicates that there is still room for further improvement of catering service facilities in Beijing.

5.3. The Future of SJ Theory: Spatial Analysis and Big Data

The spatial analysis method is an effective tool for understanding SJ, which broadens the application field of SJ theory. For example, in environmental justice, Ripley's K spatial analysis method was first used to evaluate the spatial point pattern of air toxins and environmental justice in West Oakland, California [64]. For energy justice, in addition to focusing on differences in social groups, it is also necessary to assess the justice impact of spatial differences on energy poverty risks and prevalence. This will help propose spatial injustice intervention measures based on geography [31]. In terms of traffic justice, GIS helped confirm that the diesel particulate matter (DPM) in the main highway corridors of Massachusetts was significantly higher than that in surrounding areas, and a hot spot analysis showed that an increase in DPM concentration and asthma occurrence had a statistically significant clustering. Therefore, decision makers need to consider the use of pollution reduction technologies in residential areas close to traffic corridors [71].

In addition to integrating spatial analysis, an increasing number of empirical studies are using big data. Big data provide cities with the potential to obtain valuable insights from data collected from various channels and support the construction of smart cities [72].

For example, Panoramio, Foursquare, and Twitter data jointly provide digital footprints of tourists in Madrid, helping the scenic spot to formulate new public policies to improve the tourist experience [73]. Baidu heat map data and urban POI data can help identify the urban population's spatiotemporal distribution characteristics and mechanism in Xi'an, China [74] Based on information from 10.16 million traffic monitoring records, the traffic congestion modes in Beijing have been divided into weekend mode, holiday mode, weekday mode, and weekday mode B [75]. Combined with spatial analysis and carpooling data, the multi-centre spatial structure of the Beijing metropolitan area can be defined. Unlike traditional research, this method accounts for periodicity and survey theme limitations, allowing for an easier attainment of independent conclusions [76].

This paper demonstrates how the multidimensional perception of SJ for LSRs can be realised through a combination of big data and spatial analysis. POI data helped us quickly determine the spatial location of six kinds of LSRs on a micro-scale. Using the spatial analysis methods of nearest neighbour, kernel density, and standard deviation ellipse, we were able to identify the distribution characteristics of LSRs. This provides a reliable path for future research addressing the status of LSRs. Moreover, it responds to the knowledge gaps of big data applications of SJ for LSRs. Big data and spatial analysis are important components for the future of SJ theory.

5.4. Spatial Difference of LSRs: Market Mechanism and Government Power

The market mechanism is the main reason for such a pronounced spatial difference of LSRs in Beijing. First, supply is determined by the needs of surrounding residents. Densely populated areas tend to have more choices of life service facilities. This is because consumers usually want public service facilities as close to their residences as possible, reducing unnecessary travel costs [77]. Second, establishing LSRs has corresponding costs. The market often allocates public services to those who can pay for them. Due to unequal income distribution, low-income groups cannot afford related public services [78]. We find that the current coupling coordination degree between LSRs and population density in Beijing's urban centre has not yet reached an ideal level. The overall development characteristics are high in the centre and east but low in the periphery and to the west. Due to data limitations, this paper did not include complex regression models that may have revealed the driving mechanisms of spatial differences. However, the study of coupling coordination degree based on population density still confirms the power of market mechanisms at the supply and demand level.

Government power is another reason for the spatial difference in LSRs. First, most public service investments come from the government. The higher the level of economic development, the more the government can spend on public services. This is particularly obvious in areas with a backward economy. For example, in Laos from 2000 to 2011, the GDP grew at an average annual rate of about 7%, and the proportion of total public service expenditure in the GDP increased to 11.2% [79]. Laos was able to develop public services due to its growing economic strength. Second, the government shapes the spatial patterns of public service facilities. Data from poorer areas in Sweden show that policies, such as zoning laws, can forcibly allocate community resources to highly impoverished areas, improving the quality of life for the poor [80]. Our study finds that, for Beijing, government planning dominates the spatial pattern of LSRs. Policy requirements of the urban spatial structure further strengthen the contiguous and multi-centre clusters of LSRs, which led to greater north–south than east–west differences.

5.5. Limitations and Directions for Future Research

Overall, remote sensing technology provides a basic data source for this paper to explore the sustainable development of LSRs [81]. In addition, various methods of geographic information science provide tools for us to perceive the spatial justice (sustainability) of LSRs. However, we were unable to generate a description of residents' psychological characteristics or analyse choice willingness from the perspective of psychology due to limitations of the data. The advantage of spatial analysis lies in creating a spatial visualisation of life service facilities; however, justice refers to the distribution of justice and requires more discussion from a philosophical and social sciences standpoint [82]. For example, in the future, data on differences in residents' sense of justice could be collected through questionnaires. Second, POI data have limited attributes, so the driving factors and underlying mechanisms of LSRs were unable to be examined. Identifying the market mechanism and government forces behind the supply of LSRs is a crucial direction for subsequent studies. Again, future research should consider integrating methods, such as field interviews with big data, to examine these mechanisms.

6. Conclusions and Policy Recommendations

This study finds that: (1) the total LSRs are extensive and varying in type; (2) regional differences are evident, and low-level equilibrium and high-level priority development coexist; (3) LSRs are concentrated in contiguous and multi-centre clusters with a greater north–south than east–west difference; and (4) the overall level of LSRs is low, specifically for "high in the centre and low in the periphery" and "high in the east and low in the west". Our findings inform the following recommendations:

(1) The development gap needs to be reduced between the north and south. Here, Beijing's LSRs have gradually formed an axial development pattern, and the north–south difference in LSRs is more significant than the east–west. Therefore, future LSR development should pay attention to the north–south difference, increasing investment in areas south of Chang'an Avenue. Furthermore, the distribution of LSRs around subway stations and subway lines tells us that improving transportation facilities is significant for achieving fairness in resource distribution.

(2) The spatial differences for catering services need to be addressed. The number of catering services facilities in Chaoyang District is almost 10 times that of Shijingshan District. Given that catering services are a dominant type of LSRs, this paper suggests that policymakers need to provide more catering services in Shijingshan District. This does not mean that the government should directly fund the development of catering services. Still, it can help by providing more policy support for companies and individuals that provide catering services, such as providing sufficient land supply, guiding the agglomeration of catering companies, and creating a food street or plaza with scale effect.

(3) Accommodation services could be combined with reconstructing old urban areas and the new construction of satellite cities. As a cultural symbol of Beijing, Hutong has played a crucial residential function to this day. However, due to the poor conditions of the public infrastructure, this area urgently needs renovation [83]. Some residents want to move out of the crowded hutongs, while others hope to transform the hutongs into hotels. Therefore, residents, planners, companies, and the government could work together to transform the old city and the construction of the satellite city. This embodies the procedural justice in SJ. Hence, the government should guide the orderly development of accommodation service facilities by formulating industry development plans.

(4) As the old Chinese saying goes, "Loving beauty is part of human nature". To facilitate residents' access to beauty salon services, policymakers should consider establishing at least one beauty salon in each community. Currently, beauty salon services in Beijing are centralized in the junction area of Dongcheng, Chaoyang, and Fengtai Districts. As a result, a longer commute is required for residents outside of these areas to access these services. In addition, building beauty salons can increase the vitality of the urban area by increasing the diversity of consumption in the neighbourhood, which is another way to improve the life service resources.

(5) Although the total number of POIs for dyeing, appliance repair, and portrait photography accounts for less than 10% of LSRs, existing services placement should be considered when establishing new service locations. Based on the results of the

spatial pattern, there are fewer dyeing services in the northwest of the studied area. In contrast, appliance repair and portrait photography services tended to be found in the eastern region. The results of the coupling and coordination degrees in the current study reflect the unbalanced spatial distribution of these services in Beijing. Therefore, corresponding LSRs should be added in the above areas.

Author Contributions: Conceptualization, Z.X. and Z.Z.; methodology, L.N., J.H. and C.L.; writing—original draft preparation, Z.X. and L.N.; writing—review and editing, J.H., Z.L. and Y.H.; formal analysis, Z.X. and L.N.; visualization, L.N. and Y.W.; supervision, Z.Z. and X.G.; funding acquisition, Z.Z. All authors have read and agreed to the published version of the manuscript.

Funding: The paper was supported by the Key Project of National Social Science Fund (21AZD041) and National Natural Science Foundation of China (42077433).

Institutional Review Board Statement: Not applicable.

Informed Consent Statement: Not applicable.

Data Availability Statement: Not applicable.

Conflicts of Interest: The authors declare no conflict of interest.

References

1. United Nations. Transforming Our World: The 2030 Agenda for Sustainable Development. 2015. Available online: https://sustainabledevelopment.un.org/content/documents/21252030%20Agenda%20for%20Sustainable%20Development%20web.pdf (accessed on 20 April 2022).
2. United Nations. World Urbanization Prospects. Available online: https://www.un.org/development/desa/publications/2018-revision-of-world-urbanization-prospects.html (accessed on 20 April 2022).
3. Yu, H. Pay close attention to and take measures to accelerate the development of life service industry. *China Econ. Trade Her.* **2012**, *29*, 27–30.
4. State Council. Guiding Opinions on Accelerating the Development of Life Service Industry and Promoting the Upgrading of Consumption Structure. Available online: http://www.gov.cn/zhengce/content/2015-11/22/content_10336.htm (accessed on 20 April 2022).
5. Ministry of Commerce. Thirteenth Five-Year Plan for the Development of Residents' Life Service Industry. Available online: http://www.mofcom.gov.cn/article/guihua/201701/20170102495671.shtml (accessed on 20 April 2022).
6. Ministry of Finance and State Taxation Administration. Announcement on Clarifying the Policy of Adding and Deducting Value-Added Tax of Life Service Industry. Available online: http://www.chinatax.gov.cn/chinatax/n810341/n810755/c5137754/content.html (accessed on 20 April 2022).
7. National Development and Reform Commission. Notice on Several Opinions on Promoting the Life Service Industry to Make Up for Shortcomings and Improve People's Quality of Life. Available online: http://www.gov.cn/zhengce/content/2021-11/02/content_5648192.htm (accessed on 20 April 2022).
8. Weziak-Bialowolska, D. Quality of life in cities—Empirical evidence in comparative European perspective. *Cities* **2016**, *58*, 87–96. [CrossRef]
9. Zhao, Z.; Pang, R.; Wang, S. Measurement of spatial accessibility performance of large retailing facilities in Changchun. *Geogr. Res.* **2016**, *35*, 431–441.
10. De Vita, A. Inequality and poverty in global perspective. In *Freedom from Poverty as a Human Right*; Oxford University Press: New York, NY, USA, 2007; pp. 103–132.
11. Christman, B.; Russell, H. Readjusting the political thermostat: Fuel poverty and human rights in the UK. *J. Hum. Rights Commonw.* **2016**, *2*, 116–128. [CrossRef]
12. Macintyre, S. Deprivation amplification revisited; or, is it always true that poorer places have poorer access to resources for healthy diets and physical activity? *Int. J. Behav. Nutr. Phys. Act.* **2007**, *4*, 1–7. [CrossRef]
13. Reidpath, D.D.; Burns, C.; Garrard, J.; Mahoney, M.; Townsend, M. An ecological study of the relationship between social and environmental determinants of obesity. *Health Place* **2002**, *8*, 141–145. [CrossRef]
14. Wardle, H.; Keily, R.; Astbury, G.; Reith, G. "Risky Places?": Mapping gambling machine density and socio-economic deprivation. *J. Gambl. Stud.* **2014**, *30*, 201–212. [CrossRef]
15. Macdonald, L.; Olsen, J.R.; Shortt, N.K.; Ellaway, A. Do "environmental bads" such as alcohol, fast food, tobacco, and gambling outlets cluster and co-locate in more deprived areas in Glasgow City, Scotland? *Health Place* **2018**, *51*, 224–231. [CrossRef]
16. Yen, I.H.; Kaplan, G.A. Neighborhood social environment and risk of death: Multilevel evidence from the Alameda County Study. *Am. J. Epidemiol.* **1999**, *149*, 898–907. [CrossRef]

17. Anguelovski, I.; Cole, H.V.S.; O'Neill, E.; Baro, F.; Kotsila, P.; Sekulova, F.; Perez Del Pulgar, C.; Shokry, G.; Garcia-Lamarca, M.; Arguelles, L.; et al. Gentrification pathways and their health impacts on historically marginalized residents in Europe and North America: Global qualitative evidence from 14 cities. *Health Place* **2021**, *72*, 102698. [CrossRef]

18. Buzzacchi, L.; Leveque, P.; Taramino, R.; Zotteri, G. Using betweenness metrics to investigate the geographical distribution of retailers. *Environ. Plan. B-Urban Anal. City Sci.* **2021**, *48*, 2221–2238. [CrossRef]

19. Lv, Y.; Zheng, X.; Zhou, L. Relationships between street centrality and spatial distribution of functional urban land use: A case study of Beijing central city. *Geogr. Res.* **2017**, *36*, 1353–1363.

20. Wu, M.; Pei, T.; Wang, W.; Guo, S.; Song, C.; Chen, J.; Zhou, C. Roles of locational factors in the rise and fall of restaurants: A case study of Beijing with POI data. *Cities* **2021**, *113*, 103185. [CrossRef]

21. Ning, X.; Liu, Y.; Wang, H.; Hao, M.; Yang, B.; Wang, M. Research on Functional Land Division of the Main Urban Area in Beijing Based on Crowd Sourcing Geographic Data. *Geogr. Geo-Inf. Sci.* **2018**, *34*, 42–49.

22. Jiang, B.; Wang, Y.; Ye, X. Detecting development pattern of urban business facilities using reviews data. *Acta Geod. Cartogr. Sin.* **2015**, *44*, 1022–1028.

23. Wang, T.; Wang, Y.; Zhao, X.; Fu, X. Spatial distribution pattern of the customer count and satisfaction of commercial facilities based on social network review data in Beijing, China. *Comput. Environ. Urban Syst.* **2018**, *71*, 88–97. [CrossRef]

24. Lee, J.P.; Ponicki, W.; Mair, C.; Gruenewald, P.; Ghanem, L. What explains the concentration of off-premise alcohol outlets in Black neighborhoods? *SSM-Popul. Health* **2020**, *12*, 669. [CrossRef]

25. Fang, Y.; Mao, J.; Liu, Q.; Huang, J. Exploratory space data analysis of spatial patterns of large-scale retail commercial facilities: The case of Gulou District, Nanjing, China. *Front. Archit. Res.* **2021**, *10*, 17–32. [CrossRef]

26. Wang, F.; Chen, C.; Xiu, C.; Zhang, P. Location analysis of retail stores in Changchun, China: A street centrality perspective. *Cities* **2014**, *41*, 54–63. [CrossRef]

27. Pirie, G.H. On spatial justice. *Environ. Plan. A* **1983**, *15*, 465–473. [CrossRef]

28. Smith, D.M. *Geography and Social Justice*; Blackwell: Oxford, UK, 1994.

29. Israel, E.; Frenkel, A. Social justice and spatial inequality: Toward a conceptual framework. *Prog. Hum. Geogr.* **2018**, *42*, 647–665. [CrossRef]

30. Soja, E.W. The city and spatial justice. In *Justice et Injustices Spatiales*; Presses Universitaires de Paris Nanterre: Nanterre, France, 2016; pp. 56–72.

31. Bouzarovski, S.; Simcock, N. Spatializing energy justice. *Energy Policy* **2017**, *107*, 640–648. [CrossRef]

32. Sovacool, B.K.; Heffron, R.J.; McCauley, D.; Goldthau, A. Energy decisions reframed as justice and ethical concerns. *Nat. Energy* **2016**, *1*, 1–6. [CrossRef]

33. Guerra, E.; Kirschen, M. Housing Plus Transportation Affordability Indices: Uses, Opportunities, and Challenges. Available online: https://www.econstor.eu/handle/10419/173922 (accessed on 20 April 2022).

34. Jian, I.Y.; Luo, J.; Chan, E.H.W. Spatial justice in public open space planning: Accessibility and inclusivity. *Habitat Int.* **2020**, *97*, 102122. [CrossRef]

35. Deb, D.; Smith, R.M. Application of Random Forest and SHAP Tree Explainer in Exploring Spatial (In)Justice to Aid Urban Planning. *ISPRS Int. J. Geo-Inf.* **2021**, *10*, 629. [CrossRef]

36. Ewing, R.; Hamidi, S.; Grace, J.B.; Wei, Y.D. Does urban sprawl hold down upward mobility? *Landsc. Urban Plan.* **2016**, *148*, 80–88. [CrossRef]

37. Getis, A.; Mur, J.; Zoller, H.; Mur, J.; Fingleton, B.; Smith, T.; Plotnikova, M. Spatial Econometrics and Spatial Statistics. Available online: http://citeseerx.ist.psu.edu/viewdoc/download?doi=10.1.1.175.681&rep=rep1&type=pdf (accessed on 20 April 2022).

38. Lloyd, C. *Spatial Data Analysis: An Introduction for GIS Users*; Oxford University Press: New York, NY, USA, 2010.

39. Szewranski, S.; Kazak, J.; Sylla, M.; Swiader, M. Spatial data analysis with the use of ArcGIS and Tableau systems. In *The Rise of Big Spatial Data*; Springer: Cham, Switzerland, 2017; pp. 337–349.

40. Fainstein, S.S. The just city. *Int. J. Urban Sci.* **2014**, *18*, 1–18. [CrossRef]

41. Lefebvre, H.; Nicholson-Smith, D. *The Production of Space*; Blackwell: Oxford, UK, 1991; Volume 142.

42. Harvey, D. *Social Justice and the City*; University of Georgia Press: Athens, GA, USA, 2010; Volume 1.

43. Soja, E.W. *Postmodern Geographies: The Reassertion of Space in Critical Social Theory*; Verso: London, UK, 1989.

44. Maslow, A.H. A theory of human motivation. *Psychol. Rev.* **1943**, *50*, 370–396. [CrossRef]

45. Zheng, R. *On Spatial Justice: The Reconstruction of Justice and the Criticism of Spatial Production*; Shanghai Academy of Social Sciences Press: Shanghai, China, 2018.

46. Ernstson, H. The social production of ecosystem services: A framework for studying environmental justice and ecological complexity in urbanized landscapes. *Landsc. Urban Plan.* **2013**, *109*, 7–17. [CrossRef]

47. McConnachie, M.M.; Shackleton, C.M. Public green space inequality in small towns in South Africa. *Habitat Int.* **2010**, *34*, 244–248. [CrossRef]

48. Tsou, K.W.; Hung, Y.T.; Chang, Y.L. An accessibility-based integrated measure of relative spatial equity in urban public facilities. *Cities* **2005**, *22*, 424–435. [CrossRef]

49. Kronenberg, J.; Haase, A.; Laszkiewicz, E.; Antal, A.; Baravikova, A.; Biernacka, M.; Dushkova, D.; Filcak, R.; Haase, D.; Ignatieva, M.; et al. Environmental justice in the context of urban green space availability, accessibility, and attractiveness in postsocialist cities. *Cities* **2020**, *106*, 102862. [CrossRef]

50. Beiler, M.O.; Mohammed, M. Exploring transportation equity: Development and application of a transportation justice framework *Transp. Res. Part D-Transp. Environ.* **2016**, *47*, 285–298. [CrossRef]
51. Gu, C.; Kesteloot, C. Social polarisation and segregation phenomenon in Beijing. *Acta Geogr. Sin.* **1997**, *52*, 385–393.
52. Feng, J.; Zhong, Y. Restructuring of social space in Beijing from 2000 to 2010. *Acta Geogr. Sin.* **2018**, *73*, 711–737.
53. Huang, X.; Chen, Z. The retail business structure of the metro site based on the information entropy–a case study of the 15 subway site in Guangzhou. *Econ. Geogr.* **2014**, *3*, 38–44.
54. Ebdon, D. *Statistics in Geography*; Blackwell Publishing: Hoboken, NJ, USA, 1985.
55. Wenhao, Y.; Tinghua, A. The visualization and analysis of POI features under network space supported by kernel density estimation. *Acta Geod. Cartogr. Sin.* **2015**, *44*, 82.
56. Borruso, G. Network density estimation: A GIS approach for analysing point patterns in a network space. *Trans. GIS* **2008**, *12*, 377–402. [CrossRef]
57. Duan, Y.; Liu, Y.; Liu, X.; Wang, H. Identification of polycentric urban structure of central Chongqing using points of interest Big Data. *J. Nat. Resour.* **2018**, *33*, 788–800.
58. Lefever, D.W. Measuring geographic concentration by means of the standard deviational ellipse. *Am. J. Sociol.* **1926**, *32*, 88–94. [CrossRef]
59. Zhao, L.; Zhao, Z. Projecting the spatial variation of economic based on the specific ellipses in China. *Sci. Geogr. Sin.* **2014**, *34*, 979–986.
60. Wang, J.; Wang, S.; Li, S.; Feng, K. Coupling analysis of urbanization and energy-environment efficiency: Evidence from Guangdong province. *Appl. Energy* **2019**, *254*, 113650. [CrossRef]
61. Neutens, T.; Delafontaine, M.; Scott, D.M.; De Maeyera, P. A GIS-based method to identify spatiotemporal gaps in public service delivery. *Appl. Geogr.* **2012**, *32*, 253–264. [CrossRef]
62. Nordberg, K. Spatial Justice and local capability in rural areas. *J. Rural Stud.* **2020**, *78*, 47–58. [CrossRef]
63. Sun, F.; Zhang, J.; Ma, J.; Wang, C.; Hu, S.; Xu, D. Evolution of the Spatial-Temporal Pattern and Social Performance Evaluation of Community Sports and Fitness Venues in Shanghai. *Int. J. Environ. Res. Public Health* **2022**, *19*, 274. [CrossRef]
64. Fisher, J.B.; Kelly, M.; Romm, J. Scales of environmental justice: Combining GIS and spatial analysis for air toxics in West Oakland, California. *Health Place* **2006**, *12*, 701–714. [CrossRef]
65. Xu, F.; Hu, M.; La, L.; Wang, J.; Huang, C. The influence of neighbourhood environment on Airbnb: A geographically weighed regression analysis. *Tour. Geogr.* **2020**, *22*, 192–209. [CrossRef]
66. Greater London Authority. London Borough Profiles and Atlas. Available online: https://data.london.gov.uk/dataset/london-borough-profiles (accessed on 20 April 2022).
67. Lees, L.; Ferreri, M. Resisting gentrification on its final frontiers: Learning from the Heygate Estate in London (1974–2013). *Cities* **2016**, *57*, 14–24. [CrossRef]
68. Liu, Y.; Wu, F.; Liu, Y.; Li, Z. Changing neighbourhood cohesion under the impact of urban redevelopment: A case study of Guangzhou, China. *Urban Geogr.* **2017**, *38*, 266–290. [CrossRef]
69. Zhao, P. Too complex to be managed? New trends in peri-urbanisation and its planning in Beijing. *Cities* **2013**, *30*, 68–76. [CrossRef]
70. Times, G. Construction of 50 Key Villages in the Urban-Rural Fringe of Beijing. Available online: https://china.huanqiu.com/article/9CaKrnJvCIx (accessed on 20 April 2022).
71. McEntee, J.C.; Ogneva-Himmelberger, Y. Diesel particulate matter, lung cancer, and asthma incidences along major traffic corridors in MA, USA: A GIS analysis. *Health Place* **2008**, *14*, 817–828. [CrossRef] [PubMed]
72. Hashem, I.A.T.; Chang, V.; Anuar, N.B.; Adewole, K.; Yaqoob, I.; Gani, A.; Ahmed, E.; Chiroma, H. The role of big data in smart city. *Int. J. Inf. Manag.* **2016**, *36*, 748–758. [CrossRef]
73. Henar Salas-Olmedo, M.; Moya-Gomez, B.; Carlos Garcia-Palomares, J.; Gutierrez, J. Tourists' digital footprint in cities: Comparing Big Data sources. *Tour. Manag.* **2018**, *66*, 13–25. [CrossRef]
74. Li, J.; Li, J.; Yuan, Y.; Li, G. Spatiotemporal distribution characteristics and mechanism analysis of urban population density: A case of Xi'an, Shaanxi, China. *Cities* **2019**, *86*, 62–70. [CrossRef]
75. Zhao, P.; Hu, H. Geographical patterns of traffic congestion in growing megacities: Big data analytics from Beijing. *Cities* **2019**, *92*, 164–174. [CrossRef]
76. Liu, X.; Yan, X.; Wang, W.; Titheridge, H.; Wang, R.; Liu, Y. Characterizing the polycentric spatial structure of Beijing Metropolitan Region using carpooling big data. *Cities* **2021**, *109*, 103040. [CrossRef]
77. Cremer, H.; Dekerchove, A.M.; Thisse, J.F. An economic-theory of public facilities in space. *Math. Soc. Sci.* **1985**, *9*, 249–262. [CrossRef]
78. Onokerhoraye, A.G. A conceptual framework for the location of public facilities in the urban areas of developing countries: The Nigerian case. *Socio-Econ. Plan. Sci.* **1976**, *10*, 237–240. [CrossRef]
79. Warr, P.; Menon, J.; Rasphone, S. Public services and the poor in Laos. *World Dev.* **2015**, *66*, 371–382. [CrossRef]
80. Kawakami, N.; Winkleby, M.; Skog, L.; Szulkin, R.; Sundquist, K. Differences in neighborhood accessibility to health-related resources: A nationwide comparison between deprived and affluent neighborhoods in Sweden. *Health Place* **2011**, *17*, 132–139. [CrossRef]

81. Estoque, R.C. A Review of the Sustainability Concept and the State of SDG Monitoring Using Remote Sensing. *Remote Sens.* **2020**, *12*, 1770. [CrossRef]
82. Pereira, R.H.M.; Schwanen, T.; Banister, D. Distributive justice and equity in transportation. *Transp. Rev.* **2017**, *37*, 170–191. [CrossRef]
83. Zacharias, J.; Sun, Z.; Chuang, L.; Lee, F. The hutong urban development model compared with contemporary suburban development in Beijing. *Habitat Int.* **2015**, *49*, 260–265. [CrossRef]

 *remote sensing*

Article

Impacts of Urbanization on the Muthurajawela Marsh and Negombo Lagoon, Sri Lanka: Implications for Landscape Planning towards a Sustainable Urban Wetland Ecosystem

Darshana Athukorala [1,*], Ronald C. Estoque [2], Yuji Murayama [3] and Bunkei Matsushita [3]

[1] Graduate School of Life and Environmental Sciences, University of Tsukuba, 1-1-1 Tennodai, Tsukuba, Ibaraki 305-8572, Japan

[2] National Institute for Environmental Studies, Tsukuba, Ibaraki 305-8506, Japan; estoque.ronaldcanero@nies.go.jp or rons2k@yahoo.co.uk

[3] Faculty of Life and Environmental Sciences, University of Tsukuba, 1-1-1, Tennodai, Tsukuba, Ibaraki 305-8572, Japan; mura@geoenv.tsukuba.ac.jp (Y.M.); matsushita.bunkei.gn@u.tsukuba.ac.jp (B.M.)

* Correspondence: s1830207@s.tsukuba.ac.jp or darshana12594@gmail.com

Citation: Athukorala, D.; Estoque, R.C.; Murayama, Y.; Matsushita, B. Impacts of Urbanization on the Muthurajawela Marsh and Negombo Lagoon, Sri Lanka: Implications for Landscape Planning towards a Sustainable Urban Wetland Ecosystem. *Remote Sens.* **2021**, *13*, 316. https://doi.org/10.3390/rs13020316

Received: 25 November 2020
Accepted: 12 January 2021
Published: 18 January 2021

Publisher's Note: MDPI stays neutral with regard to jurisdictional claims in published maps and institutional affiliations.

Abstract: Urban wetland ecosystems (UWEs) play important social and ecological roles but are often adversely affected by urban landscape transformations. Spatio-temporal analyses to gain insights into the trajectories of landscape changes in these ecosystems are needed for better landscape planning towards sustainable UWEs. In this study, we examined the impacts of urbanization on the Muthurajawela Marsh and Negombo Lagoon (MMNL), an important UWE in Sri Lanka that provides valuable ecosystem services. We used remote sensing data to detect changes in the land use/cover (LUC) of the MMNL over a two-decade period (1997–2017) and spatial metrics to characterize changes in landscape composition and configuration. The results revealed that the spatial and socio-economic elements of rapid urbanization of the MMNL had been the main driver of transformation of its natural environment over the past 20 years. This is indicated by a substantial expansion of settlements (+68%) and a considerable decrease of marshland and mangrove cover (−41% and −21%, respectively). A statistical analysis revealed a significant relationship between the change in population density and the loss of wetland due to settlement expansion at the Grama Niladhari division level ($n = 99$) (where wetland includes marshland, mangrove, and water) (1997–2007: $R^2 = 0.435$, $p = 0.000$; 2007–2017: $R^2 = 0.343$, $p = 0.000$). The findings also revealed that most of the observed LUC changes occurred in areas close to roads and growth nodes (viz. Negombo, Ja-Ela, Wattala, and Katana), which resulted in both landscape fragmentation and infill urban expansion. We conclude that, in order to ensure the sustainability of the MMNL, there is an urgent need for forward-looking landscape and urban planning to promote environmentally conscious urban development in the area which is a highly valuable UWE.

Keywords: wetland; muthurajawela marsh and negombo lagoon; socio-ecological; spatio-temporal analysis; urban ecology; remote sensing

---

1. Introduction

Although wetlands account only for 4–6% of the world's surface area [1], they are regarded as one of the most productive ecosystems [2,3]. A wetland ecosystem includes marsh, fen, peatland, shallow water areas, as well as natural and human-made areas with evidence of intermittent and permanent waterlogged areas between natural wet aquatic habitats and dry terrestrial ecosystems [4,5]. Wetlands provide valuable social and ecological benefits, e.g., coastal protection, flood control, carbon sequestration and biodiversity conservation, among other ecosystem services [6–9]. As such, wetlands play important roles in the context of the United Nations' sustainable development goals (SDGs) and targets [10–12]. Unfortunately, almost 64–71% of the world's wetlands have been transformed, degraded or have disappeared in recent decades as a result of anthropogenic activities, including industrialization, agriculture, and urbanization [1,13–15].

In 2018, more than 55% of the world's population lived in urban areas, and this proportion is projected to reach 68% by 2050 [16]. Such a development would exert tremendous pressure on the natural environment across urban areas in the world as large areas transform to impervious surfaces. Studies have shown that rapid, uncontrolled, and unplanned urbanization has been impacting the quality of urban ecological environments across the world [17–22], including urban wetland ecosystems (UWEs) [23–27].

Advances in geospatial technology, including geographic information systems (GIS) and remote sensing, have greatly improved the monitoring of landscape changes over space and time. Today, information derived from these advancements provides important input to landscape planning and decision-making in many contexts, biodiversity conservation [28–31], and sustainable urbanization [32–34]. In fact, Earth observation technologies, particularly remote sensing, play important roles in the monitoring of various social and ecological indicators related to the United Nations' SDGs and targets [35], including those that are associated with wetlands [11,36,37].

With changes in land use/cover (LUC) due to urbanization, natural landscapes suffer from irreversible transformation [38]. The monitoring of landscape status over space and time is hence an essential endeavor. Scholars have shown the usefulness of geospatial techniques for characterizing landscape patterns, including those of UWEs [27,39,40], and their changes over time [41–43]. Such information that facilitates impact analysis on ecosystem services and biodiversity [44–46] can be used to direct landscape and urban development planning towards sustainable UWEs [45,47,48].

The Muthurajawela Marsh and Negombo Lagoon (MMNL), the biggest coastal saltwater peat bog in Sri Lanka, is located on the western coastal belt between the Kelani River and Negombo Lagoon lying inland to Katana, Wattala, Ja-Ela, and Negombo in the Gampaha District of the Western Province. The Muthurajawela Marsh, together with the lagoon, creates an integrated coastal wetland ecosystem. The complex development of this landscape during the Holocene period (Circa 6000–5000 years) progressed after the final glacial period (Figure 1) [49–51]. The MMNL has been, and is still today, an important UWE in the country. Its estimated monetary value is around Rs 726.5 million per year, including benefits from flood prevention, treatment of wastewater, and shallow coastal fisheries [52].

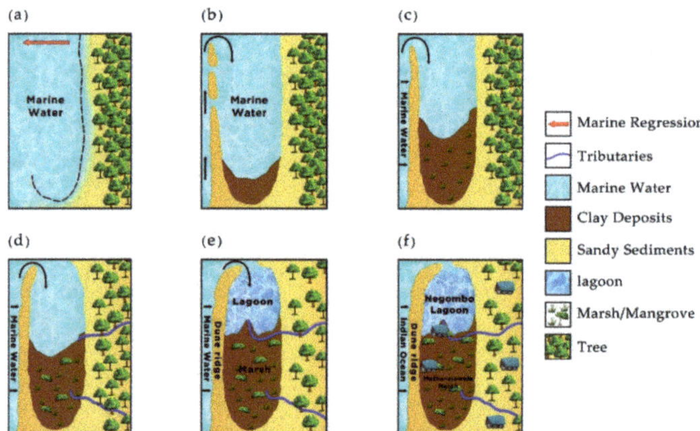

Figure 1. Geological evolution of the Muthurajawela Marsh and Negombo Lagoon (MMNL), Sri Lanka, from the Holocene period to the present. (**a**) The MMNL in the mid-Holocene period, marked with a marine regression, exposing a wider coastal area; (**b**) Wetland started to form (c. 7000); (**c**) Formation of wetland from south to north and clay deposits continued (c. 6500); (**d**) Marshland and lagoon started to form as an interdependent ecological system (c. 6500–6300); (**e**) Formation of marshland and lagoon continued at the final stage, forming one contiguous wetland (c. 6000); and (**f**) The present condition in which the MMNL is an important urban wetland ecosystem (UWE) in the Colombo Metropolitan Region with high socio-ecological significance. The images were sourced from the MMNL's master plan and MMNL's conservation management plan [49,50].

The Colombo Metropolitan Region (CMR), the country's capital and the location of the MMNL, has grown very rapidly over the recent decades. For example, CMR's population grew from 3.9 million in 1981 to 5.8 million in 2012 [53]. Consequently, the area of built-up lands in the CMR also expanded dramatically from 11,165 ha in 1992 to 35,876 ha in 2014 [54]. There have been several studies on MMNL [55,56], but a study that focuses on the impacts of urbanization on this highly valuable UWE is still lacking.

Hence, the focus in this study is specifically on detecting changes in the LUC and examining the landscape composition and configuration of the MMNL over the past two decades (1997–2017) by using remote sensing data and spatial metrics respectively. The loss of wetland (water, mangrove, and marshland) due to urbanization (settlement expansion) was quantified and the influence of population growth to this LUC transition was investigated. The implications of the findings for landscape and urban planning towards sustainable UWEs are discussed.

2. Materials and Methods

2.1. The MMNL, Sri Lanka

The MMNL, extending over roughly 134 km^2 (Figure 2), consists of the Gampaha District's four Divisional Secretariats (DS). This area has been experiencing rapid urbanization and economic growth over the past three decades. The landscape of the MMNL encompasses various land surface features, including the lagoon and marsh and mangrove areas, as well as some highly and moderately urbanized lands.

The MMNL receives freshwater from the eastward direction via two channels (viz. lower Aththanagalu Oya and lower Kelani Ganga). The area has a gentle slope, with an elevation range of 0–30 m. According to the geological timetable, the MMNL belongs to the Quaternary soil group, which is composed of soil deposits from wind-blown sands, river deposits, and lacustrine sediments [57]. The area has a tropical monsoon climate as per the Koppen classification [58]. The wet season is from May to September, and the dry season is from December to early March. Mean annual rainfall is between 2000 mm and 2500 mm, with mean annual daytime temperatures ranging from 22.5 °C to 25.0 °C [59].

Demographically, the MMNL contains urban and rural settlements. However, due to a rising population and rapid urban expansion, many parts of this valuable UWE have become highly vulnerable to the impacts of urbanization, including the Muthurajawela Marsh, which contains two protected areas (Figure 2). The lagoon is a shallow-water coastal water body and a highly productive fish area [52]. It is joined to the Indian Ocean by a single narrow opening in the north. This, too, will continue to be affected if the rapid urbanization of the area is not controlled and carefully planned.

In fact, the lagoon had initially been utilized by the fishing industry and the neighboring area had been occupied by settlements and industries [51,60]. However, over the past 60 years, parts of the lagoon have been reclaimed for various purposes, including illegitimate settlements that have extended to intertidal sands along the channel segments of the estuary [51,52]. Unlawful activities, such as illegal settlements, illegal fishing, and illegal cutting of trees, as well as waste dumping and water pollution, are among the important current concerns with regard to the management of the MMNL [51,52].

2.2. LUC Mapping

We used three Landsat images for this study, viz. two TM (Thematic Mapper) images captured in 1997-02-07 and 2007-01-02, and one OLI/TIRS (Operational Land Imager/ Thermal Infrared Sensor) image captured in 2017-01-31. They were sourced from the USGS (https://earthexplorer.usgs.gov/). The images have a spatial resolution of 30 m. Only one Landsat scene was needed for the study area. Temporal consistency and cloud cover were considered in the selection of the images. All the images were acquired during the cloud-free, dry season.

Before LUC classification, we first created a wetland classification scheme comprising four classes: settlement, marshland, mangrove, and water (Table 1). These LUC classes reflect the physical characteristics of the study area and have been widely used in previous studies [61–63].

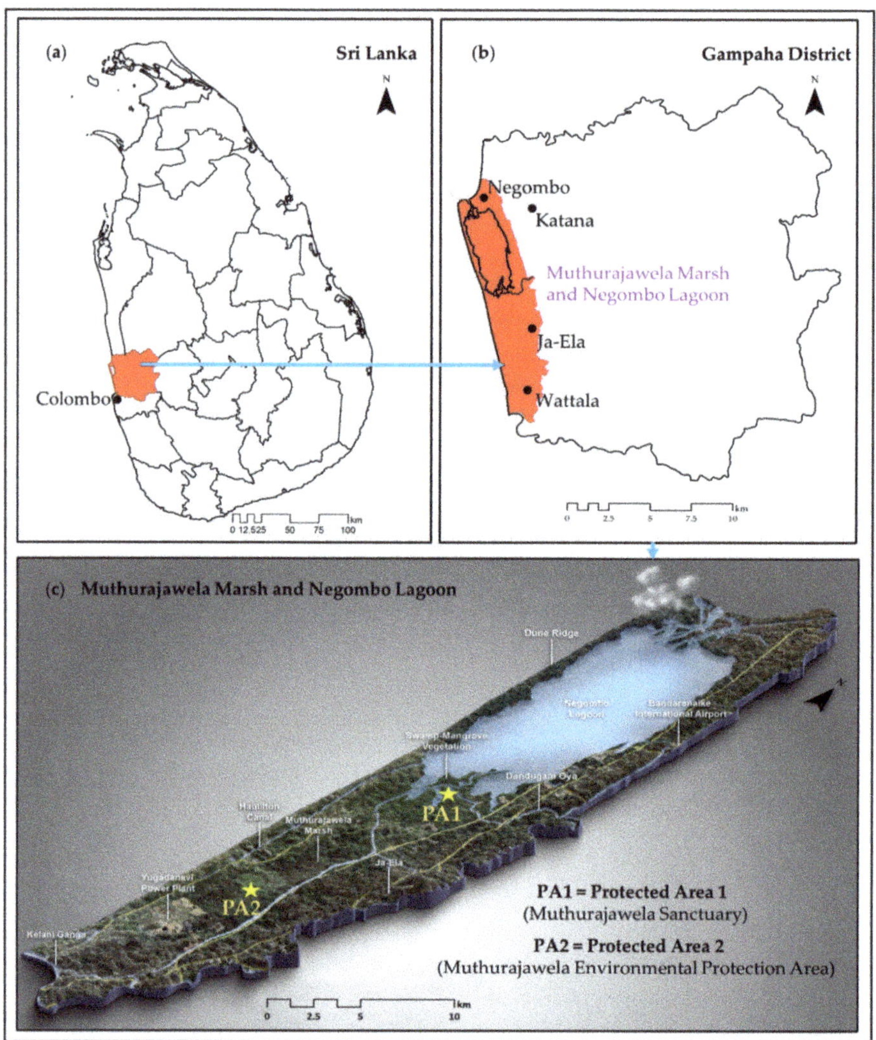

Figure 2. Location of the MMNL. (**a**) Map of Sri Lanka, (**b**) Gampaha District, and (**c**) a 3D map of the MMNL produced using a 30 m digital elevation model (ASTER). The Google Earth image was acquired on 17 February 2017.

The three Landsat images were classified using a hybrid classification method, i.e., a combination of unsupervised and supervised classification techniques [64,65], which was performed in ArcGIS 10.6. First, we used the ISODATA clustering algorithm, an unsupervised classification algorithm, to produce 14 clusters for each image. We used bands 5, 4, and 3 for the TM images and bands 6, 5, and 4 for the OLI/TIRS image. Second, we performed a supervised classification using the maximum likelihood algorithm and the result of the first step as input. For the training sites, we digitized 15 training samples for each class, where the number of pixels per sample ranged from 20 to 782. In total, we digitized 60 samples per image. We assessed the accuracy of the classified 1997, 2007, and 2017 LUC maps using 400 random points. Google Earth was the source of reference data for 2007 and 2017. For 1997, we used topographic maps from the Survey Department of Sri Lanka [66].

Table 1. Wetland classification scheme used in this study.

LUC Class	Land Surface Features
Settlement	Urban, residential, industrial, transportation (roads, train lines), communications and utilities infrastructure, airports, home gardens, concrete structures, power plants, and asphalt areas.
Marshland	Seasonally flooded areas with abandoned paddy fields, intermittently flooded areas with agriculture, marsh plant communities, trees, scrub and grassland, peat soil, bog soil and back swamp, and other cultivated areas.
Mangrove	Seasonally flooded areas with mangroves, intermittently flooded areas with mangroves and mangrove pneumatophore areas.
Water	Lagoon, streams, canals, and ponds.

2.3. Assessment of LUC Changes in the MMNL

We calculated the loss and gain areas and rates for each LUC class using Equations (1) and (2), respectively [67].

$$L/G \ area \ = A_b - A_a \tag{1}$$

$$L/G \ rate \ (\%) \ = (A_b - A_a)/A_a \times 100 \tag{2}$$

where L/G area refers to the area that each class lost or gained (ha) between two time points. L/G rate refers to the percentage of loss or gain (%) of each class area. A_a and A_b are the beginning and the end values of each class, respectively.

2.4. Assessment of Wetland Loss Across Grama Niladhari (GN) Divisions

Due to rapid urbanization, there is a high likelihood of the wetland area and its surrounding areas being converted to urban land use. Here, we identified the top wetland-losing GNs susceptible to urbanization (settlement expansion) over the study period. Currently, there are 99 GN divisions in the study area. To do this, we calculated the density of wetland loss due to urbanization across the GNs during the first time period (TP1) (1997–2007) and second time period (TP2) (2007–2017) using Equations (3) and (4), respectively.

$$WLD_{TP1}(\%) = \frac{WL_{TP1}}{A} \times 100 \tag{3}$$

$$WLD_{TP2}(\%) = \frac{WL_{TP2}}{A} \times 100 \tag{4}$$

where $WLD_{TP1}(\%)$ and $WLD_{TP2}(\%)$ refer to the density of wetland loss density in a particular GN due to urbanization (settlement expansion) during the two time periods (first and second, respectively). WL_{TP1} and WL_{TP2} are the areas of wetland loss in a particular GN due to urbanization during the two time periods (first and second, respectively). A refers to the area of a particular GN. Wetland includes water, mangrove, and marshland.

2.5. Relationship between Urbanization and Wetland Loss

We examined the relationship between urbanization as proxied by the change in population density (CPD) (Equations (5) and (6)) and density of wetland loss due to settlement expansion (WLD) at the GN division level. Since the 1997, 2007, and 2017 population data at the GN division level were not available, we extrapolated such data based on the average population growth rate (APGR) of Katana Wattala, Ja-Ela, and Negombo DS divisions for 1981, 2001, and 2012 census years. Each of these DS divisions consisted of a number of GN divisions. Thus, for those GN divisions that belonged to a particular DS division, only one APGR was used. Finally, scatter plots were produced between CPD (x) and WLD (y) for both periods (TP1 and TP2) to examine the relationship between urbanization and wetland loss.

$$CPD_{TP1}(\%) = \frac{PD_{2007} - PD_{1997}}{PD_{1997}} \times 100 \tag{5}$$

$$CPD_{TP2}(\%) = \frac{PD_{2017} - PD_{2007}}{PD_{2007}} \times 100 \tag{6}$$

where $CPD_{TP1}(\%)$ and $CPD_{TP2}(\%)$ refer to the change in population density in a particular GN during the two time periods (first and second, respectively). PD refers to the population density of a particular GN.

2.6. Landscape Pattern Analysis

Many scholars have discussed the usefulness of spatial metrics for landscape pattern analysis [68,69]. Landscape-level metrics provide general information about landscape patterns in the study area. On the other hand, class-level metrics include more detailed descriptions of landscape patterns based on class-level information. Using FRAGSTATS V4.2, landscape-level and class-level metrics were computed to gain insights into the changes in the landscape pattern of the MMNL (Table 2).

Table 2. Class and landscape-level spatial metrics [70].

Metric	Equation	Unit	Definition
Number of Patches (NP)	$NP = n_i$	None	Reflects the number of patches of the similar patch type or LUC class; a simple measure of the degree of fragmentation
Patch Density (PD)	$PD = \frac{n_i}{A}(10,000)(100)$	No. per 100 ha	Equal to the number of patches at each LUC class per unit area. A limited, yet important feature of the landscaping
Edge Density (ED)	$ED = \frac{\sum_{k=1}^{m} e_{ik}}{A}(10,000)$	Meters per ha	Measures based on edge length of a specific LUC class per unit area
Largest Patch Index (LPI)	$LPI = \dfrac{\substack{n \\ \max \\ j=i}(a_{ij})}{A}(100)$	Percent	Quantifies the percentage of total landscape area taken up by the largest patch at the class level. It is a simple indicator of dominance
Landscape Shape Index (LSI)	$LSI = \frac{0.25\sum_{k=1}^{m} e_{ik}^{*}}{\sqrt{A}}$	None	A measure of the total edge or edge density within the landscape divided by the total landscape.
Cohesion (COHESION)	$COHESION = \left[1 - \frac{\sum_{j=1}^{n} P_{ij}^{*}}{\sum_{j=1}^{n} P_{ij}^{*}\sqrt{a_{ij}^{*}}}\right] \cdot \left[1 - \frac{1}{\sqrt{Z}}\right]^{-1} \cdot (100)$	None 0–100	The physical connectivity of the corresponding patch type of LUC class. Rises with more clustering of the patch type in its configuration, resulting in a more physical combination.
Shannon's Diversity Index (SHDI)	$SHDI = -\sum_{i-1}^{m}(P_i \,{}^{\circ} InP_i)$	Information	Reflects the landscape heterogeneity and compares various landscapes or the same landscape at different times as a relative index.
Shannon's Evenness Index (SHEI)	$SHEI = \frac{-\sum_{i=1}^{m}\left(P_i \,{}^{\circ} InP_i\right)}{Inm}$	None	Maximum evenness of the area, reflecting a clear trend among the patch types at the landscape level.

Where: i = any LUC patch; n_i = number of patches of LUC category i; A = total area of LUC (m^2); e_{ik} *is sum of* edge total (m) in LUC class I—counting landscape boundary and segments; j = 1,2, 3, . . . , n sum of the specific patch area; a_{ij} = patch area ij in number of the pixel ; p_i = proportion of the i—any LUC area of the total landscape; p_{ij} = circumference of patch ij regarding the sum of cell surface; Z = total pixel in the landscape; m = total patch in the entire area, without landscape border. Patch number was determined based on the eight-cell neighborhood rule.

3. Results

3.1. Classification Accuracy and LUC Changes

The overall accuracy of the three classified LUC maps was 83.0% in 1997, 84.5% in 2007, and 84.8% in 2017 (Appendix A Table A1). The primary cause of the classification errors was spectral confusion because some of the pixels of the LUC classes had similar spectral reflectance due to soil moisture levels and vegetation types [71,72]. We found that some mangrove pixels were misclassified as marsh, and some marsh pixels were misclassified as settlements and mangroves, and so on (Appendix A Table A1). Nevertheless, the accuracy levels of the classified LUC maps of the MMNL are adequate for this study. Other related studies have reported overall accuracies ranging from 69% to 82% [73–75].

The LUC change analysis revealed that marshland in the MMNL had been drastically shrinking due primarily to the expansion of settlements. For example, of the 1767 ha and 2282 ha total loss of the marshland class in TP1 and TP2, respectively, 820.89 ha (43%) and 691.92 ha (42%) were lost from the mangrove class (Figure 3 and Table 3). During TP1, the settlement class gained a total area of 1464 ha, whereas during TP2, it gained a total area 1880.91 ha.

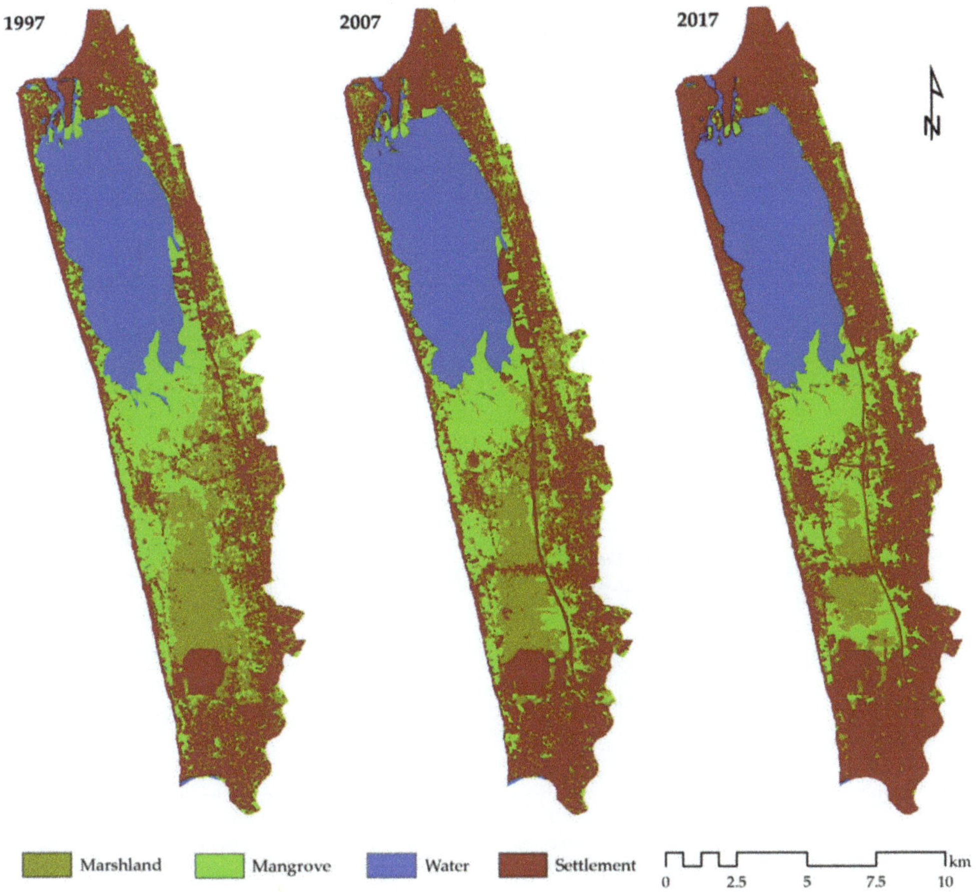

Figure 3. LUC maps of the MMNL in 1997, 2007, and 2017 derived from Landsat data (see Methods).

Table 3. LUC transitions in the MMNL (ha).

1997	2007				Total	Loss
	Marshland	**Mangrove**	**Water**	**Settlement**		
	(a) 1997–2007					
Marshland	3007.89	519.93	2.43	1244.88	4775.13	1767.24
Mangrove	615.42	1081.44	2.79	202.68	1902.33	820.89
Water	65.07	1.44	3118.41	16.83	3201.75	83.34
Settlement	547.29	33.12	1.08	2950.38	3531.87	581.49
Total	4235.67	1635.93	3124.71	4414.77	13,411.08	
Gain	1227.78	554.49	6.3	1464.39		

2007	2017				Total	Loss
	Marshland	**Mangrove**	**Water**	**Settlement**		
	(b) 2007–2017					
Marshland	1953.45	538.2	30.51	1713.51	4235.67	2282.22
Mangrove	541.71	944.01	0.9	149.31	1635.93	691.92
Water	10.62	3.15	3092.85	18.09	3124.71	31.86
Settlement	324.09	24.48	4.32	4061.88	4414.77	352.89
Total	2829.87	1509.84	3128.58	5942.79	13,411.08	
Gain	876.42	565.83	35.73	1880.91		

The mangrove class experienced considerable losses to both the marshland and settlement classes in the two periods. There were some gains in the area of the mangrove class, but its total loss outweighed its total gain, resulting in net losses during both periods. Nevertheless, the gain of mangrove from marshland in both periods (520 ha in TP1 and 538 ha in TP2) is a positive sign. This could have been due to the government's efforts to conserve the MMN by conducting mangrove reforestation activities in previous years.

Table 4 shows the L/G of the LUC classes in the MMNL in terms of area and rate. The results revealed that the mangrove class had a net decrease of 266 ha (14%) and 126 ha (8%) in TP1 and TP2, respectively. The marshland class had a net reduction of 539 ha (11%) and 1406 ha (33%). By contrast, the settlement class had a net increase of 883 ha (25%) and 1528 ha (35%).

Table 4. Losses and gains of the LUC classes in the MMNL.

	L/G Area (ha)	**L/G Rate (%)**
1997–2007		
Marshland	−539.46	−11.30
Mangrove	−266.40	−14.00
Water	−77.04	−2.41
Settlement	882.90	25.00
2007–2017		
Marshland	−1405.80	−33.19
Mangrove	−126.09	−7.71
Water	3.87	0.12
Settlement	1528.02	34.61

Figure 4 shows the spatial distribution of wetland loss due to urbanization (settlement expansion) in both periods, i.e., the loss of marshland, mangrove, and water. The results revealed that the central part (south of the lagoon) and the east part of the MMNL became more fragmented due to road construction. During the 2007–2017 period, the area exhibited a ribbon type of development. Another pattern that emerged is the settlement cluster in the middle, western part of the MMNL. In general, settlements consumed mostly the marshland and mangrove areas in the northern, eastern, and southern parts of the MMNL.

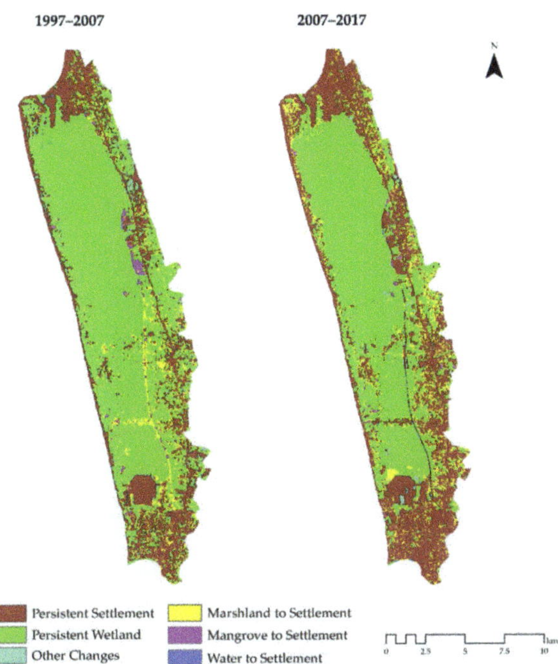

Figure 4. Spatial distribution of wetland loss in the MMNL due to urbanization (settlement expansion). (**left**) 1997 to 2007; and (**right**) 2007 to 2017. Wetland includes water, mangrove, and marshland.

3.2. Change in Population Density and Loss of Wetland Due to Urbanization

The maps of the GN-level change in population density (CPD) and density of wetland loss due to urbanization (settlement expansion) (WLD) are shown in Figure 5. The top five GNs in terms of WLD during TP1 were Thimbirigasyaya, Nayakakanda South, Kurunduhena, Welikadamulla, Telangapatha; those during TP2 were Thalahena, Palliyawatta South, Maha Pamunugama, Pitipana North, Udammita South. A more exhaustive list of the top GNs in terms of WLD is given in Appendix A Table A2.

The statistical analysis revealed a positive, significant correlation between CPD and WLD in both time periods (TP1: $R^2 = 0.435$, $p = 0.000$; TP2: $R^2 = 0.343$, $p = 0.000$) (Figure 6), indicating that as the CPD increased, the WLD also increased. This suggests that urbanization was, indeed, an important factor or a driver of wetland loss in the MMNL.

3.3. Changes in Landscape Composition and Configuration

At the class-level, the results revealed that between 1997 and 2017, the marshland class had become more fragmented, as indicated by the overall increase in its number of patches (NP), patch density (PD), edge density (ED), and the decline in its largest patch index (LPI), landscape shape index (LSI), and COHESION due to conversions to settlement marshland and water classes (Figure 7). On the other hand, the mangrove class had become less fragmented, as indicated by the overall decrease in its NP, PD, and ED. While its COHESION decreased during the first period, it increased during the second period. This suggests that the mangrove class, with the loss of some of its fragmented patches, had become more contiguous from 2007 to 2017. The decrease in LPI and LSI during the 1997–2017 period was due to mangrove gains, especially those resulting from the conversion of marshland, suggesting the development of more regular shapes at the edges of the mangrove. Conversely, the analysis results showed that LPI and LSI in marshland had fewer regular shapes at the edges, and the patches were more adjoining compared to the mangrove class. Therefore, the overall result of the mangrove showed less fragmentation than the marshland class. The settlement class had also become less fragmented, as indicated by the overall decrease in its NP, PD, and ED, and the overall increase in its COHESION. The expansion of settlements to adjacent areas resulted in an infilling pattern of urban growth, and eventually in a more contiguous and aggregated configuration of the settlement class. This observation is also supported by the increasing pattern of the settlement class' LPI.

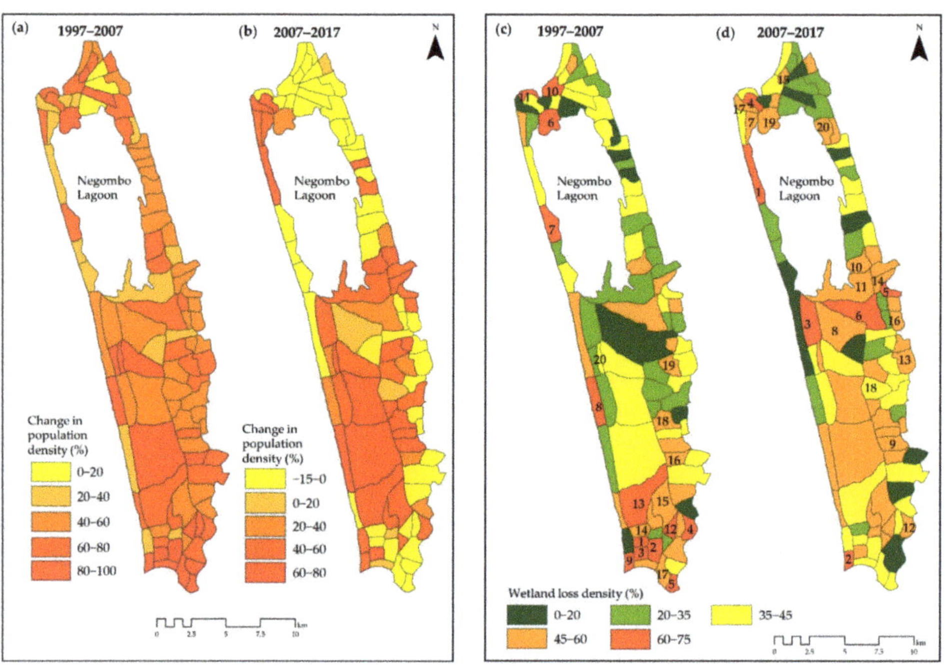

Figure 5. GN-level change in population density (CPD) (**a,b**) and density of wetland loss due to urbanization (settlement expansion) (WLD) (c and d) in the MMNL. Wetland includes water, mangrove, and marshland. The numbers on maps (**c,d**) refer to the numbers of the GNs in Appendix A Table A2.

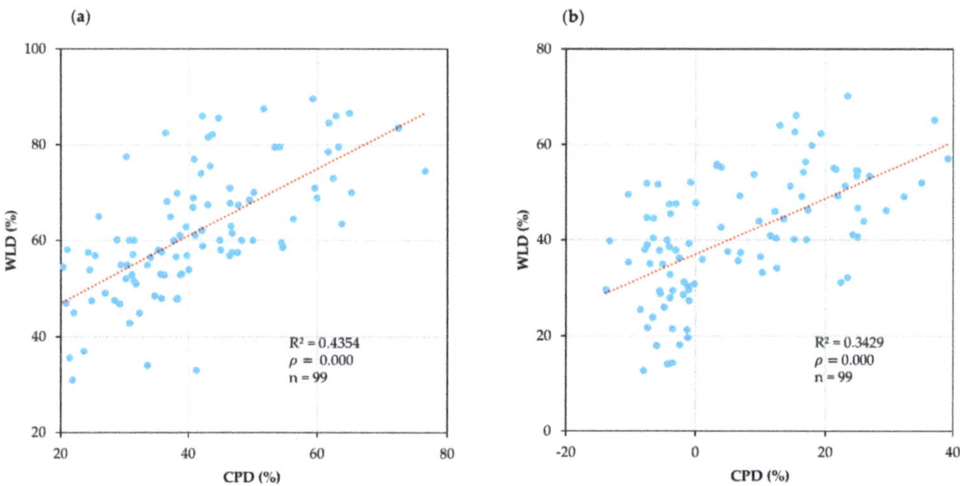

Figure 6. Relationship between change in population density (CPD) and density of wetland loss due to settlement expansion (WLD) in the MMNL during (**a**) 1997–2007 (TP1) and (**b**) 2007–2017 (TP1). Each point is a GN division.

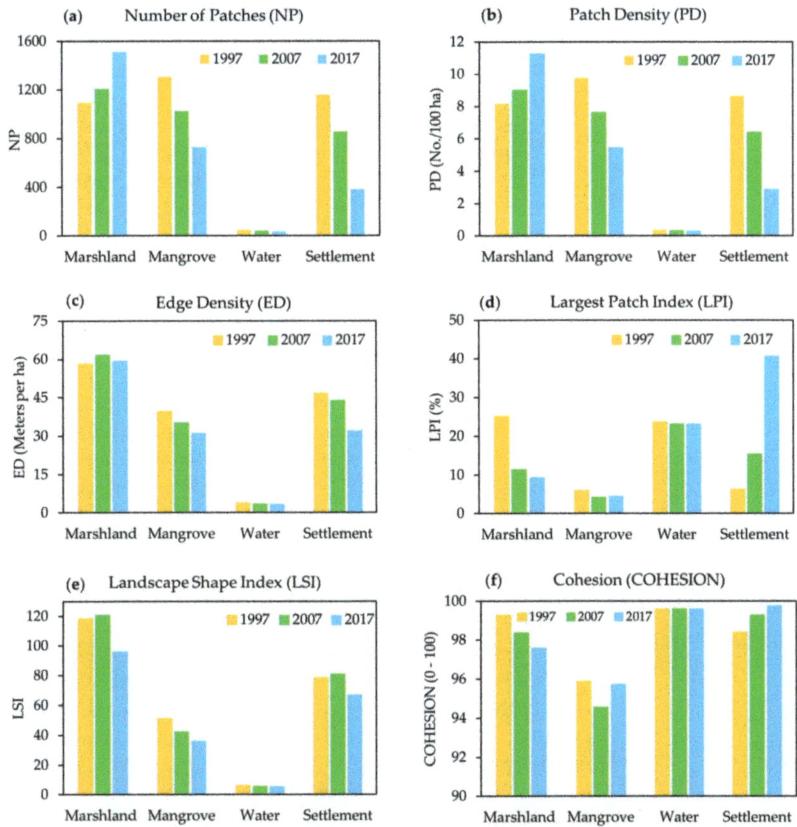

Figure 7. Class-level spatial metrics for the MMNL (1997–2017).

At the landscape level, the results revealed that SHDI and SHEI had gradually declined from 1997 to 2017 (Table 5). This indicates that the landscape of MMNL had tended to be less fragmented, clustering, and aggregating. Moreover, the overall decrease in patch richness at the class-level had resulted in an overall decrease in SHDI and SHEI at the landscape-level in the study area.

Table 5. Landscape-level spatial metrics for the MMNL (1997–2017).

Year	SHDI	SHEI
1997	1.3383	0.9654
2007	1.3263	0.9567
2017	1.2832	0.9256

4. Discussion

4.1. Landscape Transformation of the MMNL

The MMNL is an important UWE in Sri Lanka owing to its biodiverse ecosystem that is home to numerous wildlife, water habitat species, and migratory birds [50], besides the various ecosystem services it provides [52]. Our findings showed that the landscape of this highly valuable UWE had been transformed dramatically over the past two decades, losing considerable expanses of its marshland and mangrove cover due to rapid, unplanned and uncontrolled urbanization (settlement expansion) (Figures 3 and 4; Table 3).

Urbanization, led by socio-economic and biophysical factors, has altered and is still altering the MMNL landscape. If this wetland change trend continues, it may adversely impact the ecosystem services, biodiversity and aesthetic value of the area. There are indications of an infilling urban growth pattern in the MMNL (Figure 3) and clear signs of illegal settlements inside the wetland area. The uncontrolled urban expansion of the CMR and its effects on landscape changes have caused many socio-economic and ecological problems, as well as an overall degradation of the natural environment in the study area.

Today, the MMNL has been fragmented into four parts owing to settlements, the construction of the main road and the Colombo-Katunayake Expressway [76], and the area experiencing a ribbon type-development during the 2007–2017 period. This expressway runs along the marshland, and a small piece of the Negombo lagoon can be clearly identified in our classified maps, especially in 2007 and 2017 (Figures 3 and 4). Using urban wetland modelling, Zubair et al. (2017) [77] found that two of the main watersheds had increased, but subsequently decreased in one due to urban expansion. These findings generally support our results on the effects of human intervention, as indicated in previous research [78].

Availability and reclamation of natural wetland areas according to environment-friendly policies and enforcement of regulations are crucial to the protection and conservation of the MMNL. Restoration of wetland vegetation is vital, particularly in the highly populated areas of the GN divisions. A top-to-bottom approach should be adopted to ensure judicious use of wetland to ensure its protection and sustainability. Generally, wetland areas play an essential role in mitigating the urban heat island effect [79]. The MMNL is situated in the CMR which covers a considerable area. Conserving this highly valuable wetland will promote the cooling effect for better living conditions for the city dwellers of the CMR. Therefore, the protection and sustainability of the wetland should be promoted systematically by policymakers and urban planners.

In this study, we used a hybrid method (unsupervised and supervised) to classify the LUC of the MMNL from Landsat imagery (see Section 2.2). The method minimized classification errors. Overall, this hybrid classification provides comprehensive classifications of natural plant vegetation and soil moisture levels in urban wetland areas [80]. The overall accuracy of our three classified LUC maps was 83.0% in 1997, 84.5% in 2007, and 84.8% in 2017. Similar findings were reported by reference [64] in their small wetlands mapping in Kenya and Tanzania, where using unsupervised and supervised approaches, the overall classification accuracy was 83%. Lane et al. (2014) [81] reports an overall accuracy of 86.5% in wetland classification using eight-band high-resolution satellite data and a hybrid mapping approach in the Selenga River Delta in southeastern Siberia, Russia.

In general, settlement expansion can be correlated with rapid population growth in the MMNL. It is important to note that the MMNL is located in the Gampaha District of Sri Lanka, the second most populous district in Sri Lanka after the Colombo District. Rural-urban migration due to the establishment of Export Processing Zones (viz. Biyagama, Katunayake) in the Gampaha District [82] contributed to the higher population growth during the 1990s. Job opportunities in these Export Processing Zones provided better living conditions for migrants. Given the decline in agricultural productivity in the country's dry regions, the government encouraged rural-urban migration to reduce poverty [83]. In particular, post-war policies and development projects in the CMR resulted in the country's industrial capital becoming an important driver of rapid urban growth of the CMR after 2009 [84].

Figure 8 projects continuous growth in four DS divisions in the study area from 1981 to 2051. The dramatic increase in the urban population of four DS divisions in the study area is expected to continue in the future. From 1997 to 2017, the population of the study area increased by 15.51%. The population density of the Wattala and Ja-Ela DS divisions was higher than that of Katana and Negombo, indicating high urban pressure radiating from the capital of Colombo and the core of the Gampaha District (Figures 2, 3 and 5). However, the Negombo DS division should not be ignored because this DS division has a significant effect on the wetland's northern part (Figure 2), which has been impacted by rapid population growth leading to residential (including illegal settlements) and non-residential developments in industrial and commercial sectors.

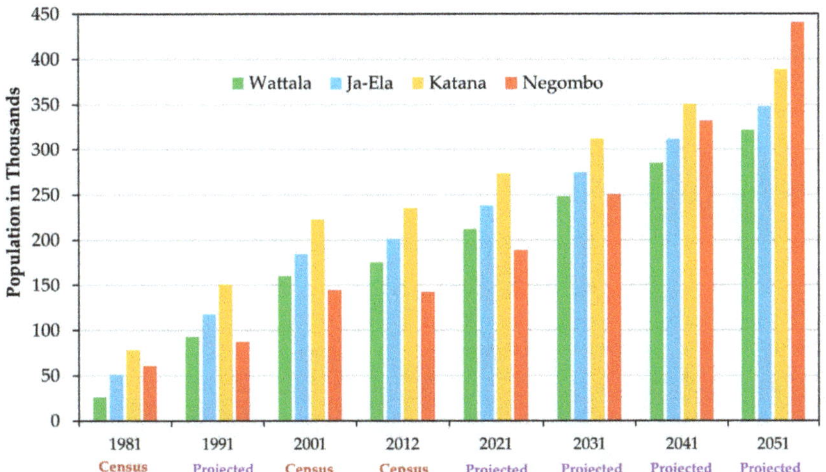

Figure 8. Projected population trend for Wattala, Ja-Ela, Katana and Negombo DSDs of Gampaha District: The population data for 1981, 2001 and 2012 were sourced from the Department of Census and Statistics, Sri Lanka [85]. The 1991 population was projected using growth rates of 5.1%, 2.6%, 2.1%, and 6.9% for Wattala, Ja-Ela, Katana, and Negombo, respectively. From 2021 to 2051, growth rates were projected using rates of 0.87%, 0.83%, 0.51%, and −0.13% for Wattala, Ja-Ela, Katana, and Negombo, respectively.

In wetland change analyses, researchers have identified some implications of wetland landscape pattern changes due to urbanization [86,87]. There is evidence that environmental degradation is very much related to economic growth, as reflected particularly in per capita income [88]. A spatial metrics analysis shows that the MMNL settlement areas have become less fragmented (Figure 7) due to the decrease in the number of settlement patches along the roads and the dispersion of newly established patches from existing settlement areas around growth nodes (i.e., Negombo, Ja Ela, and Wattala) (Figure 2) and administrative centers. Furthermore, current settlement patches have become more extensive, and the gaps between settlement patches have diminished. While the expansion process has led to the development of settlement patches, there is a reduction in the distance between settlement patches due to the impact of dispersion in the MMNL. This rapid urban development of the MMNL and its subsequent wetland landscape changes have created many socio-ecological problems (Figure 9).

The MMNL has already been encroached upon by the urban sprawl; further infilling patterns are anticipated. Several factors can cause further fragmentation of the MMNL's landscape and dispersed growth. The MMNL is located in the western coastal plain of Sri Lanka (Figure 2), and there are no significant physical restrictions, such as high altitudes and steep slopes, which promote fragmentation of the landscape. In the past two decades, a significant expansion of road networks (Figure 3) has shortened travel time from the urban center (capital city of Colombo and the core of the Gampaha District) to the suburbs (Negombo, Katana, Wattala, and Ja Ela) and Katunayake Bandaranaike International Airport. While people moving to the suburbs facilitate the fragmentation of the wetland landscape, the lagoon fishing industry and its settlements, as well as illegal wetland reclamation, also contribute to wetland degradation and land fragmentation. Wetland areas along the roads have progressively been turned into settlements (Figure 4). Moreover, push factors, such as high housing costs and service value and land prices in urban core areas have facilitated the fragmentation of wetland areas, including marshland and mangrove areas, over the last two decades in the MMNL. Accordingly, such issues concerning future wetland landscape and urban planning activities in and around the MMNL must be addressed.

Figure 9. Some fieldwork photos of the MMNL: (**a**) Human encroachment in Negombo Lagoon area, (**b**) Illegal settlements of the study area, (**c**) Dumping of garbage into the lagoon, and (**d**) Human activities inside the Muthurajawela marshland. Source: First author (D. A.), 2017.

4.2. Implications for Wetland Sustainability

A variety of environmental and socio-economic problems have arisen as a result of the rapid urbanization process in the MMNL, such as extensive degradation of wetland ecosystems and loss of important wetland ecosystem services that affect flood control, carbon sequestration, wetland productivity, the quantity of fish in the lagoon, urban poverty, and slum development. Besides the ongoing urbanization process that has led to the loss of wetland areas, current policies and regulations have not adequately addressed increasing land fragmentation and scattered settlements in the MMNL. Land fragmentation is threatening the sustainability of this urban wetland.

This study was performed on 4 DS divisions and 99 GN divisions in the Gampaha District (Figures 2 and 5). Local government authorities should consider stricter enforcement of protection and conservation measures for wetland areas that have become increasingly degraded. It would be useful for the Department of Wildlife Conservation, the Ministry of Environment, the Central Environmental Authority, and the Urban Development Authority to undertake wetland conservation projects, in particular, wetland restoration projects and community-based approaches with judicious use of wetland. While local NGOs and communities are also involved in implementing the Muthurajawela Wetland Management Plan [49,50,89], the aims of their management plan have yet to be achieved. Time is of the essence. Therefore, it is strongly recommended that city planners and local policymakers implement new laws and legislation without further delay to protect this wetland.

RS data and GIS techniques are useful for spatial analysis [90]. The above study could be useful in determining how much of the future landscape changes (e.g., the spatial extent of urban areas towards natural environment and urban wetland areas) could be compensated by comprehensive land-use practices and environmental rehabilitation activities such as reforestation programs in mangrove and also natural areas. The recovery of the wetland has significant potential to reduce the pressure on the carrying capacity of the MMNL by human activity. Once the MMNL has been changed into a settlement, it would be difficult to reverse the conversion.

Against this backdrop, it is evident that scientifically sound knowledge is needed to help urban landscape planners and policymakers tackle socio-ecological issues and achieve the essential SDGs [35] to ensure wise and sustainable use of the MMNL. Indeed, wetlands provide multiple services and benefits to people and are vital to attaining the SDGs [91] and Aichi Targets of the Convention on Biological Diversity [92]. The fourth Strategic Plan (2016–2024) of the Ramsar Convention has identified four major objectives and 19 specific objectives in this regard. Therefore, the authorities should aim to implement the Ramsar Convention's Strategic Plans and attain the Aichi Targets so as to mitigate socio-ecological problems associated with urbanization and protect wetland landscapes.

4.3. Limitations and Prospects for Future Wetland Study

This study used three satellite images captured in 1997, 2007, and 2017. More satellite images taken at different time points would provide valuable details, leading to a better spatial-temporal analysis. We need to point out that this investigation was limited by the lack of other clear and accessible satellite images temporarily compatible with the three satellite images employed. Another limitation was the unavailability of actual socio-economic data of the study area. Simultaneous and direct measurements are required to understand effective wetland conservation and protection strategies. Direct observations of urban wetlands can provide valuable knowledge for potential wetland management over the next few decades. We propose that future studies use much higher-resolution remote sensing images to investigate wetlands (at least for Sri Lanka). We need contemporary and long-term observations to appreciate the challenges in ensuring urban wetland sustainability.

5. Conclusions

We examined the impacts of urbanization on the Muthurajawela Marsh and Negombo Lagoon (MMNL), an important urban wetland ecosystem (UWE) in Sri Lanka owing to the valuable ecosystem services it provides. We found a substantial expansion of its settlements (+68%) and a considerable decrease in the extent of its marshland and mangrove forests (−41% and −21%, respectively). A statistical analysis revealed a positive, significant relationship between the change in population density and the loss of wetland due to settlement expansion, indicating that urbanization had indeed played a major role in the landscape transformation of the MMNL. The findings also revealed that most of the observed LUC changes occurred in areas close to roads and growth nodes (viz. Negombo, Ja-Ela, Wattala, and Katana), resulting in landscape fragmentation and infill urban expansion. The results indicated that the spatial and socio-economic elements of rapid urbanization of the MMNL had been the main driver of the transformation of its natural environment over the past 20 years. The study also showed that a hybrid mapping approach (unsupervised and supervised) can improve urban wetland mapping accuracy from remote sensing satellite imagery.

Overall, we conclude that, in order to ensure the sustainability of the MMNL, which is a highly valuable UWE, there is an urgent need for forward-looking landscape and urban planning that could promote environmentally-conscious urban development in the area.

Author Contributions: Conceptualization, D.A.; methodology, D.A. and R.C.E.; software, D.A. and R.C.E.; validation D.A., R.C.E., and Y.M.; formal analysis, D.A. and R.C.E.; investigation, D.A. and R.C.E.; resources, D.A.; data curation, D.A.; writing—original draft preparation, D.A.; writing—review and editing, R.C.E., Y.M., and B.M.; visualization, D.A.; supervision, Y.M. and B.M.; project administration, Y.M. All authors have read and agreed to the published version of the manuscript.

Funding: This research received no external funding.

Institutional Review Board Statement: Not applicable.

Informed Consent Statement: Not applicable.

Data Availability Statement: The data that support the findings of this study are available from the corresponding author on reasonable request.

Conflicts of Interest: The authors declare no conflict of interest.

Abbreviations

UWEs	Urban Wetland Ecosystems
MMNL	Muthurajawela Marsh and Negombo Lagoon
SDGs	Sustainable Development Goals
GIS	Geographic Information Systems
LUC	Land Use/Cover
CMR	Colombo Metropolitan Region
DS	Divisional Secretariat
TM	Thematic Mapper
OLI/TIRS	Operational Land Imager/Thermal Infrared Sensor
ETM+	Enhanced Thematic Mapper Plus
GN	Grama Niladhari
WLD	Wetland Loss Density
WL	Wetland Loss
CPD	Change in Population Density
APGR	Average Population Growth Rate
PD	Population Density
NP	Number of Patches
PD	Patch Density
ED	Edge Density
LPI	Largest Patch Index
LSI	Landscape Shape Index
COHESION	Cohesion
SHDI	Shannon's Diversity Index
SHEI	Shannon's Evenness Index
TP	Time Point
RS	Remote Sensing

Appendix A

Table A1. Confusion matrices of the classified LUC maps of the MMNL.

Classified Data	Reference Data				Total	User's Accuracy (%)
	Marshland	Mangrove	Water	Settlement		
(a) 1997						
Marshland	108	17	3	6	134	80.86
Mangrove	15	105	4	5	129	81.40
Water	7	3	42	1	53	79.25
Settlement	3	4	0	77	84	91.67
Total	133	129	49	89	400	
Producer's accuracy (%)	81.20	81.40	85.71	86.52		
Overall accuracy (%) = 83.01						
(b) 2007						
Marshland	112	16	5	4	137	81.75
Mangrove	11	98	4	3	116	84.48
Water	6	1	47	1	55	85.45
Settlement	5	5	1	81	92	88.04
Total	134	120	57	89	400	
Producer's accuracy (%)	83.58	81.67	82.46	91.01		
Overall accuracy (%) = 84.50						
(c) 2017						
Marshland	93	7	7	9	116	80.17
Mangrove	12	108	4	3	127	85.04
Water	5	3	42	1	51	82.35
Settlement	4	5	1	96	106	90.57
Total	114	123	54	109	400	
Producer's accuracy (%)	81.58	87.80	77.78	88.07		
Overall accuracy (%) = 84.75						

Table A2. Top 20 GNs in the MMNL in terms of density of wetland loss due to urbanization (settlement expansion) (WLD) (%) (see Equations (3) and (4)).

Rank	GNs	1997–2007 (TP1)	GNs	2007–2017 (TP2)
1	Thimbirigasyaya	76.47	Thalahena	70.24
2	Nayakakanda South	72.37	Palliyawatta South	66.13
3	Kurunduhena	64.86	Maha Pamunugama	66.11
4	Welikadamulla	63.69	Pitipana North	65.22
5	Telangapatha	63.21	Udammita South	64.06
6	Siriwardana Pedesa	62.82	Dandugama	61.13
7	Dungalpitiya	62.36	Pitipana South East	56.51
8	Paranambalama	61.74	Delathura West	56.01
9	Palliyawatta South	61.63	Nagoda	55.18
10	Munnakkarai North	59.94	Bandarawatta West	53.51
11	Doowa	59.57	Ambalammulla	52.41
12	Hendala North	59.26	Welikadamulla	52.35
13	Balagala	56.3	Ja-Ela	52.18
14	Elakanda	54.68	Alawathupitiya	52.12
15	Kerawalapitiya	54.47	Wella Weediya South	52.02
16	Welisara	54.20	Kudahakapola South	51.72
17	Galwetiya	53.42	Pitipana Central	51.46
18	Nedurupitiya	51.71	Indivitiya	51.37
19	Weligampitiya North	50.19	Siriwardana Pedesa	51.21
20	Bopitiya	50.01	Kurana West	50.13

References

1. Gardner, R.C.; Barchiesi, S.; Beltrame, C.; Finlayson, C.; Galewski, T.; Harrison, I.; Paganini, M.; Perennou, C.; Pritchard, D.; Rosenqvist, A.; et al. *Ramsar Convention State of the World's Wetlands and Their Services to People: A Compilation of Recent Analyses*; Ramsar Briefing note 7: Gland, Switzerland, 2015.
2. Costanza, R.; D'Arge, R.; De Groot, R.; Farber, S.; Grasso, M.; Hannon, B.; Limburg, K.; Naeem, S.; O'Neill, R.V.; Paruelo, J.; et al. The value of the world's ecosystem services and natural capital. *Nature* **1997**, *387*, 253–260. [CrossRef]
3. Dronova, I.; Gong, P.; Wang, L. Object-based analysis and change detection of major wetland cover types and their classification uncertainty during the low water period at Poyang Lake, China. *Remote Sens. Environ.* **2011**, *115*, 3220–3236. [CrossRef]
4. Ramsar. RAMSAR Homepage. Available online: https://www.ramsar.org/ (accessed on 10 October 2020).
5. Ramsar Convention. *An Introduction to the Convention on Wetlands*, 5th ed.; Ramsar Convention Secretariat: Gland, Switzerland, 2016.
6. Penatti, C.N.; Almeida, T.I.R.D.; Ferreira, L.G.; Arantes, E.A.; Coe, M.T. Satellite-based hydrological dynamics of the world's largest continuous wetland. *Remote Sens. Environ.* **2015**, *170*, 1–13. [CrossRef]
7. Reiss, C.K.; Hernandez, E.; Brown, M.T. Application of the landscape development intensity (LDI) index in wetland mitigation banking. *Ecol. Modell.* **2014**, *271*, 83–89. [CrossRef]
8. Gedan, K.B.; Kirwan, M.L.; Wolanski, E.; Barbier, E.B.; Silliman, B.R. The present and future role of coastal wetland vegetation in protecting shorelines: Answering recent challenges to the paradigm. *Clim. Chang.* **2011**, 7–29. [CrossRef]
9. Dabrowska-Zielinska, K.; Budzynska, M.; Tomaszewska, M.; Bartold, M.; Gatkowska, M.; Malek, I.; Turlej, K.; Napiorkowska, M. Monitoring Wetlands Ecosystems Using ALOS PALSAR (L-Band, HV) Supplemented by Optical Data: A Case Study of Biebrza Wetlands in Northeast Poland. *Remote Sens.* **2014**, 1605. [CrossRef]
10. Ramsar Convention. *The Fourth Ramsar Strategic Plan 2016–2024*, 5th ed.; Ramsar Convention Secretariat: Gland, Switzerland, 2016.
11. Weise, K.; Höfer, R.; Franke, J.; Guelmami, A.; Simonson, W.; Muro, J.; O'Connor, B.; Strauch, A.; Flink, S.; Eberle, J.; et al. Wetland extent tools for SDG 6.6.1 reporting from the Satellite-based Wetland Observation Service (SWOS). *Remote Sens. Environ.* **2020**, *247*, 111892. [CrossRef]
12. United Nations. Sustainable Development Goals. Available online: https://www.un.org/sustainabledevelopment/sustainable-development-goals/ (accessed on 10 October 2020).
13. MA (Millennium Ecosystem Assessment). *Ecosystems and Human Well-Being: Synthesis*; Island Press: Washington, DC, USA, 2005.
14. Euliss, N.H.; Mushet, D.M.; Newton, W.E.; Otto, C.R.V.; Nelson, R.D.; Labaugh, J.W.; Scherff, E.J.; Rosenberry, D.O. Placing prairie pothole wetlands along spatial and temporal continua to improve integration of wetland function in ecological investigations. *J. Hydrol.* **2014**, *513*, 490–503. [CrossRef]
15. Bouahim, S.; Rhazi, L.; Ernoul, L.; Mathevet, R.; Amami, B.; Er-riyahi, S.; Muller, S.D.; Grillas, P. Combining vulnerability analysis and perceptions of ecosystem services in sensitive landscapes: A case from western Moroccan temporary wetlands. *J. Nat. Conserv.* **2015**, *27*, 1–9. [CrossRef]
16. United Nations, Department of Economic and Social Affairs, Population Division. *World Urbanization Prospects: The 2018 Revision*; United Nations: New York, NY, USA, 2019.

17. Estoque, R.C.; Murayama, Y. Quantifying landscape pattern and ecosystem service value changes in four rapidly urbanizing hill stations of Southeast Asia. *Landsc. Ecol.* **2016**, *31*, 1481–1507. [CrossRef]
18. Michishita, R.; Jiang, Z.; Xu, B. Monitoring two decades of urbanization in the Poyang Lake area, China through spectral unmixing. *Remote Sens. Environ.* **2012**, *117*, 3–18. [CrossRef]
19. Zhou, D.; Zhao, S.; Zhang, L.; Liu, S. Remotely sensed assessment of urbanization effects on vegetation phenology in China's 32 major cities. *Remote Sens. Environ.* **2016**, *176*, 272–281. [CrossRef]
20. Zhang, F.; Yang, X. Improving land cover classification in an urbanized coastal area by random forests: The role of variable selection. *Remote Sens. Environ.* **2020**, *251*, 112105. [CrossRef]
21. Gold, A.C.; Thompson, S.P.; Magel, C.L.; Piehler, M.F. Urbanization alters coastal plain stream carbon export and dissolved oxygen dynamics. *Sci. Total Environ.* **2020**, *747*, 141132. [CrossRef]
22. Gillies, R.R.; Box, J.B.; Symanzik, J.; Rodemaker, E.J. Effects of urbanization on the aquatic fauna of the Line Creek watershed, Atlanta-A satellite perspective. *Remote Sens. Environ.* **2003**, *86*, 411–422. [CrossRef]
23. Mondal, B.; Dolui, G.; Pramanik, M.; Maity, S.; Biswas, S.S.; Pal, R. Urban expansion and wetland shrinkage estimation using a GIS-based model in the East Kolkata Wetland, India. *Ecol. Indic.* **2017**, *83*, 62–73. [CrossRef]
24. Das, A.; Basu, T. Assessment of peri-urban wetland ecological degradation through importance-performance analysis (IPA): A study on Chatra Wetland, India. *Ecol. Indic.* **2020**, *114*, 106274. [CrossRef]
25. Hou, X.; Feng, L.; Tang, J.; Song, X.P.; Liu, J.; Zhang, Y.; Wang, J.; Xu, Y.; Dai, Y.; Zheng, Y.; et al. Anthropogenic transformation of Yangtze Plain freshwater lakes: Patterns, drivers and impacts. *Remote Sens. Environ.* **2020**, *248*, 111998. [CrossRef]
26. Li, Z.; Jiang, W.; Wang, W.; Chen, Z.; Ling, Z.; Lv, J. Ecological risk assessment of the wetlands in Beijing-Tianjin-Hebei urban agglomeration. *Ecol. Indic.* **2020**, *117*, 106677. [CrossRef]
27. Lin, W.; Cen, J.; Xu, D.; Du, S.; Gao, J. Wetland landscape pattern changes over a period of rapid development (1985–2015) in the ZhouShan Islands of Zhejiang province, China. *Estuar. Coast. Shelf Sci.* **2018**, *213*, 148–159. [CrossRef]
28. Nagendra, H.; Lucas, R.; Honrado, J.P.; Jongman, R.H.G.; Tarantino, C.; Adamo, M.; Mairota, P. Remote sensing for conservation monitoring: Assessing protected areas, habitat extent, habitat condition, species diversity, and threats. *Ecol. Indic.* **2013**, *33*, 45–59. [CrossRef]
29. Yang, C.; Zhang, C.; Li, Q.; Liu, H.; Gao, W.; Shi, T.; Liu, X.; Wu, G. Rapid urbanization and policy variation greatly drive ecological quality evolution in Guangdong-Hong Kong-Macau Greater Bay Area of China: A remote sensing perspective. *Ecol. Indic.* **2020**, *115*, 106373. [CrossRef]
30. Wiens, J.; Sutter, R.; Anderson, M.; Blanchard, J.; Barnett, A.; Aguilar-Amuchastegui, N.; Avery, C.; Laine, S. Selecting and conserving lands for biodiversity: The role of remote sensing. *Remote Sens. Environ.* **2009**, *113*, 1370–1381. [CrossRef]
31. Wabnitz, C.C.; Andréfouët, S.; Torres-Pulliza, D.; Müller-Karger, F.E.; Kramer, P.A. Regional-scale seagrass habitat mapping in the Wider Caribbean region using Landsat sensors: Applications to conservation and ecology. *Remote Sens. Environ.* **2008**, *112*, 3455–3467. [CrossRef]
32. Zhang, D.; Xu, J.; Zhang, Y.; Wang, J.; He, S.; Zhou, X. Study on sustainable urbanization literature based on Web of Science, Scopus, and China national knowledge infrastructure: A scientometric analysis in CiteSpace. *J. Clean. Prod.* **2020**, *264*, 121537. [CrossRef]
33. Dewan, A.M.; Yamaguchi, Y. Land use and land cover change in Greater Dhaka, Bangladesh: Using remote sensing to promote sustainable urbanization. *Appl. Geogr.* **2009**, *29*, 390–401. [CrossRef]
34. Zhu, Z.; Zhou, Y.; Seto, K.C.; Stokes, E.C.; Deng, C.; Pickett, S.T.A.; Taubenböck, H. Understanding an urbanizing planet: Strategic directions for remote sensing. *Remote Sens. Environ.* **2019**, *228*, 164–182. [CrossRef]
35. Estoque, R.C. A review of the sustainability concept and the state of SDG monitoring using remote sensing. *Remote Sens.* **2020**, *12*. [CrossRef]
36. Jaramillo, F.; Desormeaux, A.; Hedlund, J.; Jawitz, J.W.; Clerici, N.; Piemontese, L.; Rodríguez-Rodriguez, J.A.; Anaya, J.A.; Blanco-Libreros, J.F.; Borja, S.; et al. Priorities and interactions of Sustainable Development Goals (SDGs) with focus on wetlands. *Water* **2019**, *11*. [CrossRef]
37. Fitoka, E.; Tompoulidou, M.; Hatziiordanou, L.; Apostolakis, A.; Höfer, R.; Weise, K.; Ververis, C. Water-related ecosystems' mapping and assessment based on remote sensing techniques and geospatial analysis: The SWOS national service case of the Greek Ramsar sites and their catchments. *Remote Sens. Environ.* **2020**, *245*, 111795. [CrossRef]
38. Estoque, R.C.; Murayama, Y. Measuring Sustainability Based Upon Various Perspectives: A Case Study of a Hill Station in Southeast Asia. *Ambio* **2014**, *43*, 943–956. [CrossRef] [PubMed]
39. Liu, A.J.; Cameron, G.N. Analysis of landscape patterns in coastal wetlands of Galveston Bay, Texas (USA). *Landsc. Ecol.* **2001**, *16*, 581–595. [CrossRef]
40. Festus, O.; Ji, W.; Zubair, O.A. Characterizing the Landscape Structure of Urban Wetlands Using Terrain and Landscape Indices. *Land* **2020**, *9*, 29. [CrossRef]
41. Hassan, M.M. Monitoring land use/land cover change, urban growth dynamics and landscape pattern analysis in five fastest urbanized cities in Bangladesh. *Remote Sens. Appl. Soc. Environ.* **2017**, *7*, 69–83. [CrossRef]
42. Aguilera, F.; Valenzuela, L.M.; Botequilha-Leitão, A. Landscape metrics in the analysis of urban land use patterns: A case study in a Spanish metropolitan area. *Landsc. Urban Plan.* **2011**, *99*, 226–238. [CrossRef]

43. Japelaghi, M.; Gholamalifard, M.; Shayesteh, K. Spatio-temporal analysis and prediction of landscape patterns and change processes in the Central Zagros region, Iran. *Remote Sens. Appl. Soc. Environ.* **2019**, *15*, 100244. [CrossRef]
44. Su, S.; Xiao, R.; Jiang, Z.; Zhang, Y. Characterizing landscape pattern and ecosystem service value changes for urbanization impacts at an eco-regional scale. *Appl. Geogr.* **2012**, *34*, 295–305. [CrossRef]
45. Liu, G.; Zhang, L.; Zhang, Q.; Musyimi, Z.; Jiang, Q. Spatio-temporal dynamics of wetland landscape patterns based on remote sensing in yellow river delta, China. *Wetlands* **2014**, *34*, 787–801. [CrossRef]
46. Haas, J.; Ban, Y. Urban growth and environmental impacts in Jing-Jin-Ji, the Yangtze, River Delta and the Pearl River Delta. *Int. J. Appl. Earth Obs. Geoinf.* **2014**, *30*, 42–55. [CrossRef]
47. Li, Y.; Zhu, X.; Sun, X.; Wang, F. Landscape effects of environmental impact on bay-area wetlands under rapid urban expansion and development policy: A case study of Lianyungang, China. *Landsc. Urban Plan.* **2010**, *94*, 218–227. [CrossRef]
48. McInnes, R. *Urban Development, Biodiversity and Wetland Management*; UN HABITAT: Bioscan (UK) Ltd: Oxford, UK, 2010.
49. Greater Colombo Economic Commission, Euroconsult. *Master Plan of Muthurajawela and Negombo Lagoon*; Gunaratne offset Ltd.: Colombo, Sri Lanka, 1991.
50. Central Environmental Authority, Euroconsult. *Conservation Master Plan, Muthurajawela Marsh and Negombo Lagoon*; Gunaratne offset Ltd.: Colombo, Sri Lanka, 1994.
51. Bambaradeniya, C.N.B.; Ekanayake, S.P.; Kekulandala, L.D.C.B.; Samarawickrama, V.A.P.; Ratnayake, N.D.; Fernando, R.H.S.S. *An Assessment of the Status of Biodiversity in the Muthurajawela Wetland Sanctuary*; IUCN: Colombo, Sri Lanka, 2002; ISBN 9558177172.
52. Central Environmental Authority (CEA). *National Wetland Directory of Sri Lanka*; CEA: Colombo, Sri Lanka, 2006; ISBN 9558177547.
53. The World Bank. *Turning Sri Lanka's Urban Vision into Policy and Action*; The World Bank: Washington, DC, USA, 2012; ISBN 9789558908440.
54. Subasinghe, S.; Estoque, R.; Murayama, Y. Spatiotemporal Analysis of Urban Growth Using GIS and Remote Sensing: A Case Study of the Colombo Metropolitan Area, Sri Lanka. *ISPRS Int. J. Geo-Inf.* **2016**, *5*, 197. [CrossRef]
55. Jayathilake, M.B.; Chandrasekara, W.U. Variation of avifaunal diversity in relation to land-use modifications around a tropical estuary, the Negombo estuary in Sri Lanka. *J. Asia-Pac. Biodivers.* **2015**, *8*, 72–82. [CrossRef]
56. Rebelo, L.M.; Finlayson, C.M.; Nagabhatla, N. Remote sensing and GIS for wetland inventory, mapping and change analysis. *J. Environ. Manag.* **2009**, *90*, 2144–2153. [CrossRef] [PubMed]
57. Cooray, P.G. *An Introduction to the Geology of Sri Lanka (Ceylon)*, 2nd ed.; Colombo National Museums Publication: Colombo, Sri Lanka, 1984.
58. Köppen, W. Klassifikation der Klimate nach Temperatur, Niederschlag und Jahresablauf (Classification of climates according to temperature, precipitation and seasonal cycle). *Petermanns Geogr. Mitt.* **1918**, *64*, 193–203.
59. Department of Meteorology, Sri Lanka. Weather Forecasts. Available online: http://www.meteo.gov.lk/index.php?lang=en (accessed on 10 October 2020).
60. Greater Colombo Economic Commission, Euroconsult. *Environmental Profile of Muthurajawela and Negombo Lagoon*; Gunaratne offset Ltd.: Colombo, Sri Lanka, 1991.
61. Mao, D.; Wang, Z.; Du, B.; Li, L.; Tian, Y.; Jia, M.; Zeng, Y.; Song, K.; Jiang, M.; Wang, Y. National wetland mapping in China: A new product resulting from object-based and hierarchical classification of Landsat 8 OLI images. *ISPRS J. Photogramm. Remote Sens.* **2020**, *164*, 11–25. [CrossRef]
62. Reschke, J.; Hüttich, C. Continuous field mapping of Mediterranean wetlands using sub-pixel spectral signatures and multi-temporal Landsat data. *Int. J. Appl. Earth Obs. Geoinf.* **2014**, *28*, 220–229. [CrossRef]
63. Ghosh, S.; Das, A. Urban expansion induced vulnerability assessment of East Kolkata Wetland using Fuzzy MCDM method. *Remote Sens. Appl. Soc. Environ.* **2019**, *13*, 191–203. [CrossRef]
64. Mwita, E.; Menz, G.; Misana, S.; Becker, M.; Kisanga, D.; Boehme, B. Mapping small wetlands of Kenya and Tanzania using remote sensing techniques. *Int. J. Appl. Earth Obs. Geoinf.* **2012**, *21*, 173–183. [CrossRef]
65. Shanmugam, P.; Ahn, Y.H.; Sanjeevi, S. A comparison of the classification of wetland characteristics by linear spectral mixture modelling and traditional hard classifiers on multispectral remotely sensed imagery in southern India. *Ecol. Modell.* **2006**, *194*, 379–394. [CrossRef]
66. Survey Department of Sri Lanka. Available online: https://www.survey.gov.lk/ (accessed on 2 May 2020).
67. Liu, X.; Dong, G.; Wang, X.; Xue, Z.; Jiang, M.; Lu, X.; Zhang, Y. Characterizing the spatial pattern of marshlands in the Sanjiang Plain, Northeast China. *Ecol. Eng.* **2013**, *53*, 335–342. [CrossRef]
68. Estoque, R.C.; Murayama, Y. Landscape pattern and ecosystem service value changes: Implications for environmental sustainability planning for the rapidly urbanizing summer capital of the Philippines. *Landsc. Urban Plan.* **2013**, *116*, 60–72. [CrossRef]
69. Zhang, Q.; Chen, C.; Wang, J.; Yang, D.; Zhang, Y.; Wang, Z.; Gao, M. The spatial granularity effect, changing landscape patterns, and suitable landscape metrics in the Three Gorges Reservoir Area, 1995–2015. *Ecol. Indic.* **2020**, *114*, 106259. [CrossRef]
70. McGarigal, K. *Fragstats*; US Department of Agriculture, Forest Service, Pacific Northwest Research Station: Corvallis, OR, USA, 2015; pp. 1–182. [CrossRef]
71. Lu, D.; Weng, Q. Urban Classification Using Full Spectral Information of Landsat ETM+ Imagery in Marion County. *Indiana* **2005**, *71*, 1275–1284. [CrossRef]
72. Liu, Y.; Zha, Y.; Gao, J.; Ni, S. Assessment of grassland degradation near Lake Qinghai, West China, using Landsat TM and in situ reflectance spectra data. *Int. J. Remote Sens.* **2004**, *25*, 4177–4189. [CrossRef]

73. Sader, S.A.; Ahl, D.; Liou, W.S. Accuracy of landsat-TM and GIS rule-based methods for forest wetland classification in Maine. *Remote Sens. Environ.* **1995**, *53*, 133–144. [CrossRef]
74. Slagter, B.; Tsendbazar, N.-E.; Vollrath, A.; Reiche, J. Mapping wetland characteristics using temporally dense Sentinel-1 and Sentinel-2 data: A case study in the St. Lucia wetlands, South Africa. *Int. J. Appl. Earth Obs. Geoinf.* **2020**, *86*, 102009. [CrossRef]
75. Singh, P.; Javed, S.; Shashtri, S.; Singh, R.P.; Vishwakarma, C.A.; Mukherjee, S. Influence of changes in watershed landuse pattern on the wetland of Sultanpur National Park, Haryana using remote sensing techniques and hydrochemical analysis. *Remote Sens. Appl. Soc. Environ.* **2017**, *7*, 84–92. [CrossRef]
76. Expressway Operation Maintenance And Management Division Road Development Authority-Sri Lanka. Available online: http://www.exway.rda.gov.lk/index.php?page=expressway_network/e03 (accessed on 5 May 2020).
77. Zubair, O.A.; Ji, W.; Weilert, T.E. Modeling the Impact of Urban Landscape Change on Urban Wetlands Using Similarity Weighted Instance-Based Machine Learning and Markov Model. *Sustainability* **2017**, *9*, 2223. [CrossRef]
78. Wei, Z.; Jingang, J.; Yubi, Z.H.U. Change in Urban Wetlands and Their Cold Island Effects in Response to Rapid Urbanization. *Chin. Geogr. Sci.* **2015**, *25*, 462–471. [CrossRef]
79. Athukorala, D.; Murayama, Y. Spatial Variation of Land Use/Cover Composition and Impact on Surface Urban Heat Island in a Tropical Sub-Saharan City of Accra, Ghana. *Sustainability* **2020**, *12*, 7953. [CrossRef]
80. Ozesmi, S.L.; Bauer, M.E. Satellite remote sensing of wetlands. *Wetl. Ecol. Manag.* **2002**, *10*, 381–402. [CrossRef]
81. Lane, C.R.; Liu, H.; Autrey, B.C.; Anenkhonov, O.A.; Chepinoga, V.V.; Wu, Q. Improved Wetland Classification Using Eight-Band High Resolution Satellite Imagery and a Hybrid Approach. *Remote Sens.* **2014**, *6*, 12187. [CrossRef]
82. Abeywardene, J.; de Alwis, R.; Jayasena, A.; Jayaweera, S.; Sanmugam, T. *Export Processing Zones in Sri Lanka: Economic Impact and Social Issues*; Center for Women's Research: Colombo, Sri Lanka, 1994.
83. Kelegama, S.; Corea, G. *Economic Policy in Sri Lanka: Issues and Debates*; SAGE Publications: New York, NY, USA, 2004; ISBN 9780761932789.
84. Hogg, C.L. *Sri Lanka: Prospects for Reform and Reconciliation*; Chatham House: London, UK, 2011.
85. Department of Census and Statistics-Sri Lanka. Available online: http://www.statistics.gov.lk/ (accessed on 3 May 2020).
86. Ancog, R.; Ruzol, C. Urbanization adjacent to a wetland of international importance: The case of Olango Island Wildlife Sanctuary, Metro Cebu, Philippines. *Habitat Int.* **2015**, *49*, 325–332. [CrossRef]
87. Thibault, P.A.; Zippererb, W.C. Temporal changes of wetlands within an urbanizing agricultural landscape. *Landsc. Urban Plan.* **1994**, *28*, 245–251. [CrossRef]
88. Akbostanci, E.; Türüt-Aşik, S.; Tunç, G.I. The relationship between income and environment in Turkey: Is there an environmental Kuznets curve? *Energy Policy* **2009**, *37*, 861–867. [CrossRef]
89. Ramsar. Implementation of the Ramsar Convention in General, and of the Ramsar Strategic Plan 1997-2002 in Particular, during the Period Since the National Report Prepared in 1995 for Ramsar COP6 and 30 June 1998. Available online: https://www.ramsar.org/document/cop7-national-reports-sri-lanka (accessed on 14 January 2021).
90. Estoque, R.C.; Murayama, Y. Examining the potential impact of land use/cover changes on the ecosystem services of Baguio city, the Philippines: A scenario-based analysis. *Appl. Geogr.* **2012**, *35*, 316–326. [CrossRef]
91. Ramsar Convention. *Scaling up Wetland Conservation, Wise Use and Restoration to Achieve the Sustainable Development Goals*; Ramsar Convention: Ramsar, Iran, 2018; pp. 1–13.
92. United Nations Environment programme (UNEP). *The Aichi Passport*; United Nations Environment programme (UNEP): Nairobi, Kenya, 2012.

Article

Mapping Multi-Temporal Population Distribution in China from 1985 to 2010 Using Landsat Images via Deep Learning

Haoming Zhuang [1], Xiaoping Liu [1,2,*], Yuchao Yan [3], Jinpei Ou [1], Jialyu He [1] and Changjiang Wu [1]

1. Guangdong Key Laboratory for Urbanization and Geo-Simulation, School of Geography and Planning, Sun Yat-sen University, Guangzhou 510275, China; zhuanghm3@mail2.sysu.edu.cn (H.Z.); oujinpei3@mail.sysu.edu.cn (J.O.); hesysugis@foxmail.com (J.H.); wuchj9@mail.sysu.edu.cn (C.W.)
2. Southern Marine Science and Engineering Guangdong Laboratory (Zhuhai), Zhuhai 519082, China
3. Sino-French Institute for Earth System Science, College of Urban and Environmental Sciences, Peking University, Beijing 100091, China; yuchaoyan@pku.edu.cn
* Correspondence: liuxp3@mail.sysu.edu.cn

Abstract: Fine knowledge of the spatiotemporal distribution of the population is fundamental in a wide range of fields, including resource management, disaster response, public health, and urban planning. The United Nations' Sustainable Development Goals also require the accurate and timely assessment of where people live to formulate, implement, and monitor sustainable development policies. However, due to the lack of appropriate auxiliary datasets and effective methodological frameworks, there are rarely continuous multi-temporal gridded population data over a long historical period to aid in our understanding of the spatiotemporal evolution of the population. In this study, we developed a framework integrating a ResNet-N deep learning architecture, considering neighborhood effects with a vast number of Landsat-5 images from Google Earth Engine for population mapping, to overcome both the data and methodology obstacles associated with rapid multi-temporal population mapping over a long historical period at a large scale. Using this proposed framework in China, we mapped fine-scale multi-temporal gridded population data (1 km × 1 km) of China for the 1985–2010 period with a 5-year interval. The produced multi-temporal population data were validated with available census data and achieved comparable performance. By analyzing the multi-temporal population grids, we revealed the spatiotemporal evolution of population distribution from 1985 to 2010 in China with the characteristic of concentration of the population in big cities and the contraction of small- and medium-sized cities. The framework proposed in this study demonstrates the feasibility of mapping multi-temporal gridded population distribution at a large scale over a long period in a timely and low-cost manner, which is particularly useful in low-income and data-poor areas.

Keywords: population mapping; Landsat; deep learning; multi-temporal; ResNet-N; Google Earth Engine; China; SDGs

Citation: Zhuang, H.; Liu, X.; Yan, Y.; Ou, J.; He, J.; Wu, C. Mapping Multi-Temporal Population Distribution in China from 1985 to 2010 Using Landsat Images via Deep Learning. *Remote Sens.* **2021**, *13*, 3533. https://doi.org/10.3390/rs13173533

Academic Editor: Ronald C. Estoque

Received: 3 August 2021
Accepted: 3 September 2021
Published: 6 September 2021

Publisher's Note: MDPI stays neutral with regard to jurisdictional claims in published maps and institutional affiliations.

1. Introduction

Understanding the spatiotemporal distribution of the population is fundamental in a wide range of fields, including resource management [1,2], disaster response [3–6], public health [7–9], and urban planning [10,11]. The United Nations' Sustainable Development Goals (SDGs) also require the accurate and timely assessment of where people live to formulate, implement, and monitor sustainable development policies [12,13].

Census data released by an official body are authoritative and vital data about population distribution [14]. However, census data are based on administrative units; thus, they have several inherent limitations and are ill-suited to many spatial studies. Firstly, there is significant spatial heterogeneity in population distribution, which cannot be reflected by census data, which assumes a completely uniform distribution of the population within census units [15]. Secondly, the size of administrative units varies significantly in urban and rural areas, which results in the Modifiable Areal Unit Problem [16]. Thirdly, administrative boundaries may change over time and are seldom compatible with practical applications, making census data challenging to integrate with other spatial data sets, preventing interdisciplinary research and temporal dynamic analysis [17]. In order to overcome the limitations of census data, fine-grained gridded population data, which are spatially continuous, are produced to supplement census data [18,19].

Several approaches have been developed to produce fine-scale gridded population data in the past few decades, such as areal weighting [20], spatial interpolation [21–23], and dasymetric mapping [24–32]. Among them, dasymetric mapping technology [33], which uses fine-scale auxiliary variables and specific weighting schemes to re-allocate census counts to grid cells, is the most widely used and effective one [19]. Commonly adopted ancillary data include land use/cover data [34–36], nighttime light data [26,37], terrain data [38], and social sensing data (e.g., points-of-interest [39], mobile phone records [40], and social media data [41]). Multiple methods, including empirical rules [15], statistical models (e.g., linear regression [34] and geographically weighted regression [17]), and machine learning models (e.g., random forest [42], expectation-maximization [43], and neural network [44]), have been proposed to estimate the distribution weight of grid cells. Various gridded population data at regional and global scales have been produced and published using the methods mentioned above, including the Gridded Population of the World (GPW) [45], Global Human Settlement Population layer (GHS-POP) [46], Global Rural-Urban Mapping Project (GRUMP) [47], WorldPop [48], and LandScan [49].

The accuracy of gridded population data is determined by the quality of auxiliary data [19]. Numerous previous studies focused on integrating novel auxiliary variables related to population distribution to improve the quality of the produced population grids [15,50,51]. The thematic, spatial, and temporal accuracy of auxiliary data themselves is also crucial to the quality of the final gridded data [19]. For example, classification error in land use/land cover data will be propagated to the produced gridded population data. Furthermore, involving more auxiliary variables contributes more uncertainty to the final result [34]. In order to produce multi-temporal gridded population data, temporally explicit and consistent auxiliary data are essential, whose availability and sustainability are questionable, especially at a large scale, precluding the production of continuous multi-temporal data over a long historical period [34].

Remote sensing (RS) data (e.g., satellite imagery) that can capture the physical characteristics of the ground at low cost, broad coverages, and high spatiotemporal resolution are becoming increasingly available with improvements in imaging technology over time [36,52]. The physical characteristics of the ground and human activities interact with each other. Human activities can lead to distinct spatial landscapes, which inversely constrain how people live, produce, and travel, providing the possibility of consistent and sustainable population estimation with RS imagery as auxiliary data without the problems mentioned above [53]. However, the raw RS imagery is highly unstructured, and its association with population count is complex and nonlinear, making it challenging to construct a mapping from raw RS imagery to population count [54]. An emerging supervised deep learning approach, convolutional neural networks (CNN), which are capable of extracting the hidden hierarchical structures of RS images [55], have shown outstanding performance in obtaining knowledge from RS images in the domain of geography (e.g., land use classification [56], spatial interpolation [57], and poverty mapping [58]). Therefore, it is possible that CNN can form a mapping from RS imagery to population count.

A few studies have tried to estimate population counts from RS imagery directly. Doupe proposed the use of a VGG-like network to estimate population density in Tanzanian and Kenya from Landsat images and achieved remarkable performance and generalizability [54]. Robinson regarded the population estimation task as a classification problem and used a similar VGG-like network to classify RS image patches into 14 population density levels. They produced gridded population data for the United States in 2010, achieved high performance, and qualitatively explained the predictions in terms of the input RS imagery [59]. Xing proposed a Neighbor-ResNet architecture by embedding the neighbor knowledge into ResNet in order to estimate the volumes of human activity from Google imagery in 18 cities in China [53]. The attempts mentioned above verify the feasibility and superiority of integrating CNN and RS imagery for population mapping. However, the established models have not been used to map historical population distributions and understand their spatiotemporal evolution.

China is the world's most populous developing country. Fine-scale population distribution data are crucial for China's sustainable development [13]. Numerous gridded population data of China, with various spatial resolutions, have been developed [25,39]. A few studies have also used time-invariant and time-explicit auxiliary variables to produce multi-temporal gridded population distribution data [17,30,60]. However, due to the lack of appropriate auxiliary datasets and effective methodological frameworks, there are rarely continuous multi-temporal gridded population data for China over a long historical period to aid in our understanding of the spatiotemporal evolution of the population.

In this study, we developed a framework integrating a ResNet-N deep learning architecture with the consideration of neighborhood effects with a vast number of Landsat-5 images from Google Earth Engine (GEE) [61] for population mapping to overcome both the data and methodology obstacles of rapid multi-temporal population mapping over a long historical period at a large scale. Once the framework was constructed, we developed multi-temporal gridded population data with a 1 km resolution for China (excluding Taiwan, Hong Kong, and Macao) for the 1985–2010 period with a 5-year interval and analyzed its spatiotemporal evolution.

2. Materials and Methods

This study aimed to develop a framework integrating a deep learning model with Landsat-5 RS images from GEE to estimate population count. Once the framework is established, large-scale population mapping can be achieved only with easily accessible and regularly updatable RS imagery. Furthermore, we produce multi-temporal gridded population data (1 km × 1 km) of China for the 1985–2010 period with a 5-year interval and analyze the spatiotemporal evolution of the population distribution of China in this period. The flowchart of this study is illustrated in Figure 1, containing three main parts: (1) we collected ground-truth population count grid cells and corresponding Landsat-5 RS image patches as reference datasets for training, validating, and testing the developed deep learning model; (2) a ResNet-N architecture considering neighborhood effects was developed to establish the end-to-end mapping between population count and RS image patches; (3) based on the trained model, we estimated the gridded population count of China with corresponding Landsat-5 image patches from GEE as input from 1985 to 2010. Furthermore, the produced raw estimations were adjusted by available census data to acquire the final gridded population data. Finally, we validated the produced datasets and analyzed the spatiotemporal evolution of China's population distribution.

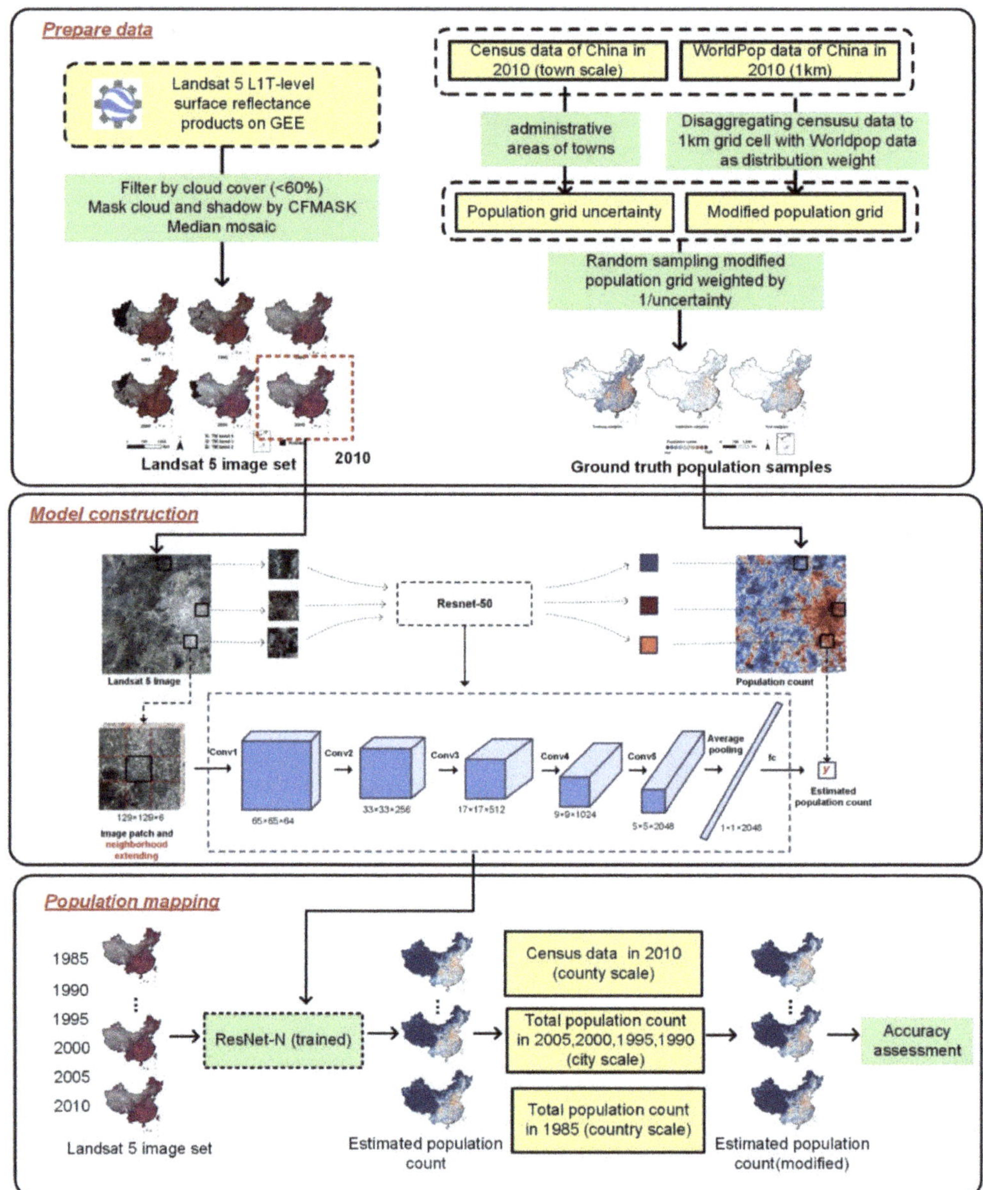

Figure 1. The flowchart of the proposed framework for mapping population distribution of China by integrating the ResNet-N model and Landsat-5 images from GEE.

2.1. Data Sources and Preprocessing

2.1.1. Ground-Truth Population Grid

In order to establish an end-to-end mapping between Landsat-5 RS image patches and population count by deep learning architecture, it is necessary to collect ground-truth population grid cells as training samples. However, the ground-truth population grid does not exist [19]. In this study, an alternative method (Figure 2) was utilized to collect the closest ground-truth population grid samples with a resolution of 1 km.

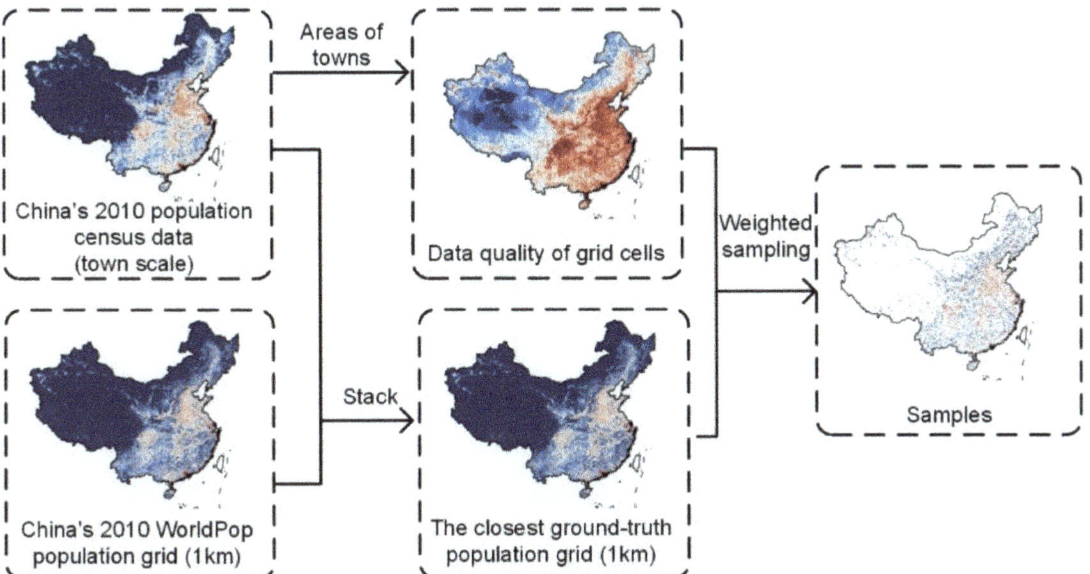

Figure 2. The flowchart of collecting the closest ground-truth population grid cell samples with a resolution of 1 km.

We obtained China's 2010 population census data at the town level (level 4 administrative unit), the finest-scale census data publicly available, from China's Sixth National Population Census. Towns are the fundamental administrative units in China, with relatively small jurisdiction areas, 58% of which are less than 100 km^2, so that the spatial heterogeneity of population distribution is tiny within towns. However, it is not adequate to use the average population density of towns as references due to heterogeneity within towns [62]. We obtained the WorldPop gridded population data [48] with a resolution of 1 km for China in 2010 to remedy this problem. The WorldPop data are produced by coupling a random forest algorithm with various auxiliary data to disaggregate county-level (level 3 administrative units) census data, recognized as some of the finest gridded population data to date [5]. In this study, we used WorldPop data in 2010 as a weighting layer to redistribute the total population count of each town to grid cells to account for the spatial heterogeneity within towns in part. Numerous towns are small in area. Therefore, this modified population map represents the closest ground-truth population grid that is available to use as training data. Finally, we sampled 100,000 grid cells from the ground-truth population grid weighted by the quality of grid cells to tradeoff the reliability and representativeness of the samples. The administrative areas of towns act as a data quality metric of grid cells [19]. Let $area^k$ represent the area of the kth town; then, the weight of selecting a grid cell inside the kth town is given as $\frac{1}{area^k}$. We discarded the grid cells with a population count of less than 10. It is unnecessary and intractable to distinguish the subtle change in population count via RS images in a 1 km^2 area [59]. The distribution of ground-truth population samples is heavily tailed, with kurtosis of 1312.41 and skewness of 26.31. To balance the dataset and ease the training of the deep learning model, the population count was logarithmized [53]. The collected samples were randomly divided into three groups: training (70%), validation (10%), and testing (20%). Figure 3 presents the spatial distribution of the collected population samples.

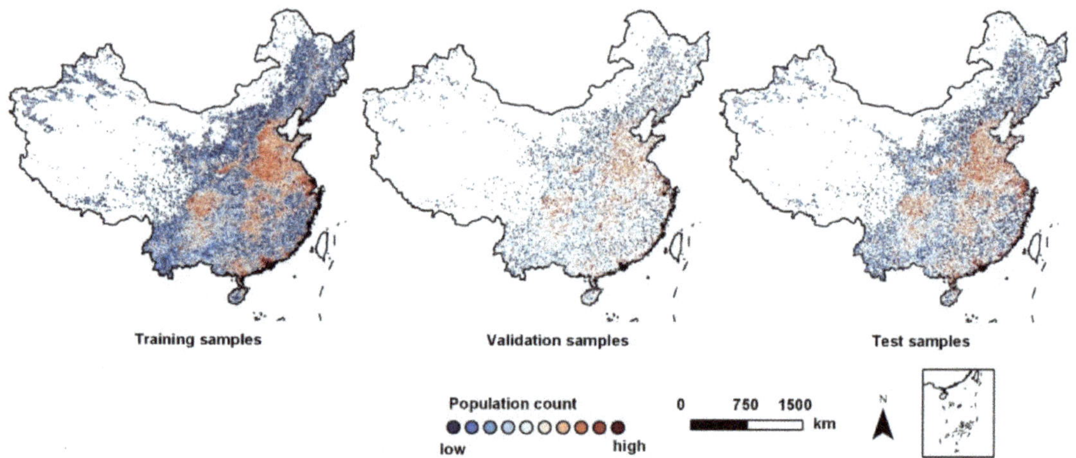

Figure 3. The spatial distribution of the ground-truth population samples.

2.1.2. Landsat-5 RS Imagery

This study used RS images from Landsat-5 collected by the Thematic Mapper (TM) sensor, covering the 1985–2010 period [52]. The full Landsat-5 L1T-level surface reflectance archive [63] covering China with a cloud score of less than 60 was preprocessed and downloaded effortlessly from GEE, a cloud-based platform for processing petabyte-scale geospatial datasets [61]. The L1T-level products have undergone geometric, radiation, and atmospheric corrections and are ready for use [64,65]. After masking clouds and shadows using Landsat quality flag information [66], a composite for a given year was produced in the form of a median mosaic of all available Landsat scenes. To address the shortage of cloud-free images, we included the Landsat scenes of the year before and the year after the target year in the composite. By referring to previous research, six bands were retrieved, i.e., Band 1 (blue), Band 2 (green), Band 3 (red), Band 4 (near-infrared), Band 5 (shortwave infrared 1), and Band 7 (shortwave infrared 2), all with a spatial resolution of 30 m [65]. Figure 4 presents the cloud-free Landsat composites with standard false-color band combination from 1985 to 2010. Due to the shortage of cloud-less images, there are missing data in western areas of China in some target years. As these areas are usually sparsely populated with slight variation, we used the valid data in the nearest adjacent year to supplement these areas.

Previous studies have revealed that the detailed characteristics of various landscapes can be well reflected by these 6 visible and invisible bands [54]. Figure 5 presents the probability density distribution of population count in the ground-truth samples and the example RS image patches that correspond to various population counts. Obviously, different magnitudes of population count correspond to distinct landscape characteristics in the RS image patches. The interplay between population count and RS images indicates the potential of estimating population count based only on RS images from Landsat-5 via a deep learning architecture.

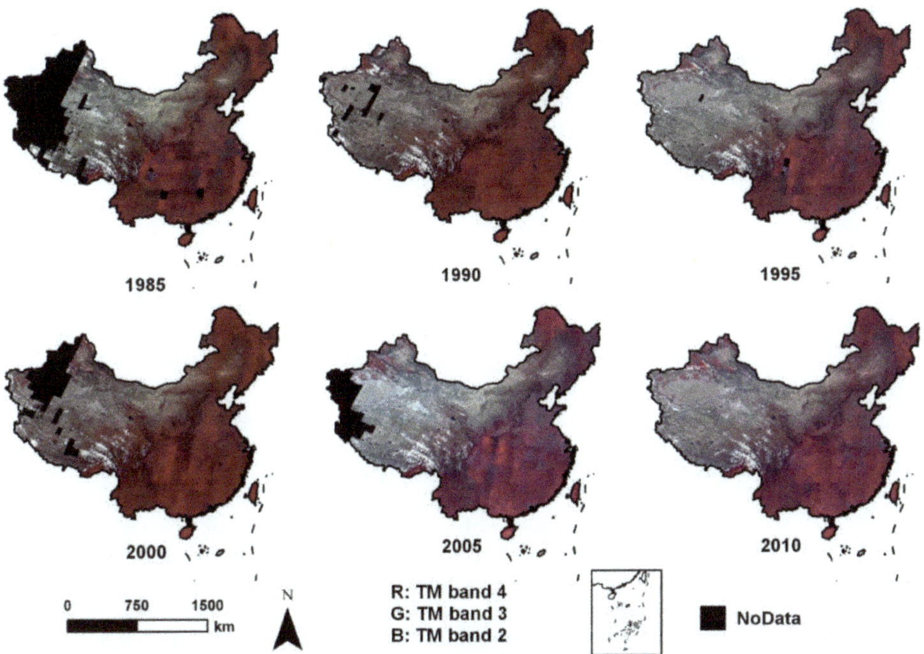

Figure 4. Cloud-free Landsat-5 composites of China from 1985 to 2010.

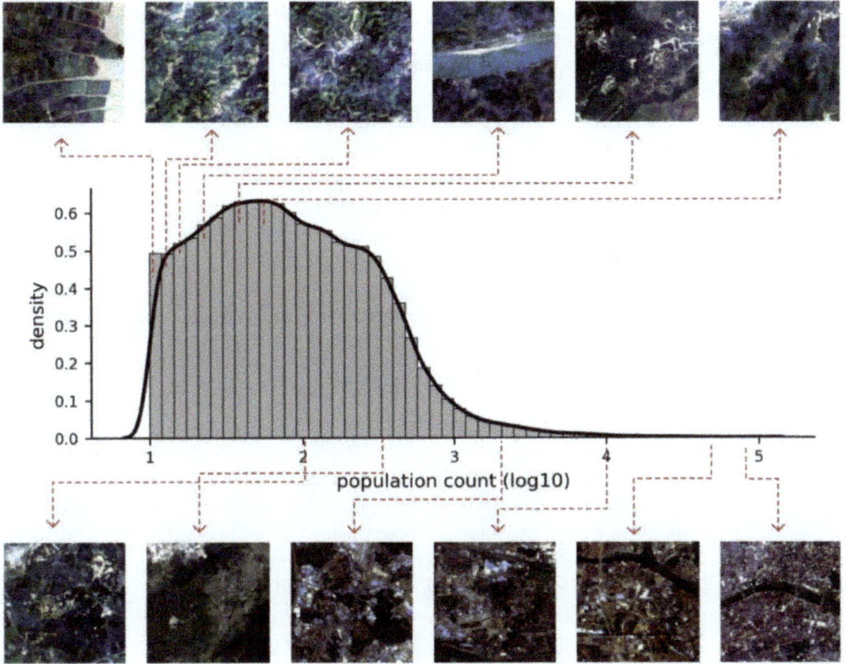

Figure 5. Probability density distribution of population count in the ground-truth samples and example RS image patches that correspond to various population counts.

2.2. Methods

2.2.1. Building a Mapping from RS Image Patches to Population Counts via ResNet-N Model

In this study, we view the gridded population estimation task as a regression problem. The method framework is shown in Figure 6. Given an image patch θ^i of the grid cell i and the corresponding logarithmized population count p_i, we express our learning task as building a mapping function:

$$p_i = f(\theta_i) \tag{1}$$

where $f(\cdot)$ is the mapping function to be learned by deep learning models. Acknowledging the highly nonlinear and complex relationship between RS images and population count, a ResNet (specifically, ResNet-50 was adopted) model considering neighborhood effects (ResNet-N) was utilized to approximate such a complex mapping relationship [53]. The ResNet model is one of the state-of-the-art CNN architectures and has been widely adopted to mine geographical knowledge from RS images [55,56]. The fundamental building block of Resnet-50 is the bottleneck, a convolution layer with an identity shortcut connection, which solves the problem of gradient vanishing [55]. As shown in Figure 6, ResNet-50 contains 7 layers (groups). Conv1 is a plain convolution layer with 64 convolution kernels of size 3×3, which slide on the RS image to extract hidden features and output 64 feature maps. Conv2 contains 3 bottleneck blocks, each with 128 convolution kernels of size 3×3, which slide on the feature maps generated by Conv1 to extract higher-level features. Likewise, Conv3 contains 4 bottleneck blocks, each with 512 convolution kernels; Conv4 contains 6 bottleneck blocks, each with 1024 convolution kernels, and Conv5 contains 3 bottleneck blocks, each with 2048 convolution kernels. In the network, deeper layers excavate more abstract and informative features related to the task from previous feature maps. Between each convolution layer (or bottleneck block), the feature map is reduced by half to aggregate information. Finally, the average pooling layer squeezes the feature map to 1 dimension, which is inputted into the fully connected layer (fc) to regress the population count. The ReLU activation function and batch normalization are used in all convolution layers to facilitate the training of networks [67]. Figure A1 illustrates how the input RS image evolves to the output population count in the network. Because of the autocorrelation of population distribution, the center grid cell population count may be affected by landscapes in the neighborhood. Hence, we constructed extended image patches by extending the center image patch to include its 3×3 neighboring patches to embed neighborhood knowledge [53,68]. Hence, the layer-wise convolutional operations of the ResNet model can extract interior and neighborhood and integrate latent features for population estimation when sliding on the extended image patches. In order to regress the population count directly, the softmax activation function in the final fully connected layer was removed. We used the log-cosh function for back-propagation training:

$$Loss(\hat{p}, p) = \sum_{i=1}^{s} log_{10}(\cos h(\hat{p}_i - p_i)) \tag{2}$$

where $Loss(\hat{p}, p)$ is the log-cosh loss function, p_i is the ground-truth population count of grid cell i, $\hat{p}_i$ is the estimated population count of grid cell i, and $\cos h(\cdot)$ is the hyperbolic cosine function [68]. The log-cosh loss is similar to the L1 loss, commonly used in regression problems, but is more tolerant of anomalous estimations and achieves better performance [68]. Hyperparameters were tuned empirically based on 1/10 of the available samples. A stochastic gradient descent (SGD) optimizer with a momentum of 0.9 and a learning rate of 10^{-4} was used for weight updating. The batch size and the maximum number of epochs were set to 32 and 1000, respectively. The framework was implemented using the Tensorflow 2.0 library on a Linux server with a 2.50 GHz Intel Xeon E5-2680 CPU, an NVIDIA GTX 2080Ti GPU, and 128GB RAM.

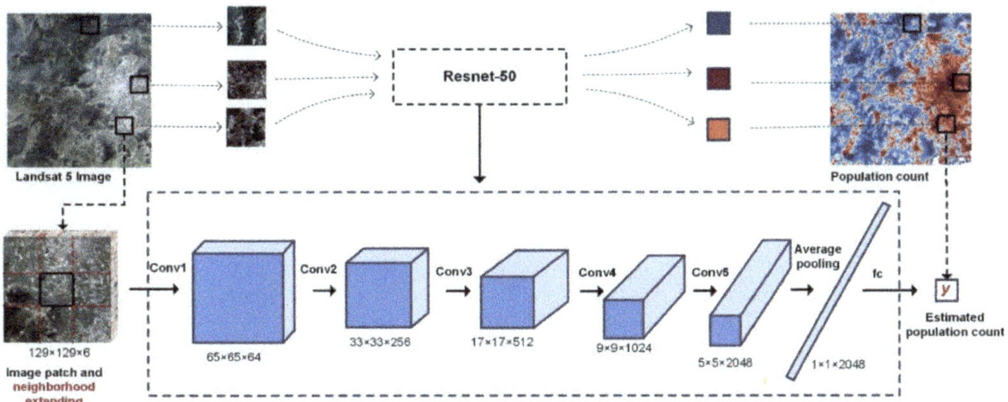

Figure 6. An end-to-end ResNet-N model to estimate population count from RS images by embedding the neighbor knowledge into ResNet.

2.2.2. Mapping Multi-Temporal Population Distributions in China via ResNet-N Model and Landsat-5 RS Images

This study aims to produce multi-temporal gridded population maps with 1 km spatial resolution for China via establishing a framework integrating a deep learning model with Landsat-5 RS images from GEE. Our research area, mainland China, is covered by a grid of 7346 × 4507 consisting of 1 km × 1 km cells. We excluded grid cells with a population count of <10 in the ground-truth population grid in 2010, which can be regarded as uninhabited areas, to reduce the computational burden, resulting in 5,508,904 grid cells being retained. A 1 km × 1 km cell in the population grid approximately covers a 34 × 34 image patch with a spatial resolution of 30 m on Landsat-5 composites. To consider the contribution of neighborhood effects on population count, we constructed extended image patches with a width and height of 102, including the center patch and its 3 × 3 neighboring patches. The extended image patch was then resized to a fixed size of 129 × 129 for inputting into the deep learning model. We obtained a centroid for each cell in the population grid and extracted a 102 × 102 image patch center around the obtained centroid from the Landsat-5 composites for each target year from 1985 to 2010. A total of 33,053,424 RS image patches were extracted and normalized to 0–1. Among them, the image patches that corresponded to the 2010 ground-truth population samples were utilized for training, evaluating, and testing the deep learning model. Once the model was trained, all RS image patches were inputted into the model to measure the population count of each position of each target year.

2.2.3. Modifying Raw Population Estimation via Census Data

Ensuring that the aggregated grid population counts at census units match the known official total population count is necessary. The dasymetric mapping method is used to achieve this goal. For a census unit s with a known official total population count p_s, the following equations are used to modify the raw population estimations:

$$w_i = \frac{p_i^r}{\sum_{i \in s} p_i^r} \tag{3}$$

$$p_i^m = p_s \times w_i \tag{4}$$

where p_i^r is the raw population count of cell i estimated by the deep learning model, w_i is the distribution weight of cell i, and p_i^m represents the modified population count of cell i. In 2010, we used county-scale census data from the National Bureau of Statistics of China to modify the estimation. Due to data limitations, we used the city-scale (level 2 administrative unit) total population count from WorldPop generated using the dasymetric method based on county-scale census data to modify estimations for 2005 and 2000 [48]. For 1995 and 1990, we used the city-scale total population count from GPWv3 produced by the areal weighting method based on official census data at the county scale to modify the estimations [45]. For 1985, due to the unavailability of census data, a single country-wide population count from the World Bank Database was used for modification. The data source and administrative unit level of the census data or total population count are summarized in Table A1.

2.2.4. Accuracy Assessment

We used six quantitative metrics to assess the performance of the proposed population mapping framework and the produced multi-temporal gridded population data, including Pearson's correlation coefficient (R), the coefficient of determination (R^2), mean absolute error (MAE), percentage mean absolute error (%MAE), root mean squared error (RMSE), and percentage root mean squared error (%RMSE):

$$R = \sum_{i=1}^{n} \frac{(p_{i,o} - \overline{p_o})(p_{i,s} - \overline{p_s})}{\sqrt{\sum_{i=1}^{n}(p_{i,o} - \overline{p_o})^2}\sqrt{\sum_{i=1}^{n}(p_{i,s} - \overline{p_s})^2}} \tag{5}$$

$$R^2 = \frac{n\sum_{i=1}^{n} p_{i,o}p_{i,s} - \sum_{i=1}^{n} p_{i,o}\sum_{i=1}^{n} p_{i,s}}{\sqrt{n\sum_{i=1}^{n} p_{i,o}^2 - \left(\sum_{i=1}^{n} p_{i,o}\right)^2}\sqrt{n\sum_{i=1}^{n} p_{i,s}^2 - \left(\sum_{i=1}^{n} p_{i,s}\right)^2}} \tag{6}$$

$$MAE = \frac{1}{n}\sum_{i=1}^{n}|p_{i,o} - p_{i,s}| \tag{7}$$

$$\%MAE = \frac{1}{n}\sum_{i=1}^{n}\frac{|p_{i,o} - p_{i,s}|}{p_{i,o}} \tag{8}$$

$$RMSE = \sqrt{\frac{1}{n}\sum_{i=1}^{n}(p_{i,o} - p_{i,s})^2} \tag{9}$$

$$\%RMSE = \frac{\sqrt{\frac{1}{n}\sum_{i=1}^{n}(p_{i,o} - p_{i,s})^2}}{\overline{p_o}} \tag{10}$$

where $p_{i,o}$ is the ground-truth population count of the ith sample, $p_{i,s}$ denotes the estimated population count of the ith sample, n represents the total number of samples, $\overline{p_o}$ is the average of the ground-truth population count, and $\overline{p_s}$ is the average of the estimated population count. The indicator R, ranging from -1 to 1, measures the linear correlation between actual values and estimated values to evaluate the relative magnitude fitting performance [62]. The indicator R^2, with a value from -infinity to 1, measures how much variance in actual values is captured by the predicted values, assessing the absolute magnitude fitting performance [53]. R and R^2 evaluate the explainability of estimated values to actual values. MAE designates the average absolute error between actual values and estimated values. In order to highlight large errors, absolute errors are squared in RMSE. Since MAE and RMSE are not as understandable, the percentage errors (%MAE and %RMSE) assessing the proportion of the error to the actual value are also presented [54]. These 4 error metrics evaluate the absolute and percentage estimation error together. The mentioned 6 metrics complement each other and provide a comprehensive assessment of the proposed framework and the produced data [69].

3. Results

3.1. Accuracy Assessment of ResNet-N Model for Population Estimation

In this study, a ResNet-N model with neighbor augmentation was utilized to establish the end-to-end mapping between Landsat-5 RS image patches and population count. The model's performance of directly estimating the population count from RS images was evaluated by the collected 20,000 testing samples. Figure 7 shows the scatterplots of ground-truth population count (p) and estimated population count ($\hat{p}$) with their probability density distributions. As shown in Figure 7, the scatterplots of p and $\hat{p}$ present a clustered distribution pattern along the 1:1 reference line, validating that the deep learning architectures can effectively establish the mapping from RS image patches to population count. The probability density distributions of p and $\hat{p}$ exhibit similar shapes and also confirm this conclusion. Compared to the ResNet model without neighbor augmentation, the ResNet-N model with neighbor augmentation used in this study displays superior performance in terms of the six evaluation metrics. ResNet-N (R = 0.84, R^2 = 0.70) exhibits higher explainability of landscape characteristics extracted from the RS images on population count compared to ResNet (R = 0.70, R^2 = 0.56). The R^2 indicates that 70% of the variance population count can be explained by the ResNet-N, compared to 56% by the ResNet. ResNet-N (%MAE = 13.63%, %RMSE = 15.91%) also has higher absolute accuracy than ResNet (%MAE = 16.06%,%RMSE = 19.35%). The %RMSE of ResNet -N is lower than that of ResNet by 21.62% and %MAE by 17.93%. The comparatively low %RMSE and %MSE of both models reveal the capacity of the deep learning model to capture the heterogeneity in population distribution from RS images, and improved estimation performance can be achieved considering neighbor effects.

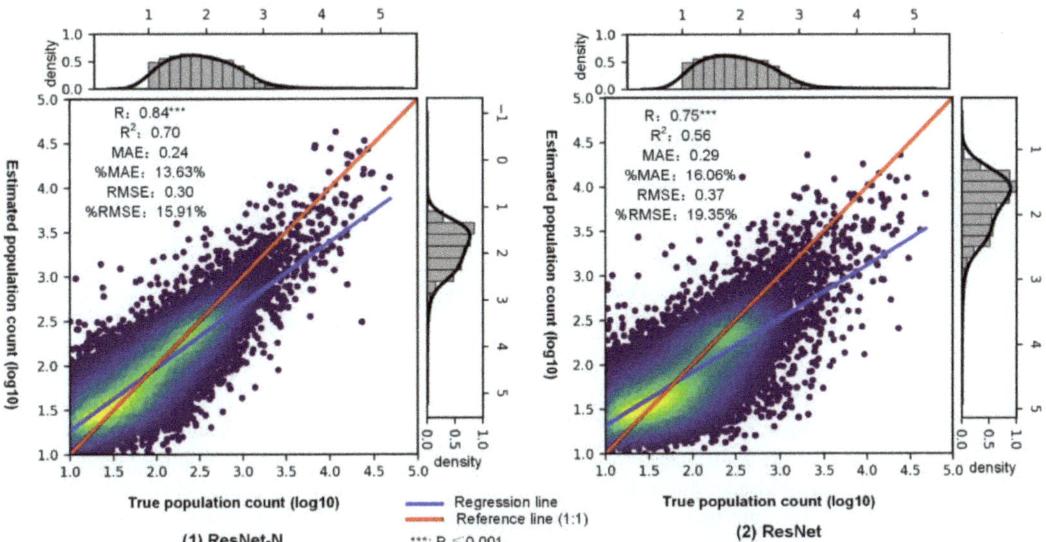

Figure 7. Scatterplots and probability density distributions of ground-truth population count and estimated population count from ResNet-N and ResNet.

For true population count (p) and estimated population count ($\hat{p}$), an investigation of the relationship between p and $\hat{p} - p$ was conducted to explore the systematic bias of estimating population count from RS images via deep learning technologies. Figure 8 shows the scatterplots of p and $\hat{p} - p$ from ResNet-N and ResNet. The results reveal that both models tend to underestimate densely populated samples and overestimate sparsely populated samples, evidenced by the significant negative correlation coefficient and the negative slope. The observed bias can be ascribed to the inherent limitations of multispectral RS images, which cannot identify the social–economic factors that affect population distribution (i.e., the high utilization efficiency of space in densely populated areas). However, consideration of neighbor effects leads to reduced biases and better estimation performance [53].

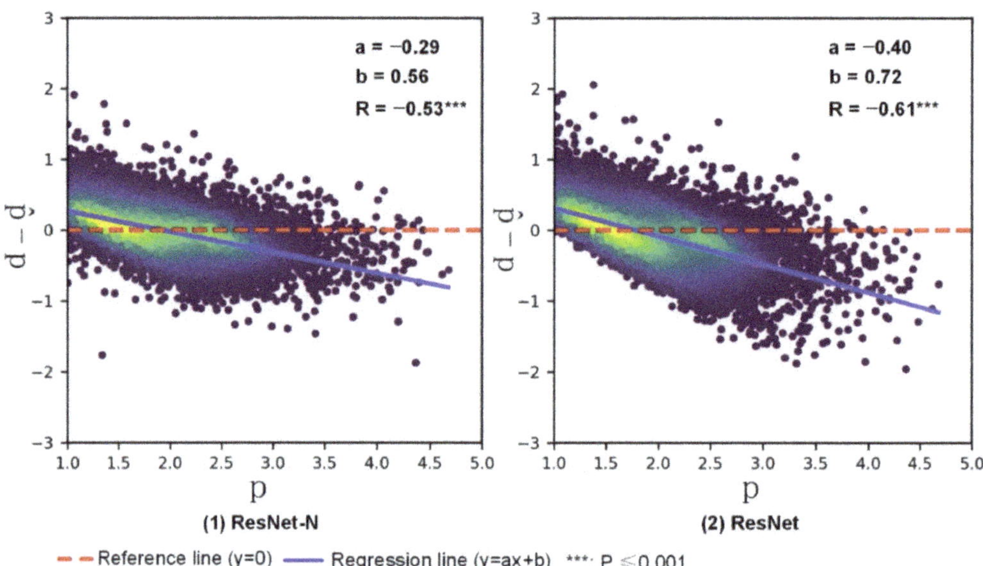

Figure 8. Scatterplots of test samples between p and $\hat{p} - p$ from ResNet-N and ResNet. (p: true population count; $\hat{p}$: estimated population count).

Interpretability is a critical aspect of a model [53,59]. A model with good interpretability usually has good performance. In this study, the ResNet-N model considers only RS images as input to estimate population count. Therefore, all estimations can be explained in terms of the landscape details from RS images. We used gradient-weighted class activation mapping (Grad-CAM), a visual explanation technology for deep learning models, to figure out what features our model learns to estimate population count [70]. Grad-CAM can output a heatmap for an RS image patch. The heat value quantifies the relative contribution of input pixels in the original patch to the estimated population count [70]. For analysis, we selected 12 typical grid cells with different magnitudes of population count. Figure 9 presents the RS image patches in the top rows and corresponding heatmaps in the bottom rows. As shown in Figure 9a, built-up areas are highlighted in heatmaps when they border natural areas. The explanation for this is that built-up areas are usually more densely populated than natural areas. Figure 9b proves the ability of our model to recognize different buildings by capturing hidden hierarchic features of RS images in the interior of the built-up area to estimate population count. As densely populated buildings (i.e., residential) and sparsely populated buildings (i.e., factories) are staggered in the built-up area, distinguish different buildings contributes to accurate population estimation. The heatmaps offer insights into how human activities interact with the underpinning physical environment and prove that our model can learn valuable features for population estimation.

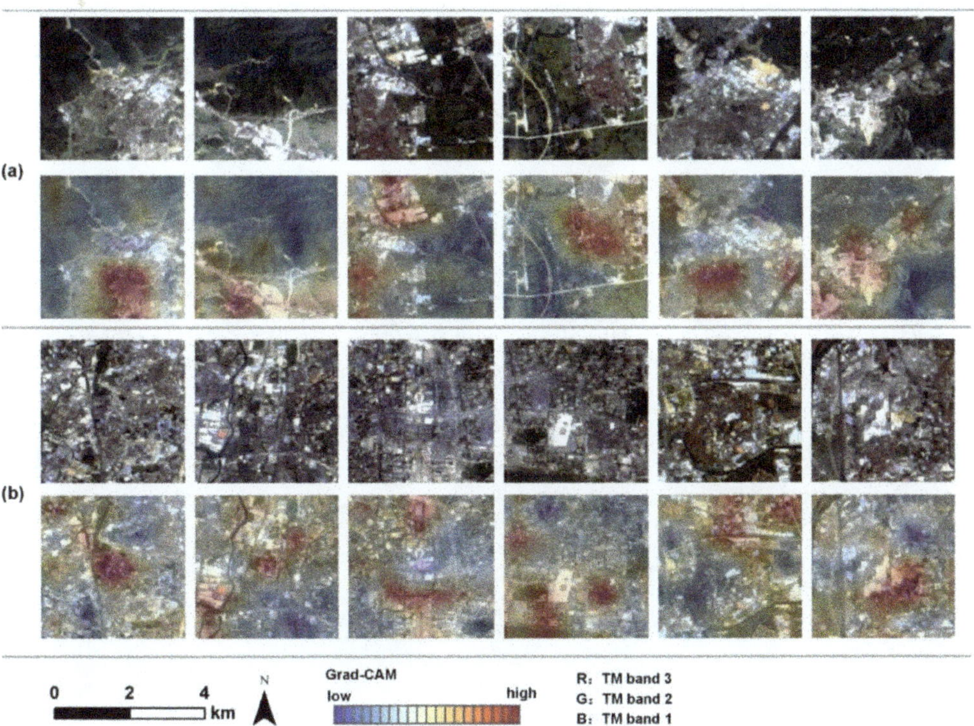

Figure 9. RS image patches (**top row**) and corresponding heatmaps (**bottom row**) produced by Grad-CAM in 12 typical grid cells. (**a**) Built-up areas border natural areas; (**b**) Interiors of built-up areas.

3.2. Validating Multi-Temporal Gridded Population Data via Census Data

A stable end-to-end mapping from RS image patches to population count was established by the ResNet-N model. It is promising that population distribution mapping can be achieved with only the formed mapping and RS images. However, it is necessary to ensure that the aggregated grid population counts at census units match the known official total population count. Furthermore, the grid cell estimation will be more accurate when scaled to match the true population value [59]. We used county-level census data to modify the raw population count estimated from RS images by the model in 2010. Validation of the modified population map was conducted using town-level census data. It is a common practice in dasymetric mapping to use census data of a finer scale to evaluate the accuracy of the produced gridded population data [39]. Two well-known gridded population datasets, WorldPop [48] and GPWv4 [45], were selected as baselines to highlight the performance of the produced data. We collected towns with a population of >100 to assess the comparative performance of the produced gridded data. As shown in Table 1, our new population map produced by coupling RS images and deep learning technologies (referred to as RSPop) achieved the best performance, with the lowest absolute and relative errors and the highest explainability and correlation with the true population count. Figure 10 presents scatterplots of the true population count and estimated population count of each town from RSPop, WorldPop, and GPWv4. Compared to other gridded population data, the scatterplot of RSPop presents a more concentrated distribution pattern along the 1:1 reference line, with the highest accuracy. In contrast, points are scattered and distributed away from the 1:1 reference line in GPWv4, which has the lowest accuracy.

Table 1. Accuracy assessment of RSPop at town scale comparing WorldPop and GPWv4.

	RSPop	WorldPop	GPWv4
R	0.89	0.87	0.82
R^2	0.77	0.69	0.61
MAE	7846.62	8138.20	9463.33
%MAE	46.21	51.19	62.48
RMSE	15,686.74	18,277.52	20,448.11
%RMSE	56.03	65.28	73.03

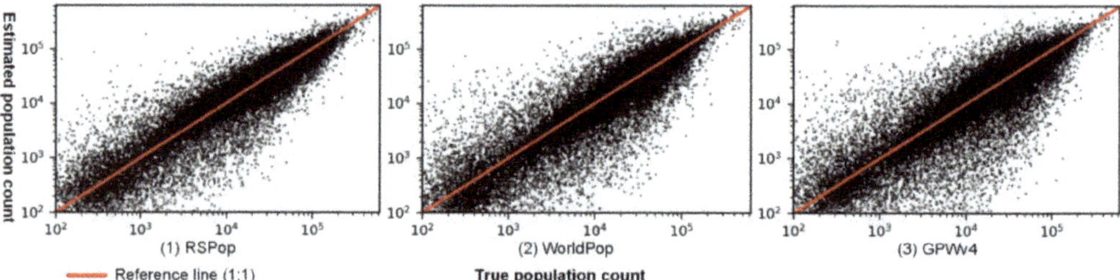

Figure 10. Scatterplots of the true population count and estimated population count from RSPop, WorldPop, and GPWv4 at town scale.

The gridded population data in 2010 produced by the proposed framework were validated and achieved the highest performance compared to other datasets. Due to the consistency of Landsat-5 images and the relative stability of human activity patterns, it can be expected that accurate gridded population data from 1985 to 2005 can be produced by the same framework, using corresponding RS images at target years as input. For the period of 1990–2005, because town-scale census data are challenging to collect, we used the city-scale total population count to modify the estimated population count and applied the county-scale total population count to verify the accuracy of the data. Total population counts at city scale and county scale in 2000 and 2005 were obtained from WorldPop, while total population counts at city scale and county scale for 1990 and 1995 were obtained from GPWv3. Both WorldPop and GPWv3 were produced based on county-scale census data [45,48]. Therefore, it would be impractical to use them for comparative analysis. Instead, as the accuracy of the gridded population data in 2010 has been verified, the population estimation in 2010 modified by city-scale census data was used for comparison at the county scale. As shown in Table 2, overall performance reductions exist for each target year in 1990–2005 compared to 2010. For example, the R^2 is reduced from 0.93 in 2010 to 0.91 in 2005, 0.88 in 2000, 0.73 in 1995, and 0.74 in 1990, with an average reduction of 12.37%. Figure 11 shows scatterplots of the true population count and estimated population count at the county level, which present clustered patterns along the 1:1 reference line. These results imply that the model trained in 2010 is generalizable to other years.

Table 2. Accuracy assessment of RSPop at county scale from 1990 to 2010.

	1990	1995	2000	2005	2010
R	0.86	0.86	0.94	0.95	0.97
R^2	0.74	0.73	0.88	0.91	0.93
MAE	93,260.92	103,362.10	83,413.22	77,668.16	69,052.32
%MAE	30.68	28.48	22.11	19.67	16.64
RMSE	163,431.57	182,624.46	127,106.24	116,733.57	105,319.00
%RMSE	38.45	40.74	27.43	24.47	21.49

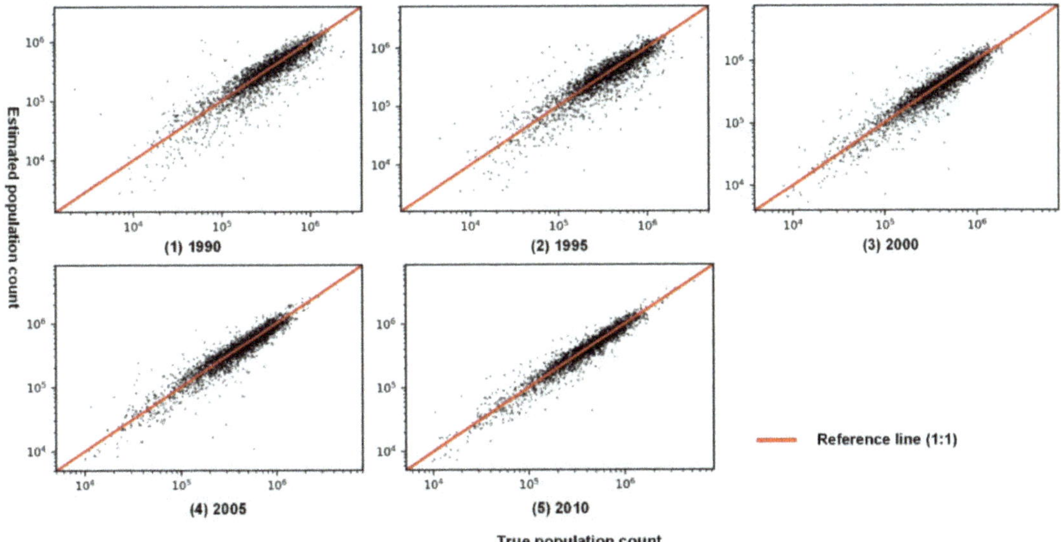

Figure 11. Scatterplots of the true population count and estimated population count at county scale from 1990 to 2010.

For 1985, as the corresponding census data were unavailable, we used a single country-wide population count from the World Bank Database to modify gridded population data and have not verified it. Due to the consistency of the proposed population mapping framework, we argue that data accuracy in 1985 is comparable to that in other years.

3.3. Accuracy Analysis of Gridded Population Data to Scales of Census Data

The availability of census data restrains the production of gridded population data, and fine-scale census data benefit accurate population mapping [48]. However, census surveys are time-consuming and labor-intensive, and, in many cases, only coarse-grained census data can be obtained [71]. Here, we utilized the population distribution in 2010 to investigate the difference in the accuracy of gridded population data based on census data of different scales. The true population and the estimated population at the town scale were compared to evaluate the accuracy of the modified data. Figure 12 shows scatterplots of the true population count and estimated population count at the town scale from gridded population data based on county-scale, city-scale, province-scale, and country-scale census data. The points of true and estimated values present clustered distribution along the 1:1 reference line at all scales, suggesting that the produced gridded population data based on all census scales can capture the heterogeneity in population distribution. Figure 13 shows the variation in the accuracy of gridded population distribution based on census data of four different scales in terms of six accuracy metrics. It is shown that with the increase in the scale of census units, data accuracy decreases. Therefore, when census data are available, it is necessary to use them to modify the raw estimations and obtain better accuracy. Comparable to GPWv4 (R^2 = 0.61, %RMSE = 73.03), based on county-census data, the R^2 and %RMSE of gridded population data based on a single country-wide population count is 0.55 and 79.14, with a difference of 9.84% and 8.37%, respectively. As the difference is relatively low and a single country-wide population count is easily accessible, it is promising that the constructed framework can generate reliable gridded population data from RS images without census data efficiently.

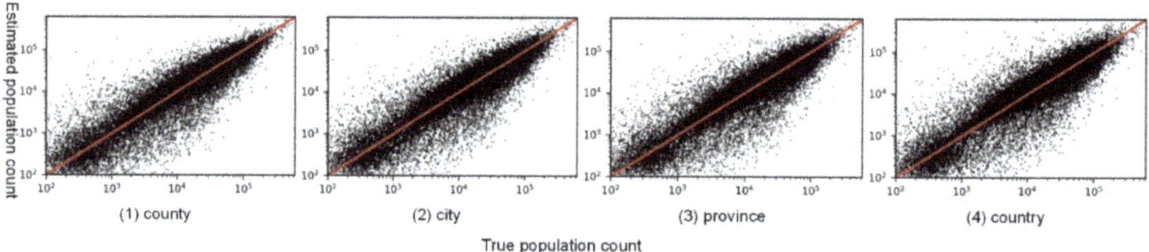

Figure 12. Scatterplots of the true population count and estimated population count at town scale based on county-scale, city-scale, province-scale, and country-scale census data.

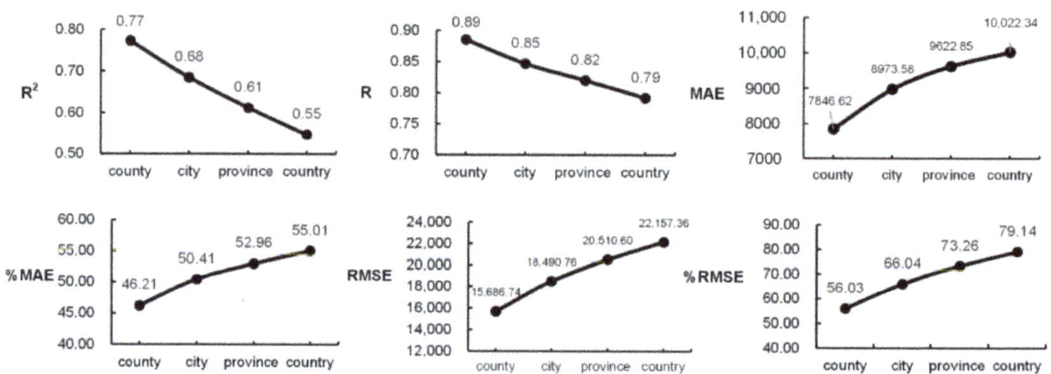

Figure 13. Variation in the accuracy of gridded population data based on county-scale, city-scale, province-scale, and country-scale census data in terms of 6 accuracy metrics.

3.4. Evolution of China's Population Distribution from 1985 to 2010

Figure 14 shows the produced gridded population maps of China with the resolution of 1 km for the years 1985, 1990, 1995, 2000, 2005, and 2010. Although the total population of China grew from 105,104,000 in 1985 to 133,770,500 in 2010, the pattern of population distribution has not changed significantly. The famous Hu-Line pattern [72], characterized by a dense population in the southeast part and a sparse population in the northwest areas of China, remains. From 1985 to 2010, the population gravity center [73] of China lay roughly at the point (113.89° E, 32.97° N), which showed a slight movement to the southeast, with a moving distance of fewer than 33 km (Figure A2), suggesting that China's population and economic center was moving towards the southeast area. In line with previous studies, China's population density is classified into eight levels in this study [34]. Among them, grid cells with a population density greater than 1500 persons/km² are regarded as high-density regions, cells with a population density between 200 and 1500 are regarded as medium-density regions, and cells with a population density less than 200 are regarded as low-density regions. Table 3 lists the percentage values of the area and population for different levels, reflecting the evolution of China's population distribution from 1985 to 2010.

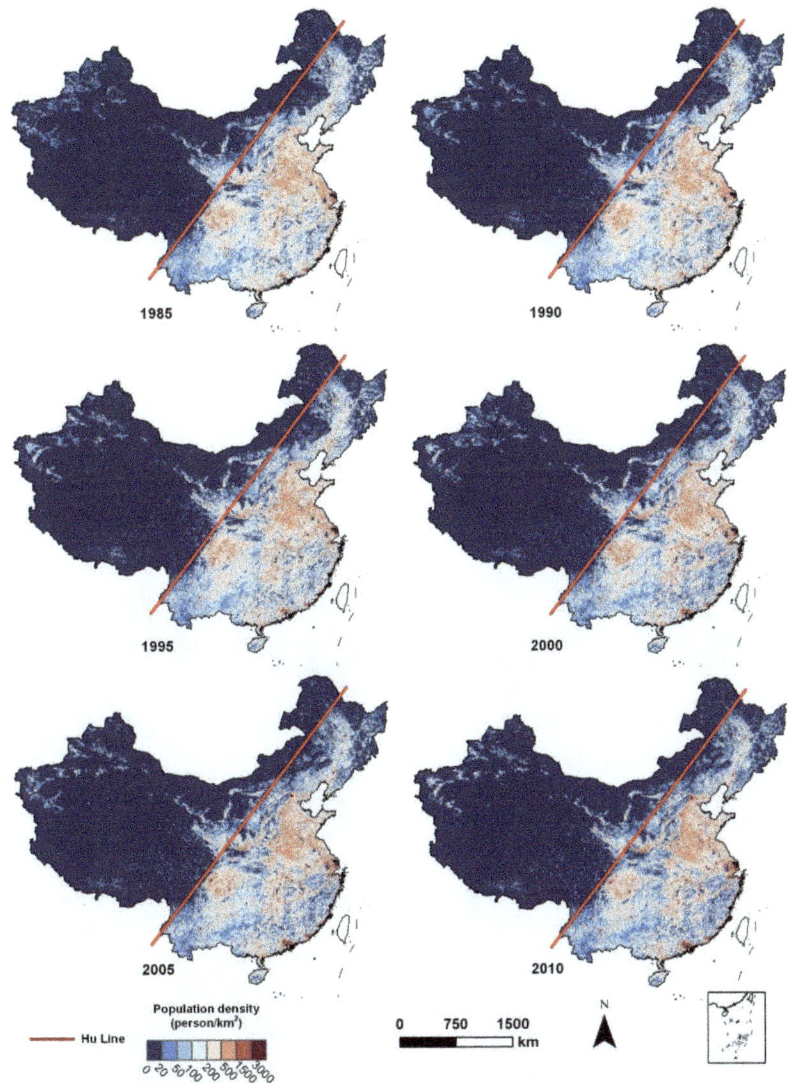

Figure 14. Gridded population data (1 km × 1 km) of China from 1985 to 2010.

Table 3. Percentage values of area and population for different density levels.

Density	1985		1990		1995		2000		2005		2010	
	Area	Population	Area	Population	Area	Population	Area	Population	Area	Population	Area	Population
Low	83.19	22.49	82.33	20.48	82.76	19.46	83.79	18.26	83.97	17.62	83.91	16.44
Medium	16.40	61.18	17.2	60.54	16.64	56.21	15.42	50.96	15.12	48.22	15.07	47.66
High	0.42	16.33	0.48	18.98	0.61	24.33	0.79	30.78	0.92	34.17	1.02	35.90

From 1985 to 2010, the area proportion of high-density regions increased from 0.42% to 1.02%, increasing by 145%, and the population proportion increased from 16.33% to 35.90%, increasing by 119.84%. Previous researches have suggested that high-density regions with a population density of >1500 persons/km^2 can be regarded as urbanized regions [34]. The expansion of regions with high population density can be ascribed to rapid urbanization and the emergence of megacities due to China's reform and opening-up policy. The area proportion of medium-density regions decreased from 16.40% in 1985 to 15.07% in 2010 a decrease of 4.46%, and the population proportion decreased from 61.18% to 47.66% decreasing by 22.09%. The expansion of megacities can explain the reduction in regions with medium population density as the concentration of the population in megacities leads to the contraction of small- and medium-sized urban regions. The area proportion of low-density regions increased from 83.19% in 1985 to 83.91% in 2010, while the population proportion decreased from 22.49% to 16.44%. The expansion of low-density regions may be attributed to immigration measures in some mountainous areas to protect the ecological environment and alleviate poverty [24]. However, with urbanization, the population becomes gradually concentrated in urban regions, leading to a reduced population in low-density regions.

Figure 15 shows the population distributions and landscape variations of three regions in large urban agglomerations in China from 1985 to 2010: (a) Beijing-Tianjin-Hebei, (b) the Yangtze River Delta, and (c) the Pearl River Delta. During this period, these areas experienced rapid urban expansion and consequent population growth, which further led to the transformation of the urban landscape. The produced continuous multi-temporal gridded population data with high spatial resolution provide support to track the co-evolvement of the human population and physical landscape.

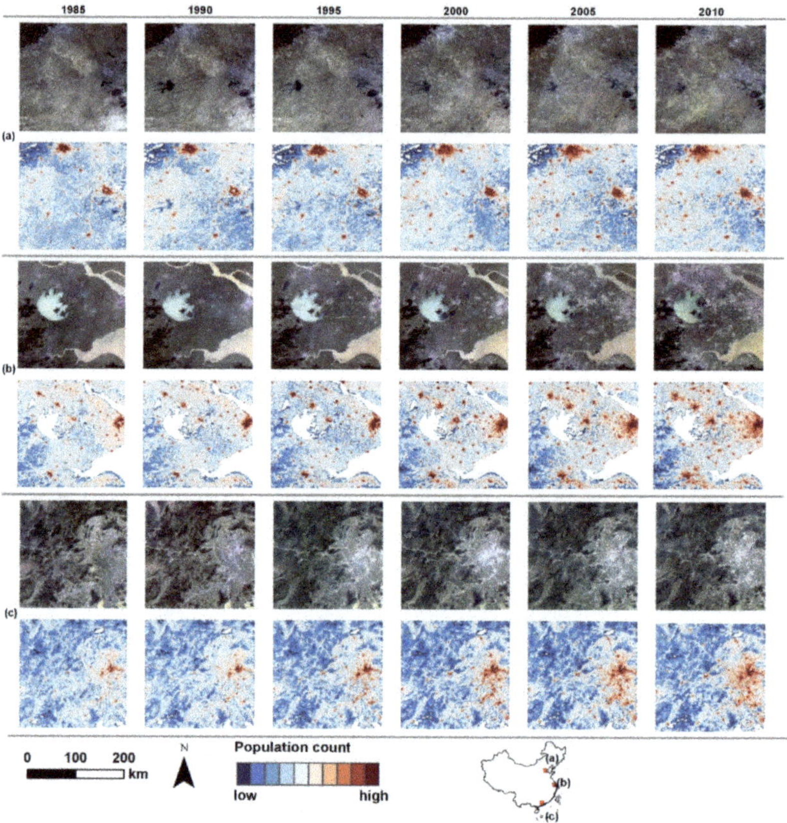

Figure 15. Population distributions (**bottom row**) and landscape variations (**top row**) of three regions in large urban agglomerations in China from 1985 to 2010. (**a**) Beijing-Tianjin-Hebei; (**b**) The Yangtze River Delta; (**c**) The Pearl River Delta.

4. Conclusions and Discussion

China, as the most populous developing country in the world, has experienced rapid economic development, population growth, and urbanization in recent decades. Fine-scale population distribution data and their dynamics are a crucial component in many fields, including resource management, disaster response, public health, urban planning, and climate change; they are also fundamental in monitoring and achieving sustainable development goals (e.g., SDG 11.6.2—annual mean levels of fine particulate matter (e.g., PM2.5 and PM10) in cities (population-weighted)) [74]. However, due to the lack of adequate methodology and appropriate data, there are rarely continuous multi-temporal gridded population data available for China over a long historical period to aid in our understanding of the evolution of population distribution.

The continuously improving remote sensing technology provides low-cost, broad-coverage, and high spatiotemporal resolution ground information, which, in conjunction with deep learning technology that can mine hidden geographical knowledge, enables continuous population distribution mapping. We introduced a framework integrating a ResNet-N deep learning architecture with the consideration of neighborhood effects with a vast number of Landsat-5 images from GEE for rapid multi-temporal population mapping over a long historical period in this study. The ResNet-N model was developed to establish the end-to-end mapping between population count and RS image patches. Based on the trained model, we estimated the gridded population count (1 km $\times$ 1 km) of China with corresponding Landsat-5 image patches from GEE as input from 1985 to 2010. The produced raw estimations were adjusted by available census data to acquire the final gridded population data.

The ResNet-N model with neighbor augmentation achieved R^2 0.70 and %RMSE 15.91%, with a better explainability and higher absolute accuracy than ResNet, which can model the interaction between the physical environment and population and capture the heterogeneity in population distribution from RS images. An interpretation analysis revealed that the constructed deep learning model could provide valuable features for population estimation since it can distinguish the differences between natural and built-up areas and between densely populated and sparsely populated buildings. The produced gridded population data in 2010 was validated via town-scale census data and showed higher accuracy than WorldPop and GPWv4. The produced gridded population data from 1990 to 2005 were validated via county-scale total population count and achieved comparable performance to data in 2010, suggesting that the produced gridded population map can analyze spatiotemporal characteristics of China's population distribution over a long period with acceptable accuracy.

The spatiotemporal analysis of multi-temporal gridded population data showed that China's population distribution pattern did not change significantly from 1985 to 2010, and the famous Hu-Line pattern remains. With China's urbanization process and the emergence of megalopolises, the high-density population regions have dramatically expanded, with the area expanding by approximately 145% and the population expanding by approximately 120%. The concentration of the population in big cities has led to the contraction of cities with medium and small sizes. China's medium-density regions have shrunk by around 4.46%, and their population has decreased by approximately 22.09%. China's low-density regions have expanded slightly with China's poverty alleviation and mountain migration strategy [24], but the population has decreased.

The coupling of deep learning technologies and easily accessible, regularly updated, and analysis-ready remote sensing data from GEE unquestionably establishes a novel avenue that promotes multi-temporal population mapping over a long period at a large scale. However, there are several limitations of this framework. First, although informative knowledge of the population distribution can be extracted from RS images directly, socioeconomic information cannot be identified. For example, the vacancy rate of buildings is difficult to capture, making it impossible to distinguish between vacant buildings and occupied buildings [54]. Especially in China, unreasonable urban expansion has led to the appearance of ghost cities characterized by high vacancy rates of buildings, which cause overestimation of the population [75,76]. Social sensing data and nighttime light data can depict multiple facets of human society, capturing related socioeconomic information [69,75]. In the future, integrating multi-source RS data and time-series social sensing data can further improve the framework [23]. Second, we produced the gridded population data for each target year independently. However, as population distribution is continuous in the time dimension, specific time-series analysis techniques are needed to stabilize temporal variation in population distribution [17]. Third, the deep learning model ResNet-N was trained based on samples collected from the entirety of China in 2010. Although the generalization performance to other years of the model trained in 2010 has been validated, further efforts are needed in considering generalization errors. As China has a large territory and exhibits significant internal variations, in the future, we will investigate whether using regionally parameterized models will improve the performance of population mapping [59].

The framework proposed in this paper demonstrates the feasibility of mapping multi-temporal gridded population distribution at a large scale over a long period in a timely and low-cost manner, which is particularly useful in low-income and data-poor regions. The framework can also be easily extended to a global scale or to map other gridded socioeconomic variables (e.g., GDP) for monitoring and assessing progress toward fulfillment of the SDGs [12].

Author Contributions: Conceptualization, H.Z. and X.L.; methodology, H.Z. and X.L.; validation, H.Z., J.H. and C.W.; formal analysis, H.Z., X.L., Y.Y. and J.O.; resources, X.L. and J.O.; writing—original draft preparation, H.Z.; writing—review and editing, H.Z., X.L., Y.Y., J.O., J.H. and C.W.; supervision, X.L.; project administration, X.L.; funding acquisition, X.L. All authors have read and agreed to the published version of the manuscript.

Funding: This research was funded by the National Key R&D Program of China (Grant No. 2019YFB2103102), the National Natural Science Foundation of China (Grant No. 41801304), and the Natural Science Foundation of Guangdong Province of China (Grant NO. 2018A030310313, 2021A1515011192).

Data Availability Statement: The data presented in this study are openly available in FigShare at https://doi.org/10.6084/m9.figshare.15095748.

Conflicts of Interest: The authors declare no conflict of interest.

Appendix A

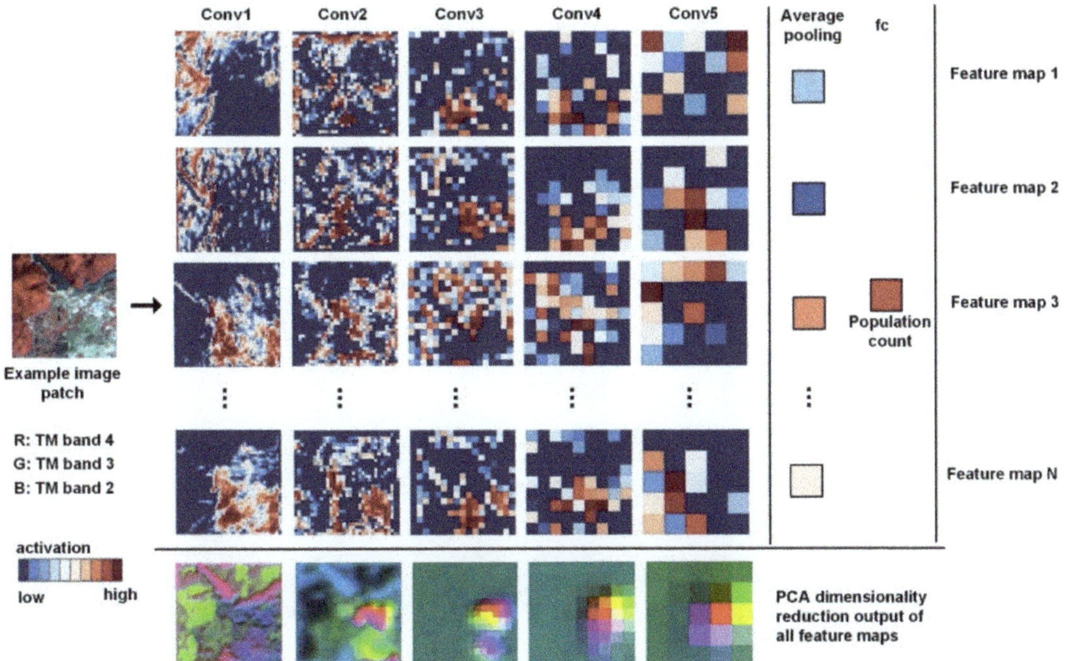

Figure A1. Illustration of how the input RS image evolves to the output population count in the ResNet-N by an example image patch. The activations of the first three and the last feature map of each network layer were visualized. The principal component analysis (PCA) dimension-reduction technique [77] was used to compress all feature maps of each layer to 3 RGB channels for visualization. It is shown that the shallow neural layers (Conv1 and Conv2) excavate concrete features such as texture, shape, and edge from natural landscapes. Then, the deep layers (Conv3, Conv4, and Conv5) extract informative abstract features based on the shallow features for population estimation.

Table A1. Source and administrative unit level of census data or total population count for modifying raw population estimation of each year.

Year	Administrative Unit Level	Source
1985	Country	World Bank Database
1990	City	GPWv3
1995	City	GPWv3
2000	City	WorldPop
2005	City	WorldPop
2010	County	National Bureau of Statistics of China

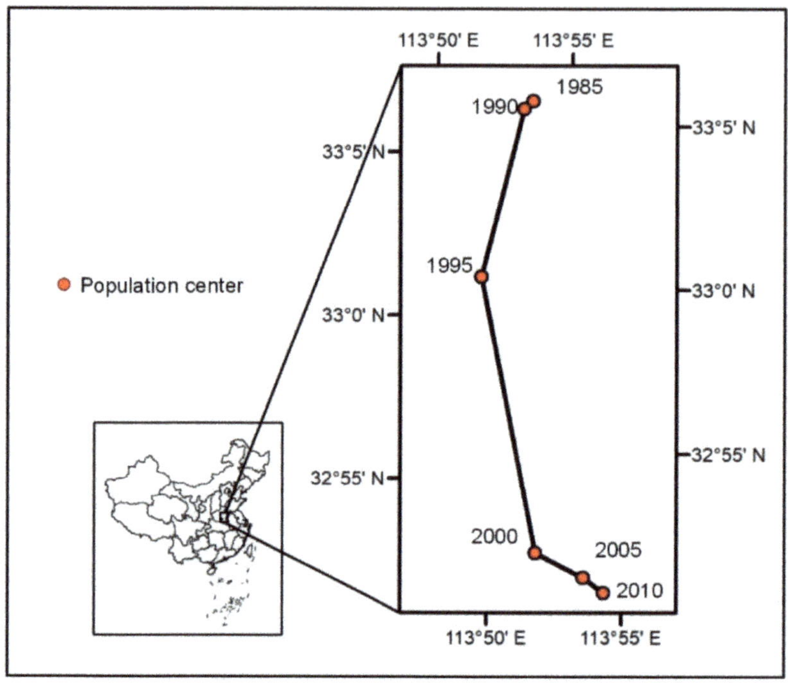

Figure A2. The movement path of population center in China from 1985 to 2010.

References

1. Parish, E.S.; Kodra, E.; Steinhaeuser, K.; Ganguly, A.R. Estimating future global per capita water availability based on changes in climate and population. *Comput. Geosci.* **2012**, *42*, 79–86. [CrossRef]
2. Deichmann, U.; Meisner, C.; Murray, S.; Wheeler, D. The economics of renewable energy expansion in rural Sub-Saharan Africa. *Energy Policy* **2011**, *39*, 215–227. [CrossRef]
3. Ehrlich, D.; Melchiorri, M.; Florczyk, A.J.; Pesaresi, M.; Kemper, T.; Corbane, C.; Freire, S.; Schiavina, M.; Siragusa, A. Remote sensing derived built-up area and population density to quantify global exposure to five natural hazards over time. *Remote Sens.* **2018**, *10*, 1378. [CrossRef]
4. Chen, Y.; Xie, W.; Xu, X. Changes of Population, Built-up Land, and Cropland Exposure to Natural Hazards in China from 1995 to 2015. *Int. J. Disaster Risk Sci.* **2019**, *10*, 557–572. [CrossRef]
5. Chen, Y.; Li, X.; Huang, K.; Luo, M.; Gao, M. High-Resolution Gridded Population Projections for China Under the Shared Socioeconomic Pathways. *Earth's Future* **2020**, *8*. [CrossRef]
6. Mohanty, M.P.; Simonovic, S.P. Understanding dynamics of population flood exposure in Canada with multiple high-resolution population datasets. *Sci. Total Environ.* **2021**, *759*, 143559. [CrossRef] [PubMed]
7. Song, Y.; Huang, B.; Cai, J.; Chen, B. Dynamic assessments of population exposure to urban greenspace using multi-source big data. *Sci. Total Environ.* **2018**, *634*, 1315–1325. [CrossRef] [PubMed]
8. Wang, H.; Li, J.; Gao, Z.; Yim, S.H.L.; Shen, H.; Ho, H.C.; Li, Z.; Zeng, Z.; Liu, C.; Li, Y.; et al. High-spatial-resolution population exposure to PM2.5 pollution based on multi-satellite retrievals: A case study of seasonal variation in the Yangtze River Delta, China in 2013. *Remote Sens.* **2019**, *11*, 2724. [CrossRef]
9. Hay, S.I.; Noor, A.M.; Nelson, A.; Tatem, A.J. The accuracy of human population maps for public health application. *Trop. Med. Int. Health* **2005**, *10*, 1073–1086. [CrossRef]
10. Song, J.; Tong, X.; Wang, L.; Zhao, C.; Prishchepov, A.V. Monitoring finer-scale population density in urban functional zones: A remote sensing data fusion approach. *Landsc. Urban Plan.* **2019**, *190*, 103580. [CrossRef]
11. Dong, N.; Yang, X.; Cai, H.; Xu, F. Research on Grid Size Suitability of Gridded Population Distribution in Urban Area: A Case Study in Urban Area of Xuanzhou District, China. *PLoS ONE* **2017**, *12*, e0170830. [CrossRef] [PubMed]
12. Estoque, R.C. A Review of the Sustainability Concept and the State of SDG Monitoring Using Remote Sensing. *Remote Sens.* **2020**, *12*, 1770. [CrossRef]
13. Wang, Y.; Huang, C.; Feng, Y.; Zhao, M.; Gu, J. Using earth observation for monitoring SDG 11.3.1-ratio of land consumption rate to population growth rate in Mainland China. *Remote Sens.* **2020**, *12*, 357. [CrossRef]

14. Zeng, C.; Zhou, Y.; Wang, S.; Yan, F.; Zhao, Q. Population spatialization in china based on night-time imagery and land use data. *Int. J. Remote Sens.* **2011**, *32*, 9599–9620. [CrossRef]

15. Huang, X.; Wang, C.; Li, Z.; Ning, H. A 100 m population grid in the CONUS by disaggregating census data with open-source Microsoft building footprints. *Big Earth Data* **2021**, *5*, 112–133. [CrossRef]

16. Fotheringham, A.S.; Wong, D.W.S. The modifiable areal unit problem in multivariate statistical analysis. *Environ. Plan. A* **1991**, *23*, 1025–1044. [CrossRef]

17. Wang, L.; Wang, S.; Zhou, Y.; Liu, W.; Hou, Y.; Zhu, J.; Wang, F. Mapping population density in China between 1990 and 2010 using remote sensing. *Remote Sens. Environ.* **2018**, *210*, 269–281. [CrossRef]

18. Mesev, V. *Remotely-Sensed Cities*; CRC Press: Boca Raton, FL, USA, 2003; ISBN 9780415260459.

19. Leyk, S.; Gaughan, A.E.; Adamo, S.B.; De Sherbinin, A.; Balk, D.; Freire, S.; Rose, A.; Stevens, F.R.; Blankespoor, B.; Frye, C.; et al. The spatial allocation of population: A review of large-scale gridded population data products and their fitness for use. *Earth Syst. Sci. Data* **2019**, *11*, 1385–1409. [CrossRef]

20. Tobler, W.; Deichmann, U.; Gottsegen, J.; Maloy, K. World population in a grid of spherical quadrilaterals. *Int. J. Popul. Geogr.* **1997**, *3*, 203–225. [CrossRef]

21. Bracken, I.; Martin, D. The generation of spatial population distributions from census centroid data. *Environ. Plan. A* **1989**, *21*, 537–543. [CrossRef]

22. Chen, Y.; Zhang, R.; Ge, Y.; Jin, Y.; Xia, Z. Downscaling Census Data for Gridded Population Mapping with Geographically Weighted Area-To-Point Regression Kriging. *IEEE Access* **2019**, *7*, 149132–149141. [CrossRef]

23. Cheng, Z.; Wang, J.; Ge, Y. Mapping monthly population distribution and variation at 1-km resolution across China. *Int. J. Geogr. Inf. Sci.* **2020**, 1–19. [CrossRef]

24. Lu, D.; Wang, Y.; Yang, Q.; Su, K.; Zhang, H.; Li, Y. Modeling spatiotemporal population changes by integrating dmsp-ols and npp-viirs nighttime light data in chongqing, china. *Remote Sens.* **2021**, *13*, 284. [CrossRef]

25. Wang, Y.; Huang, C.; Zhao, M.; Hou, J.; Zhang, Y.; Gu, J. Mapping the population density in mainland china using npp/viirs and points-of-interest data based on a random forests model. *Remote Sens.* **2020**, *12*, 3645. [CrossRef]

26. Wang, L.; Fan, H.; Wang, Y. Fine-resolution population mapping from international space station nighttime photography and multisource social sensing data based on similarity matching. *Remote Sens.* **2019**, *11*, 1900. [CrossRef]

27. He, M.; Xu, Y.; Li, N. Population spatialization in Beijing city based on machine learning and multisource remote sensing data. *Remote Sens.* **2020**, *12*, 1910. [CrossRef]

28. Luo, P.; Zhang, X.; Cheng, J.; Sun, Q. Modeling population density using a new index derived from multi-sensor image data. *Remote Sens.* **2019**, *11*, 2620. [CrossRef]

29. Zhao, Y.; Li, Q.; Zhang, Y.; Du, X. Improving the accuracy of fine-grained population mapping using population-sensitive POIs. *Remote Sens.* **2019**, *11*, 2502. [CrossRef]

30. Yu, S.; Zhang, Z.; Liu, F. Monitoring population evolution in China using time-series DMSP/OLS nightlight imagery. *Remote Sens.* **2018**, *10*, 194. [CrossRef]

31. Li, L.; Zhang, Y.; Liu, L.; Wang, Z.; Zhang, H.; Li, S.; Ding, M. Mapping changing population distribution on the qinghai–tibet plateau since 2000 with multi-temporal remote sensing and point-of-interest data. *Remote Sens.* **2020**, *12*, 4059. [CrossRef]

32. Yang, X.; Ye, T.; Zhao, N.; Chen, Q.; Yue, W.; Qi, J.; Zeng, B.; Jia, P. Population mapping with multisensor remote sensing images and point-of-interest data. *Remote Sens.* **2019**, *11*, 574. [CrossRef]

33. Eicher, C.L.; Brewer, C.A. Dasymetric mapping and areal interpolation: Implementation and evaluation. *Cartogr. Geogr. Inf. Sci.* **2001**, *28*, 125–138. [CrossRef]

34. Tan, M.; Li, X.; Li, S.; Xin, L.; Wang, X.; Li, Q.; Li, W.; Li, Y.; Xiang, W. Modeling population density based on nighttime light images and land use data in China. *Appl. Geogr.* **2018**, *90*, 239–247. [CrossRef]

35. Lo, C.P. Automated population and dwelling unit estimation from high-resolution satellite images: A GIS approach. *Int. J. Remote Sens.* **1995**, *16*, 17–34. [CrossRef]

36. Patela, N.N.; Angiuli, E.; Gamba, P.; Gaughan, A.; Lisini, G.; Stevens, F.R.; Tatem, A.J.; Trianni, G. Multitemporal settlement and population mapping from landsatusing google earth engine. *Int. J. Appl. Earth Obs. Geoinf.* **2015**, *35*, 199–208. [CrossRef]

37. Elvidge, C.D.; Baugh, K.E.; Kihn, E.A.; Kroehl, H.W.; Davis, E.R.; Davis, C.W. Relation between satellite observed visible-near infrared emissions, population, economic activity and electric power consumption. *Int. J. Remote Sens.* **1997**, *18*, 1373–1379. [CrossRef]

38. Wang, F.; Lu, W.; Zheng, J.; Li, S.; Zhang, X. Spatially explicit mapping of historical population density with random forest regression: A case study of Gansu province, China, in 1820 and 2000. *Sustainability* **2020**, *12*, 1231. [CrossRef]

39. Ye, T.; Zhao, N.; Yang, X.; Ouyang, Z.; Liu, X.; Chen, Q.; Hu, K.; Yue, W.; Qi, J.; Li, Z.; et al. Improved population mapping for China using remotely sensed and points-of-interest data within a random forests model. *Sci. Total Environ.* **2019**, *658*, 936–946. [CrossRef]

40. Deville, P.; Linard, C.; Martin, S.; Gilbert, M.; Stevens, F.R.; Gaughan, A.E.; Blondel, V.D.; Tatem, A.J. Dynamic population mapping using mobile phone data. *Proc. Natl. Acad. Sci. USA* **2014**, *111*, 15888–15893. [CrossRef]

41. Zhao, S.; Liu, Y.; Zhang, R.; Fu, B. China's population spatialization based on three machine learning models. *J. Clean. Prod.* **2020**, *256*, 120644. [CrossRef]

42. Stevens, F.R.; Gaughan, A.E.; Linard, C.; Tatem, A.J. Disaggregating census data for population mapping using Random forests with remotely-sensed and ancillary data. *PLoS ONE* **2015**, *10*, e0107042. [CrossRef]
43. Harvey, J.T. Population estimation models based on individual TM pixels. *Photogramm. Eng. Remote Sens.* **2002**, *68*, 1181–1192.
44. Cheng, L.; Wang, L.; Feng, R.; Yan, J. Remote sensing and social sensing data fusion for fine-resolution population mapping with a multi-model neural network. *IEEE J. Sel. Top. Appl. Earth Obs. Remote. Sens.* **2021**, *14*, 5973–5987. [CrossRef]
45. Doxsey-Whitfield, E.; MacManus, K.; Adamo, S.B.; Pistolesi, L.; Squires, J.; Borkovska, O.; Baptista, S.R. Taking Advantage of the Improved Availability of Census Data: A First Look at the Gridded Population of the World, Version 4. *Pap. Appl. Geogr.* **2015**, *1*, 226–234. [CrossRef]
46. Melchiorri, M.; Florczyk, A.J.; Freire, S.; Schiavina, M.; Pesaresi, M.; Kemper, T. Unveiling 25 years of planetary urbanization with remote sensing: Perspectives from the global human settlement layer. *Remote Sens.* **2018**, *10*, 768. [CrossRef]
47. Balk, D.L.; Deichmann, U.; Yetman, G.; Pozzi, F.; Hay, S.I.; Nelson, A. Determining Global Population Distribution: Methods, Applications and Data. *Adv. Parasitol.* **2006**, *62*, 119–156.
48. Tatem, A.J. WorldPop, open data for spatial demography. *Sci. Data* **2017**, *4*, 2–5. [CrossRef]
49. Dobson, J.E.; Bright, E.A.; Coleman, P.R.; Durfee, R.C.; Worley, B.A. LandScan: A global population database for estimating populations at risk. *Photogramm. Eng. Remote Sens.* **2000**, *66*, 849–857.
50. Yu, B.; Lian, T.; Huang, Y.; Yao, S.; Ye, X.; Chen, Z.; Yang, C.; Wu, J. Integration of nighttime light remote sensing images and taxi GPS tracking data for population surface enhancement. *Int. J. Geogr. Inf. Sci.* **2019**, *33*, 687–706. [CrossRef]
51. Yao, Y.; Liu, X.; Li, X.; Zhang, J.; Liang, Z.; Mai, K.; Zhang, Y. Mapping fine-scale population distributions at the building level by integrating multisource geospatial big data. *Int. J. Geogr. Inf. Sci.* **2017**, *31*, 1220–1244. [CrossRef]
52. Liu, X.; Hu, G.; Chen, Y.; Li, X.; Xu, X.; Li, S.; Pei, F.; Wang, S. High-resolution multi-temporal mapping of global urban land using Landsat images based on the Google Earth Engine Platform. *Remote Sens. Environ.* **2018**, *209*, 227–239. [CrossRef]
53. Xing, X.; Huang, Z.; Cheng, X.; Zhu, D.; Kang, C.; Zhang, F.; Liu, Y. Mapping Human Activity Volumes Through Remote Sensing Imagery. *IEEE J. Sel. Top. Appl. Earth Obs. Remote Sens.* **2020**, *13*, 5652–5668. [CrossRef]
54. Doupe, P.; Bruzelius, E.; Faghmous, J.; Ruchman, S.G. Equitable development through deep learning: The case of sub-national population density estimation. In Proceedings of the 7th Annual Symposium on Computing for Development ACM DEV-7 2016, Nairobi, Kenya, 18–20 November 2016. [CrossRef]
55. He, K.; Zhang, X.; Ren, S.; Sun, J. Deep residual learning for image recognition. In Proceedings of the IEEE Computer Society Conference on Computer Vision and Pattern Recognition, Las Vegas, NV, USA, 26–30 June 2016; pp. 770–778.
56. Helber, P.; Bischke, B.; Dengel, A.; Borth, D. Eurosat: A novel dataset and deep learning benchmark for land use and land cover classification. *IEEE J. Sel. Top. Appl. Earth Obs. Remote Sens.* **2019**, *12*, 2217–2226. [CrossRef]
57. Zhu, D.; Cheng, X.; Zhang, F.; Yao, X.; Gao, Y.; Liu, Y. Spatial interpolation using conditional generative adversarial neural networks. *Int. J. Geogr. Inf. Sci.* **2020**, *34*, 735–758. [CrossRef]
58. Jean, N.; Burke, M.; Xie, M.; Davis, W.M.; Lobell, D.B.; Ermon, S. Combining satellite imagery and machine learning to predict poverty. *Science* **2016**, *353*, 790–794. [CrossRef]
59. Robinson, C.; Hohman, F.; Dilkina, B. A deep learning approach for population estimation from satellite imagery. In Proceedings of the 1st ACM SIGSPATIAL Workshop on Geospatial Humanities, Redondo Beach, CA, USA, 7–10 November 2017; pp. 47–54.
60. Gaughan, A.E.; Stevens, F.R.; Huang, Z.; Nieves, J.J.; Sorichetta, A.; Lai, S.; Ye, X.; Linard, C.; Hornby, G.M.; Hay, S.I.; et al. Spatiotemporal patterns of population in mainland China, 1990 to 2010. *Sci. Data* **2016**, *3*, 1–11. [CrossRef]
61. Gorelick, N.; Hancher, M.; Dixon, M.; Ilyushchenko, S.; Thau, D.; Moore, R. Google Earth Engine: Planetary-scale geospatial analysis for everyone. *Remote Sens. Environ.* **2017**, *202*, 18–27. [CrossRef]
62. Hu, W.; Patel, J.H.; Robert, Z.A.; Novosad, P.; Asher, S.; Tang, Z.; Burke, M.; Lobell, D.; Ermon, S. Mapping Missing Population in Rural India: A Deep Learning Approach with Satellite Imagery. *arXiv* **2019**, arXiv:1905.02196.
63. Masek, J.G.; Vermote, E.F.; Saleous, N.E.; Wolfe, R.; Hall, F.G.; Huemmrich, K.F.; Gao, F.; Kutler, J.; Lim, T.K. A landsat surface reflectance dataset for North America, 1990–2000. *IEEE Geosci. Remote Sens. Lett.* **2006**, *3*, 68–72. [CrossRef]
64. Dwyer, J.L.; Roy, D.P.; Sauer, B.; Jenkerson, C.B.; Zhang, H.K.; Lymburner, L. Analysis ready data: Enabling analysis of the landsat archive. *Remote Sens.* **2018**, *10*, 1363. [CrossRef]
65. Liu, H.; Gong, P.; Wang, J.; Wang, X.; Ning, G.; Xu, B. Production of global daily seamless data cubes and quantification of global land cover change from 1985 to 2020—iMap World 1.0. *Remote Sens. Environ.* **2021**, *258*, 112364. [CrossRef]
66. Qiu, S.; Lin, Y.; Shang, R.; Zhang, J.; Ma, L.; Zhu, Z. Making Landsat time series consistent: Evaluating and improving Landsat analysis ready data. *Remote Sens.* **2019**, *11*, 51. [CrossRef]
67. Nguyen, G.; Dlugolinsky, S.; Bobák, M.; Tran, V.; López García, Á.; Heredia, I.; Malík, P.; Hluchý, L. Machine Learning and Deep Learning frameworks and libraries for large-scale data mining: A survey. *Artif. Intell. Rev.* **2019**, *52*, 77–124. [CrossRef]
68. Huang, X.; Zhu, D.; Zhang, F.; Liu, T.; Li, X.; Zou, L. Sensing Population Distribution from Satellite Imagery via Deep Learning: Model Selection, Neighboring Effect, and Systematic Biases. Available online: http://arxiv.org/abs/2103.02155 (accessed on 1 September 2021).
69. Yao, Y.; Zhang, J.; Hong, Y.; Liang, H.; He, J. Mapping fine-scale urban housing prices by fusing remotely sensed imagery and social media data. *Trans. GIS* **2018**, *22*, 561–581. [CrossRef]
70. Selvaraju, R.R.; Cogswell, M.; Das, A.; Vedantam, R.; Parikh, D.; Batra, D. Grad-CAM: Visual Explanations from Deep Networks via Gradient-Based Localization. *Int. J. Comput. Vis.* **2020**, *128*, 336–359. [CrossRef]

1. Wardrop, N.A.; Jochem, W.C.; Bird, T.J.; Chamberlain, H.R.; Clarke, D.; Kerr, D.; Bengtsson, L.; Juran, S.; Seaman, V.; Tatem, A.J. Spatially disaggregated population estimates in the absence of national population and housing census data. *Proc. Natl. Acad. Sci. USA* **2018**, *115*, 3529–3537. [CrossRef] [PubMed]
2. Chen, D.; Zhang, Y.; Yao, Y.; Hong, Y.; Guan, Q.; Tu, W. Exploring the spatial differentiation of urbanization on two sides of the Hu Huanyong Line—Based on nighttime light data and cellular automata. *Appl. Geogr.* **2019**, *112*, 102081. [CrossRef]
3. Liang, L.; Chen, M.; Luo, X.; Xian, Y. Changes pattern in the population and economic gravity centers since the Reform and Opening up in China: The widening gaps between the South and North. *J. Clean. Prod.* **2021**, *310*, 127379. [CrossRef]
4. UN IAEG-SDGs Global Indicator Framework for the Sustainable Development Goals and Targets of the 2030 Agenda for Sustainable Development. Available online: https://unstats.un.org/sdgs/indicators/Global%2520Indicator%2520Framework%2520after%25202020%2520review_Eng.pdf (accessed on 1 September 2021).
5. Zeng, Q.; Zhang, W. Research on the Development of "Ghost City" Based on Night Light Data: Taking Sichuan Province as an Example. *Open J. Soc. Sci.* **2019**, *7*, 176–188. [CrossRef]
6. Mingye, L. Evolution of Chinese Ghost Cities. *China Perspect.* **2017**, *2017*, 69–78. [CrossRef]
7. Jolliffe, I.T.; Cadima, J.; Cadima, J. Principal component analysis: A review and recent developments Subject Areas. *Philos. Trans. R. Soc. A* **2016**, *374*, 20150202. [CrossRef] [PubMed]

 *remote sensing*

Article

Understanding the Relationship between China's Eco-Environmental Quality and Urbanization Using Multisource Remote Sensing Data

Dong Xu [1,2,3,4,†], Jie Cheng [2,3], Shen Xu [5], Jing Geng [6], Feng Yang [7], He Fang [1,*], Jinfeng Xu [4], Sheng Wang [4], Yubai Wang [4], Jincai Huang [8], Rui Zhang [9], Manqing Liu [2,3] and Haixing Li [4,†]

[1]   Zhejiang Climate Center, Hangzhou 310052, China
[2]   State Key Laboratory of Remote Sensing Science, Faculty of Geographical Science, Beijing Normal University, Beijing 100875, China; xd@mail.bnu.edu.cn (D.X.); Jie_Cheng@bnu.edu.cn (J.C.); 202131051048@mail.bnu.edu.cn (M.L.)
[3]   Faculty of Geographical Science, Institute of Remote Sensing Science and Engineering, Beijing Normal University, Beijing 100875, China
[4]   College of Geomatics Science and Technology, Nanjing Tech University, Nanjing 211816, China; 202061223009@njtech.edu.cn (J.X.); 202061223008@njtech.edu.cn (S.W.); 202061123001@njtech.edu.cn (Y.W.); lihaixing@njtech.edu.cn (H.L.)
[5]   School of Psychological and Cognitive Sciences, Peking University, Beijing 100871, China; xs-psych@stu.pku.edu.cn
[6]   Academician Workstation of Zhai Mingguo, University of Sanya, Sanya 572000, China; jinggeng@sanyau.edu.cn
[7]   School of Environmental Science and Engineering, Southern University of Science and Technology, Shenzhen 518055, China; yftaurus@mail.bnu.edu.cn
[8]   Shenzhen Key Laboratory of Spatial Smart Sensing and Service, Shenzhen University, Shenzhen 518060, China; huangjincaiszu@szu.edu.cn
[9]   School of Geography and Information Engineering, China University of Geosciences, Wuhan 430074, China; 1202010859@cug.edu.cn
*   Correspondence: fanghe_doc@163.com
†   These authors contributed equally to this work.

Citation: Xu, D.; Cheng, J.; Xu, S.; Geng, J.; Yang, F.; Fang, H.; Xu, J.; Wang, S.; Wang, Y.; Huang, J.; et al. Understanding the Relationship between China's Eco-Environmental Quality and Urbanization Using Multisource Remote Sensing Data. *Remote Sens.* **2022**, 14, 198. https://doi.org/10.3390/rs14010198

Academic Editor: Ronald C. Estoque

Received: 31 October 2021
Accepted: 23 December 2021
Published: 2 January 2022

Publisher's Note: MDPI stays neutral with regard to jurisdictional claims in published maps and institutional affiliations.

Abstract: The rapid development of urbanization and population growth in China has posed a major threat to the green sustainable development of the ecological environment. However, the impact of urbanization on the eco-environmental quality (EEQ) in China remains to be developed. Understanding their interactive coupling mechanism is of great significance to achieve the urban sustainable development goals. By using multi-source remote sensing data and the coupling co-ordination degree model (CCDM), we intended to answer the question "What are the temporal and spatial characteristics of urbanization and EEQ in China on the pixel scale during 2000–2013, and what is the coupling mechanism between the urbanization and the EEQ?". To answer these questions, we explored the coupling mechanism between urbanization and the EEQ in China with a combined mathematical and graphics model. The results show that the urbanization and the coupling coordination degree (CCD) of the whole region continually increased from 2000 to 2013, especially in the three major urban agglomerations, with a spatial distribution pattern that was "high in the east and low in the west". Most importantly, from 2000 to 2013, the CCD type of cities in China gradually evolved from uncoordinated cities to coordinated cities. Additionally, the decisive factor affecting the CCD from 2000 to 2013 was the development of urbanization, and the degree at which urbanization had an impact on CCD was about 8.4 times larger than that of the EEQ. At the same time, the rapid urbanization that has occurred in some areas has led to a significant decline in the EEQ, thus indicating that China needs to increase its protection of the ecological environment while pursuing social and economic development in the future. This study makes up for the deficiencies in the existing literature and investigates the long-term coupling of the EEQ and urbanization in China, thereby providing a new research perspective for the sustainable development of China and even the world in the future.

Keywords: M-RSEQI; DMSP; entropy method; coupling coordination degree model; remote sensing data; urbanization

1. Introduction

Since the beginning of the current century, China has experienced a period of urbanization that is unprecedented in human history. During this period, the speed of urbanization development in China was approximately three times larger than the world average during the same period [1,2]. Rapid urbanization has led to serious environmental problems, such as the depletion of natural resources and the destruction of ecosystem services [3]. Severe ecological and environmental problems restrict China from having a healthy economy and societal development and threatens regional sustainable development [4–6]. In terms of this increasingly severe eco-environmental situation, it is increasingly important to conduct effective scientific research on monitoring the coupling of China's eco-environment quality (EEQ) and urbanization in order to ensure the realization of "China's 2030 sustainable development goals" [7].

Currently, research on EEQ assessment has made great progress [8]. In 1979, David J. Rapport and Tony Friend firstly proposed the "pressure-state-response" (PSR) framework [9], which reflects the relationship among natural, social, and economic factors from the perspective of environmental protection and economic development, and this framework has been widely used in EEQ assessment studies [8]. Based on the PSR framework, the European Environment Agency (EEA) [10], the United States Environmental Protection Agency (USEPA) [11], and the Council for Sustainable Development (CSD) [12] have all proposed improved models for EEQ assessment in different regions. With the emergence of the concept of sustainable development, research methods have gradually developed EEQ assessment models that are based on sustainable development, and these models include ecological footprint (EF) [13] and the environmental sustainability index (ESI) [14]. Through summarizing existing studies, we found that previous research had two shortcomings. First, the environmental assessments mainly focused on the assessment of ecological risks, ecological effects, and ecological fragility and lacked a comprehensive assessment of the overall EEQ of the region. Secondly, most of the studies relied heavily on panel statistical data, which led to uncertainty in the study results due to the diversity of the sources of panel statistical data. In terms of urbanization assessment, current research methods are mainly based on panel statistical data that are implemented around different assessment systems, similar to EEQ assessments [15–18]. Therefore, studies that evaluate urbanization also have the two abovementioned shortcomings.

The rapid development of remote sensing sensors and the open access of massive remote sensing data have greatly promoted Earth observation research on the regional scale [19,20]. In 2020, Estoque (2020) elaborated on the status, challenges, and opportunities of the remote sensing (RS) monitoring of the Sustainable Development Goals and pointed out that remote sensing technology is an important environmental monitoring tool that can help fill the gaps in environmental data [21]. For example, the RSEI model proposed by Xu [22] in 2013 has been widely applied to assess the EEQ in different regions [23–25]. However, these studies often overlook an important principle, i.e., the definition of EEQ should be evaluated on the basis of the entire territory of China. For example, the EEQ of the Qinghai–Tibet Plateau calculated by the RSEI index is not comparable with that of the Loess Plateau. Secondly, many studies have found that the application of the RSEI index on a large scale has larger uncertainty, especially in regions with abundant land cover types, which creates a gap in the current research on remotely sensed EEQ assessments in China [26]. Remote sensing technology has also made great progress in the quantitative assessment of urbanization. Currently, urbanization monitoring methods using remote sensing are generally divided into two categories: multispectral remote sensing and nighttime light remote sensing. For example, Li [27] proposed an urban boundary extraction algorithm

(GA-UCAT) that was based on the normalized building index (NDBI). Zhang [28] analyzed the urbanization development of India, China, Japan, and the United States from 1992 to 2000 based on DMSP nighttime light data. However, the advantages of nighttime light data in characterizing the level of urbanization have been verified [29–31].

The theories on the interaction and coupling of the EEQ and urbanization mainly include the environmental Kuznets curve (EKC) [32,33], the planetary boundaries theory [34,35], the tele-coupling theory [36,37], the footprint family theory [38,39], and the urban metabolism theory [40,41]. Related models mainly include the STIRPAT model [42], the coupling coordination degree model (CCDM) [43], and the multiagent model [44]. Among them, CCDM is the most widely used [45]. Zhao [46] studied the coupling of the EEQ and urbanization in the Yangtze River Basin from 1980 to 2013 through the EKC and dynamic CCDM and found that the EEQ and urbanization conformed to the S-shaped curve of the coupling and coordination relationship. Li [47] used the CCDM to study the coupling relationship between the EEQ and urbanization in Lianyungang and found that the coupling coordination degree (CCD) presented an inverted U-shaped curve, and Wang et al. also reached a similar conclusion [48]. Some scholars used the double exponential model and the CCDM to verify the nonlinear relationship between the EEQ and urbanization [49].

Under the context that the sustainable development goals (SDGs) are becoming a global strategic development goal, research on the coupling of the regional EEQ and urbanization has gradually become a global hot topic [45]. By reviewing previous studies, their shortcomings were summarized. Firstly, most existing studies were based on statistical data. The diverse data sources and the low temporal and spatial resolution of statistical data have led to great uncertainty in the research results. Since the sources of statistical data in China can be divided into the national and provincial levels, the same indicators published by different agencies, which contain huge human factors, are very different. Another uncertainty of statistical data comes from the timeliness. For example, with the change of the definition of an indicator, the indicator may stop updating in the future or may be replaced by another indicator, which will cause great uncertainty to the statistical data. On the other hand, studies based on statistical data may also result in low spatial resolution characteristics. Therefore, these research results usually cannot show the surface details on the pixel scale. Secondly, there is currently a lack of studies on the long-term continuous coupling of the EEQ and urbanization in China. To address these issues, this study explored the coupling mechanism between the EEQ and urbanization in China from 2000 to 2013 using the CCDM and multisource remote sensing data.

This study counterbalances the shortcomings of existing research and fills gaps in the research on the coupling mechanism between the EEQ and urbanization. Additionally, this study provides a new perspective for research on urban sustainable development in China, which can also be extended to the global research.

2. Study Area and Data

2.1. Study Area

China is located in eastern Asia, with the terrain gradually rising in elevation from east to west and a rich ecosystem. There are three forest areas in northeastern and southern China, with the Inner Mongolia Plateau constituting China's largest grassland resource and representing the largest ecosystem type in China (Figure 1). Lakes of various sizes, most of which are distributed on the Qinghai–Tibet Plateau constitute China's complex wetland ecosystem. The Qinghai–Tibet Plateau is the largest ice accumulation area in the world after the North and South poles, and this region is called the "Asian Water Tower" [50]. Therefore, the EEQ of China plays an important role in the stability of China's ecosystem.

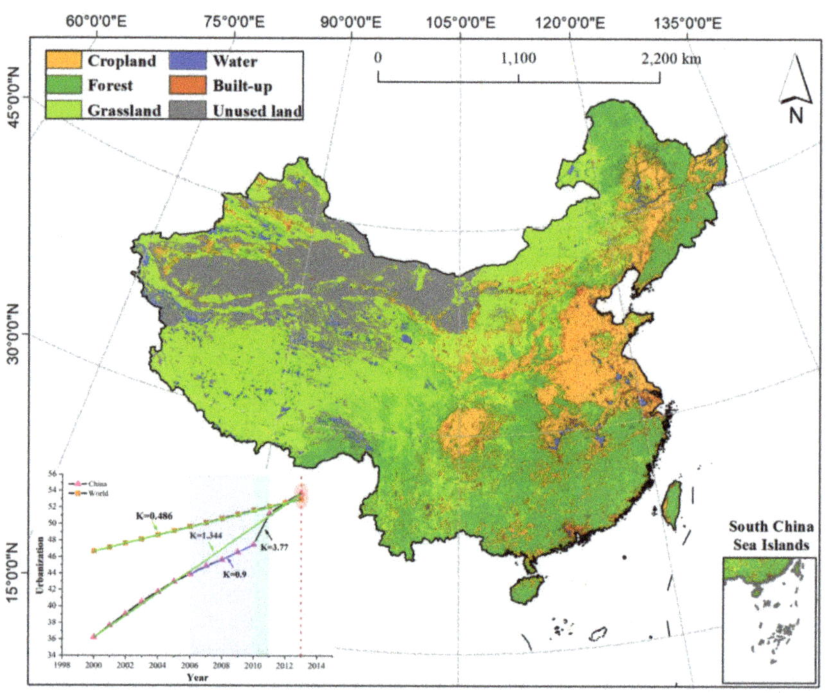

Figure 1. Land use map (MCD12Q1) of China in 2013 and line chart of the urbanization rates in China and the world from 2000 to 2013 (bottom left). The light blue shaded area represents the slow growth stage of the urbanization rate, while the light green area represents the rapid growth stage.

Since the 21st century, China has experienced the largest and fastest urbanization process in its history [51]. The urbanization rate increased from 36.22% in 2000 to 53.7% in 2013, with an annual average growth rate of 1.344, which is much higher than the world average of 0.486. As shown in Figure 1, the urbanization process in China from 2000 to 2013 experienced slow and rapid growth. From 2006 to 2010, the average growth rate of China's urbanization, with a rate of 0.9, was higher than the global growth rate of 0.498 during the same period, which was much lower than the annual average growth rate over the 14-year period. From 2010 to 2011, China's urbanization rate increased rapidly, and the growth rate reached 3.77. China's urbanization rate exceeded the world average for the first time in 2013, which is the end of the time series of this study.

2.2. Data

Multisource remote sensing data were used in this study, including precipitation (PRE), temperature (TEMP), net primary productivity (NPP), vegetation coverage (FVC), digital elevation model (DEM), land use and land cover change (LULC), population density (POP), potential evapotranspiration (PET), and DMSP (Defense Meteorological Satellite Program) nighttime light data. The data used in this study are detailed in Table 1.

Table 1. Detailed description of the data.

Data Name	Format	Spatial Resolution	Time Resolution	Source
2019QZKK0603-zgyjsl [52]	NETCDF	1000 m	Monthly	NTPDC [a]
2019QZKK0603-zgypjw [52]	NETCDF	1000 m	Monthly	NTPDC [a]
MOD17A3	HDF	1000 m	Monthly	NASA [b]
MOD13A3	HDF	250 m	Monthly	NASA [b]
SRTM DEM	TIFF	250 m	—	USGS [c]
MCD12Q1	HDF	1000 m	Seasonal	NASA [b]
Landscan POP	TIFF	1000 m	Annual	ORNL [d]
MOD16A3	HDF	1000 m	Monthly	NASA [b]
LUCC2000	TIFF	1000 m	Annual	CAS [e]
DMSP nighttime light	TIFF	1000 m	Annual	NOAA [f]
Anthropogenic heat flux (AHF)	TIFF	1000 m	Annual	Article [53]

[a] NTPDC: National Tibetan Plateau Data Center. [b] NASA: National Aeronautics and Space Administration. [c] USGS: United States Geological Survey. [d] ORNL: Oak Ridge National Laboratory. [e] CAS: Chinese Academy of Sciences. [f] NOAA: National Oceanic and Atmospheric Administration.

3. Methods

The methods that were used in this study are divided into three parts: (1) the production of the Multisource Remote Sensing EEQ Index (M-RSEQI) (Section 3.1), (2) the processing of the quantitative calculation of urbanization (Section 3.2), and (3) the evaluation of the coupling processes between the M-RSEQI and urbanization (Section 3.3). The flowchart for this article is shown in Figure 2 and includes two parts: the flowchart of study methods (Figure 2a) and an example for 2000 (Figure 2b). The main process for Figure 2a is as follows: First, the annual mean value of eight indicators representing the EEQ was calculated and standardized, and then the dimension reduction of the eight indicators was carried out to calculate the M-RSEQI. Secondly, the DMSP nighttime light data were corrected and standardized. Finally, the processed DMSP nighttime light data and M-RSEQI were input into the CCDM to calculate the CCD.

3.1. M-RSEQI

This study followed the principles for the selection of assessment indicators proposed by the China National Environmental Monitoring Centre [54], namely comprehensive principles, representative principles, scientific principles, comparability principles, and operability principles. Eight indicators were selected, as shown in Table 1.

To eliminate the influence of different index dimensions on the results, dimensionless treatment was needed for all of the indexes. The range standardization method was used to standardize the eight indicators that were used in this research. The selected indicators were divided into positive and negative indicators. Positive indicators included PRE, NPP, NDVI, DEM, and LUCC, which promoted the EEQ, while the negative indicators had negative effects on the EEQ, and included TEMP, POP, and PET. The standardized equations are as follows:

$$\text{Positive indicator}: \quad r_s^+ = (I_j - I_{\min})/(I_{\max} - I_{\min}) \tag{1}$$

$$\text{Negative indicator}: \quad r_s^- = (I_{\max} - I_j)/(I_{\max} - I_{\min}) \tag{2}$$

where r_s^+ is the standardized value of the jth indicator, I_j is the original value of the jth indicator, and $I_{\min}$ and $I_{\max}$ are the minimum and maximum values of the jth indicator.

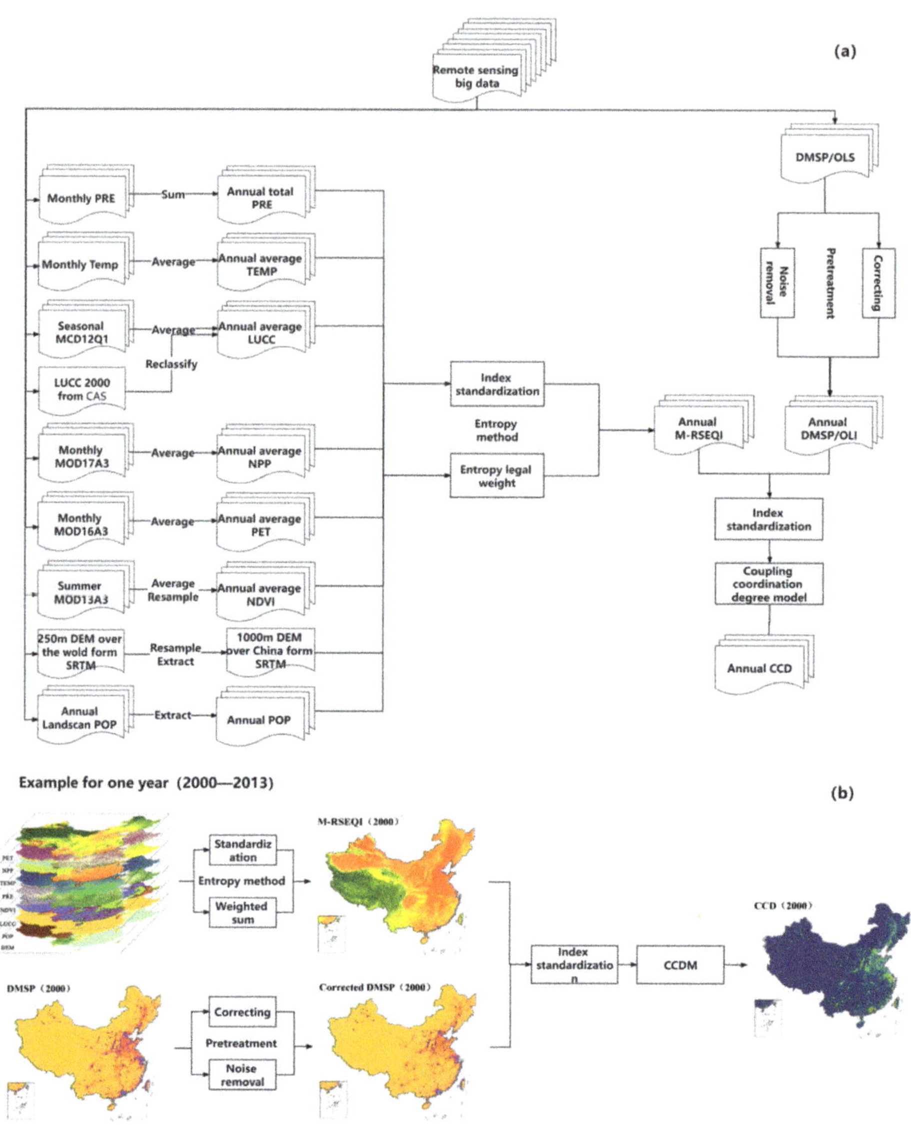

Figure 2. Flowchart of the methods. (**a**) The flowchart of data processing methods; (**b**) An example for one year (2000).

Another key point of EEQ assessment is the assessment model. Currently, the main assessment methods include objective methods and subjective methods. Common objective methods include the entropy method (EM) and principal component analysis (PCA), and common subjective methods include grey relation analysis (GRA) and the analytic hierarchy process (AHP) [55]. This study used the EM to calculate the weights of the eight indicators in order to eliminate the influence of subjective human factors on the study results. The calculation equations of the EM are as follows:

$$w_{ij} = \frac{r_{ij}^+}{\sum_{i=1}^n r_{ij}^+} \text{ or } w_{ij} = \frac{r_{ij}^-}{\sum_{i=1}^n r_{ij}^-} \tag{3}$$

$$e_j = -k\sum_{i=1}^n w_{ij} \ln w_{ij}, k = (\ln n)^{-1} \tag{4}$$

$$f_j = 1 - e_j \tag{5}$$

$$w_j = \frac{f_j}{\sum_{j=1}^m f_j} \tag{6}$$

where w_{ij} is the weight corresponding to the jth indicator of the ith city, e_j is the entropy value of the jth indicator, f_j is the redundancy coefficient of the jth indicator, and w_j is the weight corresponding to the jth indicator. In this study, i and j have values of 366 and 8, respectively.

3.2. Urbanization Assessment Based on DMSP

Cao's research [56,57] was referred to for the processing of the DMSP nighttime light data, including noise removal and correction. The correction process included the desaturation and continuous correction of different sensors (F15, F16, and F18). Figure 3 shows the change in the total nighttime light value (TNLV) before and after the DMSP nighttime light data correction of the different sensors. The corrected DMSP nighttime light data have good continuity.

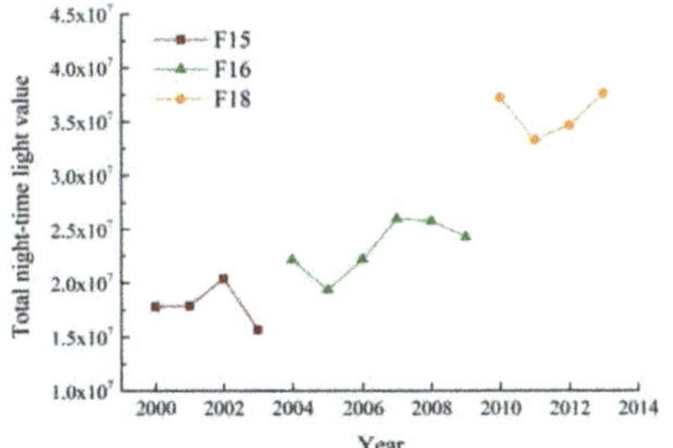

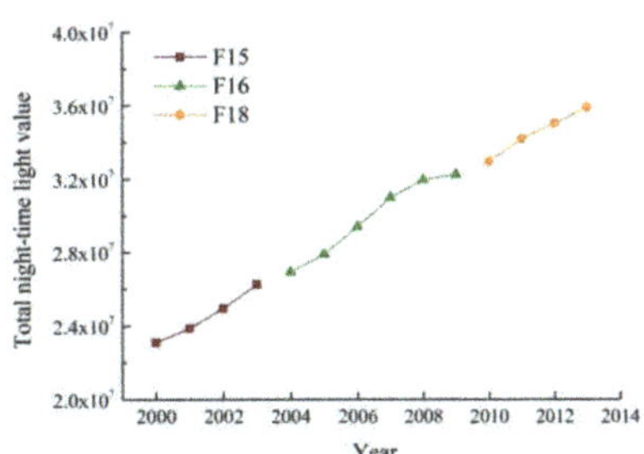

Figure 3. Total nighttime light value (TNLV) of noncorrected DMSP nighttime light data across China (**left**) and that of corrected DMSP nighttime light data (**right**).

3.3. Coupling Coordination Degree Model (CCDM)

The phenomenon in which two or more systems interact with each other is called coupling, and the coordination degree is the degree of harmony in the interaction between systems [58,59]. The CCDM was used to construct the coupling relationship between the EEQ and urbanization in China from 2000 to 2013. The coupling degree model (CDM) is expressed as

$$C = \left\{ \frac{U \times E}{[(U + E)/2]^2} \right\}^{\frac{1}{2}} \tag{7}$$

where C is the coupling degree; the larger C indicates a more coordinated relationship of the EEQ and urbanization; U is designated as urbanization; and E is M-RSEQI.

The coupling degree model is a representation of the coordination between the two subsystems. In fact, the size of the subsystem value does not affect the degree of coordination. Therefore, we introduced the CCDM based on the CDM, which avoids these issues. The formula of the CCDM is as follows:

$$D = \sqrt{C \times T}, \; T = \alpha U + \beta E \tag{8}$$

where D is the CCD and ranges from 0 to 1; T is the comprehensive assessment index; and α and β are undetermined coefficients, where $\alpha + \beta = 1$. In this study, α and β were both assumed to be 0.5.

The CCD was divided into three categories and six subcategories according to the classification criteria in Table 2.

Table 2. Classification standard of the CCD between the EEQ and urbanization based on the classification breakpoints.

CCD	Progression Stage	Comparison	Subcategories
$0 \leq D < 0.2$	Uncoordinated	U < E	Urbanization lags behind
		U > E	Eco-environment lags behind
$0.2 \leq D < 0.4$	Primary coordinated	U < E	Urbanization lags behind
		U > E	Eco-environment lags behind
$0.4 \leq D \leq 1$	Coordinated	U < E	Urbanization lags behind
		U > E	Eco-environment lags behind

3.4. Trend Analysis

This study used the univariate linear regression model to calculate the annual change trend of the M-RSEQI, urbanization, and the CCD [60–62]. The formula is as follows (the trend of M-RSEQI):

$$slope = \frac{n \times \sum\limits_{i=1}^{n} (i \times MRSEQI_i) - \sum\limits_{i=1}^{n} i \sum\limits_{i=1}^{n} MRSEQI_i}{n \times \sum\limits_{i=1}^{n} i^2 - \sum\limits_{i=1}^{n} i} \tag{9}$$

where *slope* is the slope of the linear regression equation between the *M-RSEQI* and year in China from 2000 to 2013, i is the time variable, n is the year and is set as 14 in this study, and $MRSEQI_i$ is the *M-RSEQI* of the *ith* year. When A < 0, it indicates that the EEQ is decreasing; when A > 0, it indicates that the EEQ is increasing. The absolute value represents the rate of slope change in the *M-RSEQI*.

4. Results

4.1. M-RSEQI

4.1.1. Rationality of M-RSEQI

The collinear diagnostic indicators in linear regression were used to explore the redundancy of the selected indicators in this study. The variance inflation factor (VIF) and tolerance (TOL) are reciprocal to each other and are commonly used indicators for collinearity diagnosis. When VIF < 10 or TOL > 0.1, it indicates that the selected assessment indicators are reasonable [63]. In this study, the pixel values of all eight indicators were extracted and analyzed for redundancy. The diagnosis results are shown in Table 3. The diagnosis results show that the eight indicators that were selected in this study are reasonable, and there is no information redundancy between them.

4.1.2. Spatiotemporal Changes in M-RSEQI

As shown in Figure 4a, the areas with obvious improvements in the EEQ were mainly located on the Loess Plateau and on the Greater Khingan Rangeg. The provinces with the most rapid growth trends in the M-RSEQI were Ningxia, Qinghai, and Shaanxi, which had rates of 1.295 $(10^{-3}/a)$, 1.146 $(10^{-3}/a)$, and 1.105 $(10^{-3}/a)$, respectively (Figure 4b).

Coincidently, the areas where the EEQ significantly decreased were mainly located in southern China, especially in the southwest region. Yunnan was the province with the fastest decreasing M-RSEQI trend among all of the provinces, with a trend of -1.311 (10^{-3}/a), followed by Taiwan and Fujian. The areas where the M-RSEQI remained relatively unchanged were located in the central and northern parts of Xinjiang, the central and western parts of Inner Mongolia, and the Hexi Corridor.

Table 3. Collinearity diagnosis results.

Indicator	VIF	TOL
PRE	8.097	0.123
TEMP	9.411	0.106
NPP	5.758	0.174
NDVI	5.922	0.169
DEM	2.975	0.336
LUCC	4.095	0.244
POP	1.349	0.741
PET	2.634	0.380

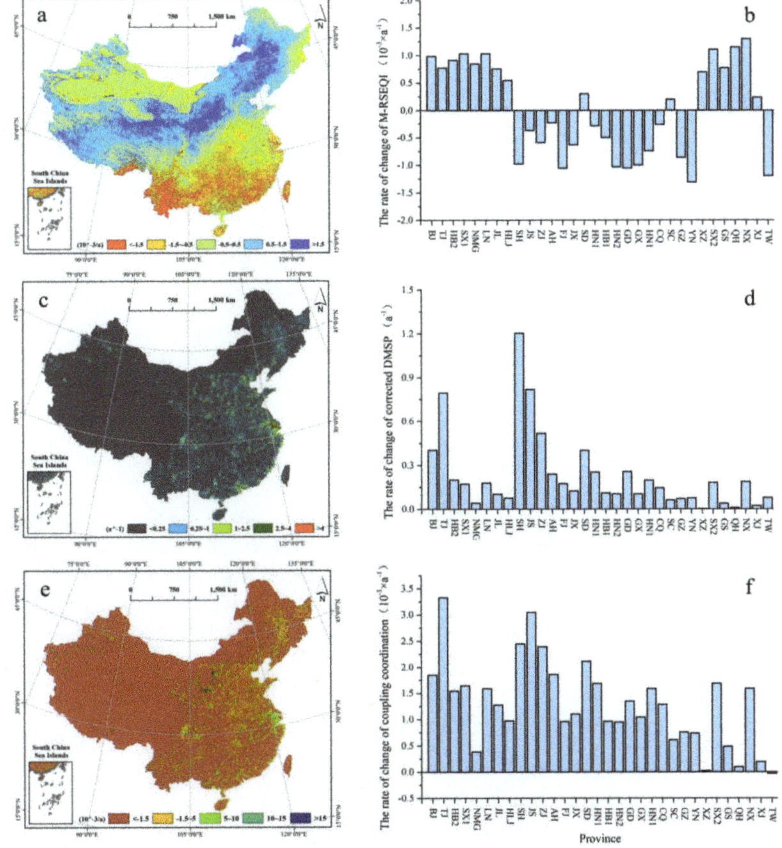

Figure 4. The change trend distribution of M-RSEQI (**a**), DMSP (**c**), and CCD (**e**) in China from 2000 to 2013 and the average trend value of M-RSEQI (**b**), DMSP (**d**), and CCD (**f**) for each province.

4.1.3. M-RSEQI in Different Ecosystems

Figure 5 shows that China's EEQ showed a fluctuating change but tended to remain stable in different ecosystems from 2000 to 2013. Among the six ecosystems, the M-RSEQI of the water area was the highest (Figure 5a), reaching 0.346 and indicating that the eco-environment of the water area was the best. The overall average value of the built-up ecosystem was the lowest (0.200), which indicates that the development of urbanization would inevitably affect the eco-environment. For all of the ecosystems, the M-RSEQI in 2007 was the highest among all of the years. We speculated that the reasons leading to the growth of China's M-RSEI in 2007 could be diversified. This could be related to the environmental protection policies that were formulated by the Chinese government to welcome the arrival of the Olympic Games. As shown in Figure 4a, the areas where the EEQ in China has improved are mainly concentrated in the Loess Plateau, and the effectiveness of China's environmental governance on the Loess Plateau in recent years is prominent.

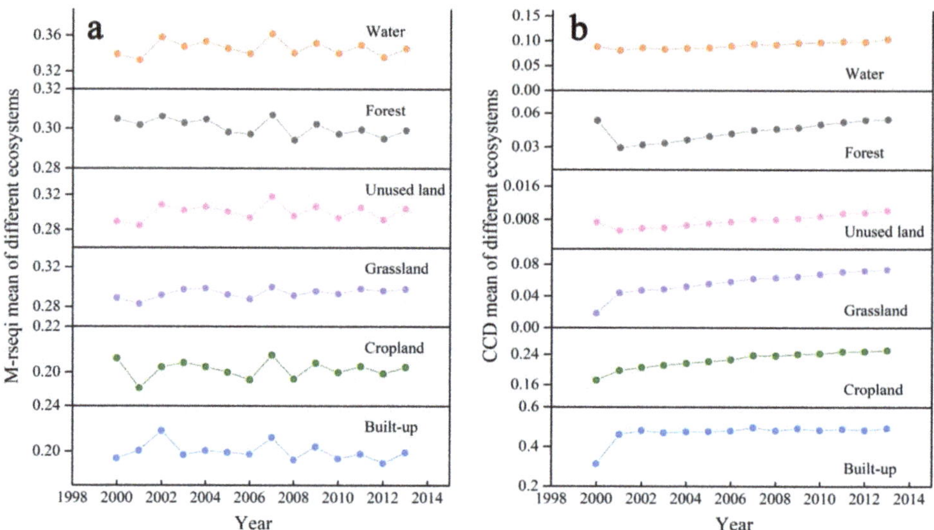

Figure 5. Annual changes in the (**a**) M-RSEQI and (**b**) CCD in China in different ecosystems from 2000 to 2013.

Figure 6a shows that the change in the M-RSEQI of the water ecosystem was the largest from 2000 to 2013, ranging from 0.333 to 0.362, which shows that the eco-environment of water in China greatly improved during this period. The M-RSEQI standard deviation for water was 0.135, which was the largest among all of the ecosystems (Figure 6, right). The interior of the water ecosystem is relatively complex and includes lakes, rivers, reservoirs, ponds, glaciers, and wetlands [64]. The standard deviations of the cropland and built-up areas were small, with values of 0.035 and 0.040, respectively, indicating that the EEQ was relatively stable.

4.2. Urbanization

4.2.1. Validity of Corrected DMSP

Corrected DMSP data were fitted with AHF [53] (anthropogenic heat flux) data to explore the validity of the corrected DMSP nighttime light data in this article (Figure 7). From the four fitting equations, it can be seen that the light intensity was positively correlated with AHF, and that R^2 was higher than 0.6, indicating a larger light intensity with a larger number of people, which was consistent with the actual situation. Therefore, the DMSP nighttime light data that were corrected in this study had high validity.

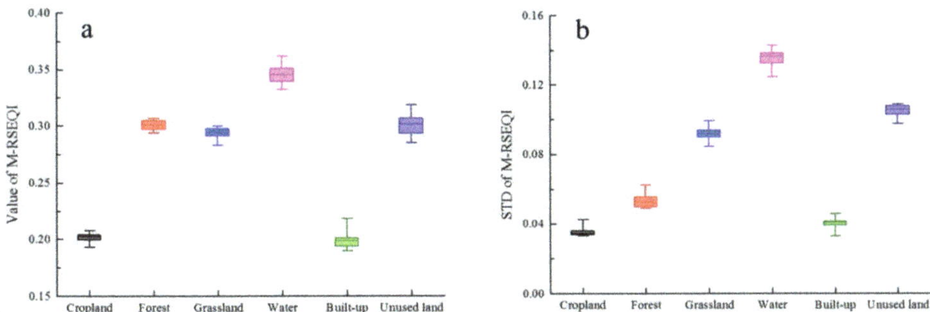

Figure 6. The mean (**a**) and standard deviation (**b**) of the M-RSEQI in different ecosystems in China from 2000 to 2013.

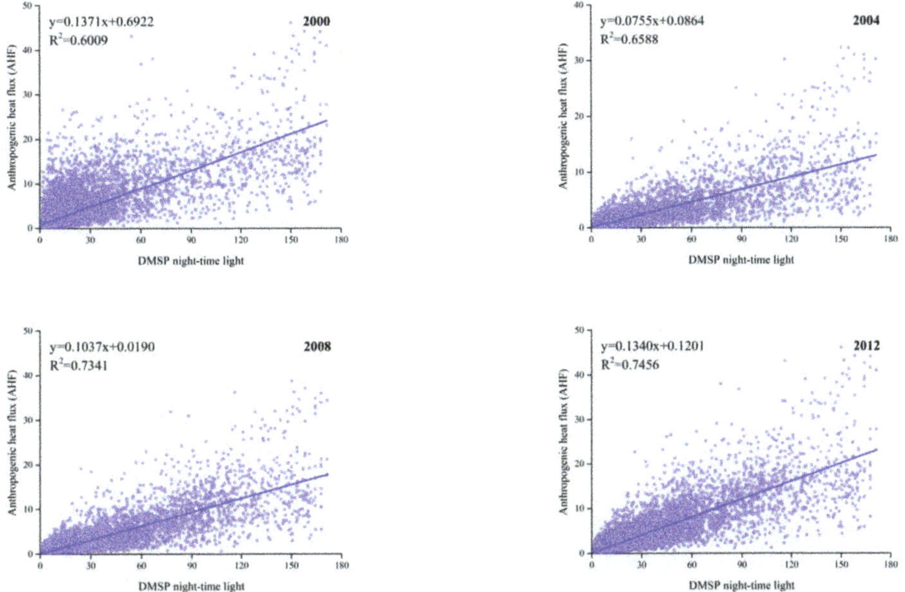

Figure 7. Scatter plot of corrected DMSP nighttime light data and AHF data in 2000, 2004, 2008, and 2012. The blue line represents the relationship obtained by fitting DMSP and AHF.

Figure 8 shows the corrected DMSP night light data from 2001 to 2013 in China. As seen in Figure 8, China's nighttime light showed a spatial distribution pattern of "high in the East and low in the west", and the nighttime light of China's three major urban agglomerations (Beijing Tianjin Hebei Urban Agglomeration, Yangtze River Delta Urban Agglomeration, and Pearl River Delta Urban Agglomeration) were the brightest. The dense road network in central and eastern China has greatly promoted the urbanization process in the region. As of 2019, China's central and eastern region is home to 97.59% of China's total population, and it contains 87.27% of China's GDP [65]. Secondly, from 2001 to 2013, the nighttime light in China was gradually brightened, which shows that China's urbanization has continued to develop over the past 14 years, which is consistent with the conclusions seen in Figure 7.

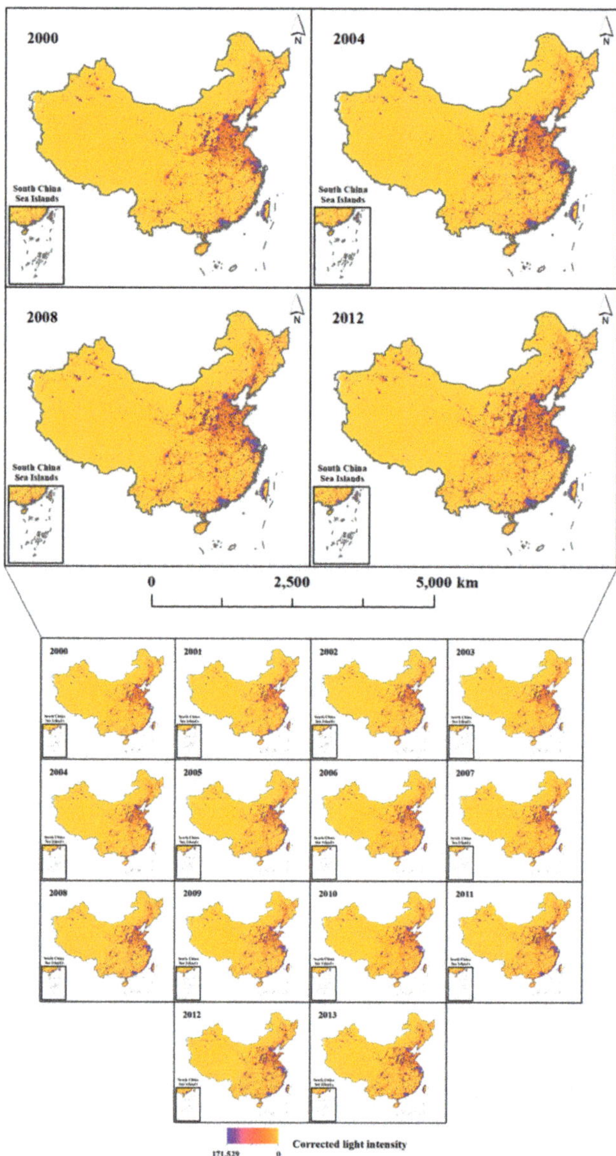

Figure 8. Spatial distribution map of corrected DMSP nighttime light data in China from 2000 to 2013.

4.2.2. Spatiotemporal Changes in Urbanization

As shown in Figure 4c, urbanization development in China from 2000 to 2013 shows the characteristics of "fast in the east and slow in the west". Among all of the provinces (including municipalities directly under the Central Government), Shanghai had the fastest urbanization development process, with a rate of 1.203 (10^{-3}/a), followed by Jiangsu (0.819) and Tianjin (0.794) (Figure 4d). During this period, the speed of urbanization in the Tibet Autonomous Region (0.002) was the slowest, followed by Qinghai (0.008) and Xinjiang (0.022), which was strongly related to geography. Poor living conditions and rare population centers hindered the urbanization development of these provinces. In contrast,

Shanghai and Tianjin, China's two major ports, are at the forefront of economic reform and opening up. With the support of national policies and advanced technology, urbanization has developed rapidly. Jiangsu Province, which is located on the Yangtze River Delta, is the only province where all of the prefecture-level cities have entered the top 100 in China.

4.3. CCD

4.3.1. Spatiotemporal Changes in CCD

Figure 9 shows the spatial distribution of CCD in China from 2001 to 2013. It can be seen that similar to DMSP nighttime light, China's CCD presented a spatial distribution pattern of "high in the east and low in the west", and the three major urban agglomerations had the highest CCD. Secondly, it can be seen from Figure 9 that China's CCD has gradually increased from 2001 to 2013, which is consistent with the conclusion in Section 4.2.1.

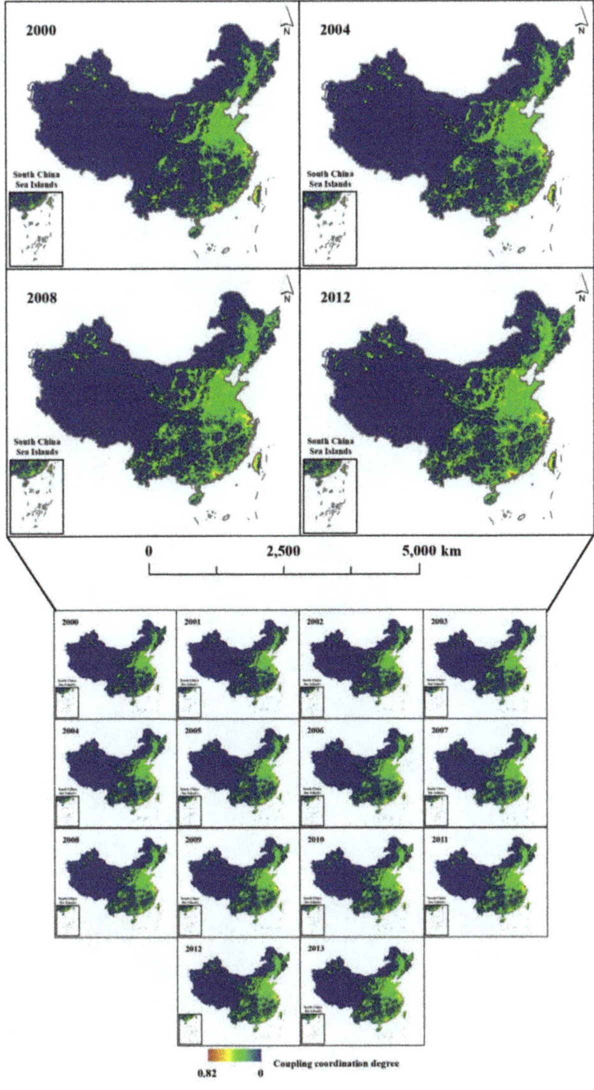

Figure 9. Spatial distribution map of CCD data in China from 2000 to 2013.

Figure 4e shows that the trends for the CCD between the EEQ and urbanization had the characteristic of "fast in the east and slow in the west". This is similar to the characteristics that were observed in the urbanization trends. This indicates that the CCD was greatly affected by urbanization development. Among all of the provinces, Tianjin had the fastest growth in CCD, with a trend of 3.326 (10^{-3}/a) (Figure 4f). This was closely related to the industrial structure adjustment policy of Tianjin. As shown in Figure 5, from 2000 to 2013, the M-RSEQI of Tianjin increased significantly, and it was also one of the provinces with the fastest trend of urbanization development. Among all of provinces, Taiwan Province showed a downward trend in CCD changes, which was likely caused by the continuous decline in its urbanization rate. The main reason was that from 2000 to 2013, the urbanization of Taiwan Province was in a stagnation state, while its EEQ showed a downward trend (Figure 4b).

As shown in Figure 10, the CCD of provincial capital cities, including municipalities directly under the central government, increased significantly from 2000 to 2013. Although Figure 10 presents that the CCD of almost all cities has shown an increasing trend, it also reflects that there are significant spatial differences in their changes. Among them, the CCD of cities in central and eastern China, represented by Beijing, Shanghai, Guangzhou, and Hangzhou, have grown rapidly, while the cities of western China, represented by Lhasa, Urumqi, and Xining, have experienced slower growth, which is consistent with the conclusion obtained from Figure 4e. Secondly, we also found an interesting phenomenon. Taiwan Province, as a typical eastern coastal area of China, demonstrates rare CCD growth. Based on the above conclusion that rapid urbanization has promoted regional CCD growth, we speculate that the main reason is the slow development of urbanization in Taiwan Province in recent years, which has also inhibited Taiwan's CCD growth.

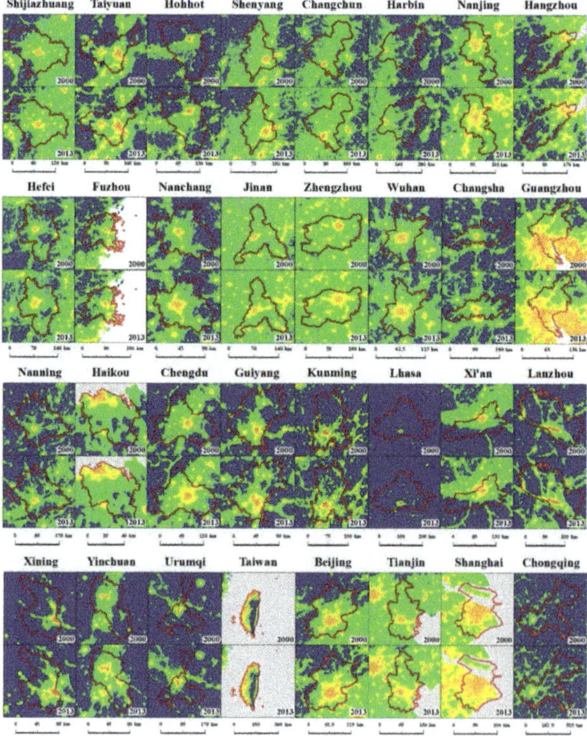

Figure 10. CCD of 28 provincial capital cities and 4 municipalities directly under the central government in 2000 and 2013.

According to the grading standards in Table 2, the CCD of all of the cities in China from 2000 to 2013 were classified. Figure 11 shows that cities with higher CCD in China were mainly distributed in the North China Plain and in the southeastern coastal areas. From 2000 to 2013, most cities in China were in an uncoordinated state, which was concentrated in the central and western regions of China. In addition, there were almost no uncoordinated or primary coordinated cities with EEQ that lagged behind. This indicates that there was a large imbalance in the spatial distribution of urbanization development, which extended to most of China's uncoordinated cities from 2000 to 2013.

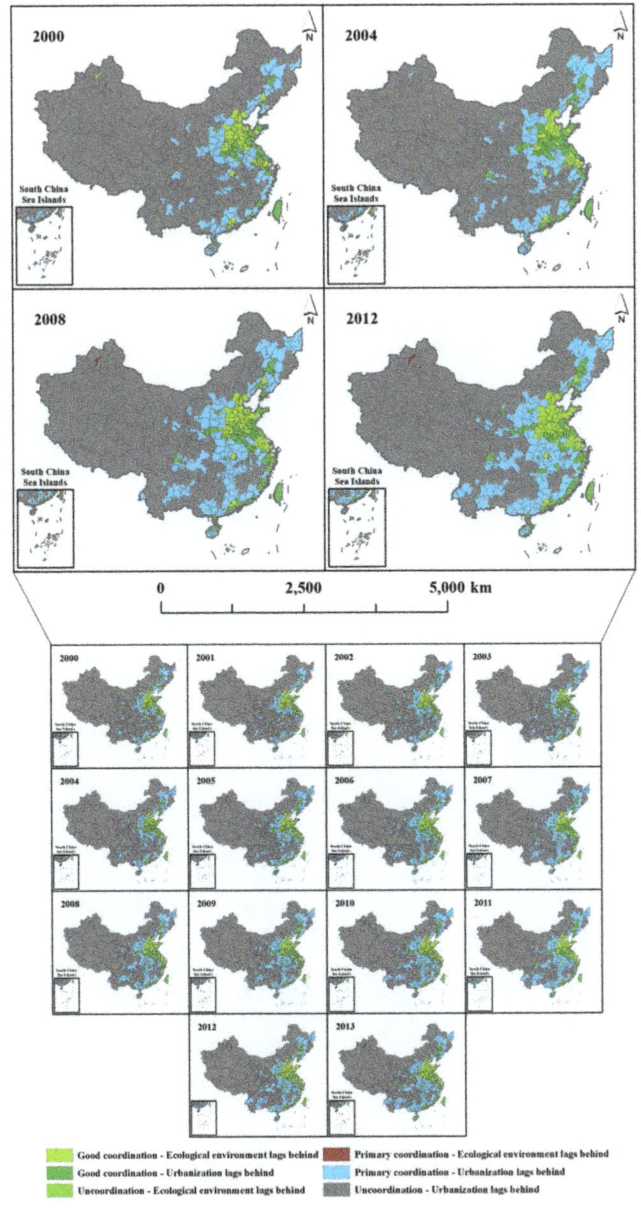

Figure 11. China's CCD classification map from 2000 to 2013.

Figure 12 shows that the number of uncoordinated cities decreased each year, while the number of primary coordinated and coordinated cities increased. From 2000 to 2013 the number of coordinated cities increased from 83 to 125. Simultaneously, the number of uncoordinated cities decreased from 181 to 110. With the continuous development of China's urbanization, the pace of urbanization gradually matched the EEQ, while the CCD also constantly improved. Since 2000, the number of Subtype VI (uncoordinated urbanization lags behind) cities was the largest, whilst the number of Subtype IV (primary coordinated: urbanization lags behind) b cities became the largest in 2013. This shows that the level of urbanization in China was lower than the EEQ from 2000 to 2013. In summary, the coupling law of the EEQ and urbanization in China was "uncoordinated to primary coordinated. primary coordinated to coordinated, and with the characteristic of "urbanization lags behind EEQ" evolving into "EEQ lags behind urbanization". In the future, the CCD of cities in China will go through the process of "Type III to Type II to Type I". When the CCD between the EEQ and urbanization of most cities in China reaches Type I, the entire CCD development process will be in a stable or circular development stage.

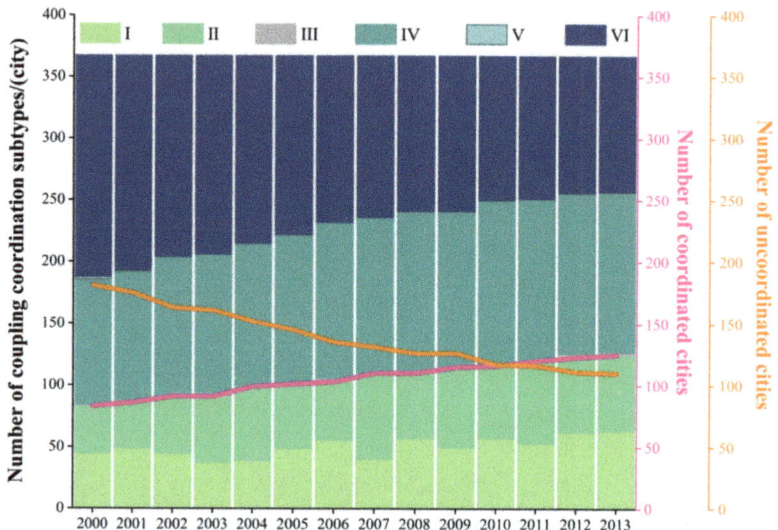

Figure 12. The statistical graph of the number of cities with different CCD subtypes. (I) Coordinated: EEQ lags behind. (II) Coordinated: urbanization lags behind. (III) Primary coordinated: EEQ lags behind. (IV) Primary coordinated: urbanization lags behind. (V) Uncoordinated: EEQ lags behind. (VI) Uncoordinated: urbanization lags behind.

4.3.2. Change Characteristics of the CCD in Different Ecosystems

Figure 5b shows that built-up ecosystems (0.470) had the highest CCD among the six ecosystems. The CCD value of unused land (0.008) was the lowest. However, the CCD values of forest, grassland, and water that had higher EEQ levels were not higher, which indicates that the decisive factor of the CCD came from urbanization. This result is consistent with the conclusion in Section 4.3.1. In addition, the CCD values of the six ecosystems maintained a steady increasing trend from 2000 to 2013, which was closely related to the continuous development of urbanization in China during this period.

5. Discussion

5.1. Coupling Mechanism Analysis

The above analysis of the results shows that at this stage, the development of urbanization has a decisive influence on the CCD. To explain this phenomenon, the coupling mechanism between China's urbanization and EEQ was explored using a combination of mathematics and geometry. The change rates of the tangent plane in the two dimensions of urbanization and M-RSEQI were used to characterize the degree of influence of urbanization and the EEQ in terms of coupling and the coordination degree. The tangent plane is calculated by

$$CCD - CCD_0 = F'_U(U - U_0) + F'_E(E - E_0) \tag{10}$$

where CCD is the coupling coordination degree; F'_U and F'_E are the partial derivatives of urbanization and RSEI-2, respectively; CCD_0, U_0, and E_0 are the mean values of the CCD, urbanization, and EEQ in 2013, which are 0.445, 0.026, and 0.216, respectively.

Figure 13a shows the spatial distribution of the CCD trends in China from 2001 to 2013. The blue and red lines represent the average trend values of nighttime light and CCD in latitude and longitude, respectively. We can see that at the scale of latitude and longitude, the trends of CCD and nighttime light have a high degree of consistency in terms of the spatial distribution, which shows that the changes in urbanization have profoundly affected the changes in CCD. Figure 13b,c show the correlation between the average trend value of CCD and nighttime light in latitude and longitude, respectively. It can be seen that there is a significant correlation ($p < 0.001$) between the two values at the latitude and longitude scale.

Urban areas are often used to evaluate a region's level of urbanization [66,67]. Therefore, we extracted the land use and land cover (LULC) [68] data for urban areas in China in 2001 and 2013, calculated the urban sprawl index (USI) of all cities in China, divided all of the cities into five categories according to the classification standard in Figure 13f, and finally counted the annual CCD values of these five types of cities and the 371 cities in China. It can be seen from Figure 13f that the CCD value of most cities showed a continuous upward trend (gray broken line). Most importantly, we found that the city with the highest USI had the highest CCD growth rate, reaching 0.494 (10^{-3}/a). On the contrary, the city with the slowest USI had the lowest CCD growth rate, which was 0.183 (10^{-3}/a). This also confirms the abovementioned speculation that urbanization has promoted the development of CCD.

Figure 13d,e show that in 2013, the growth rate of the CCD in the urbanization component was 2.659, while the increase in the M-RSEQI component was only 0.316, indicating that in 2013, the degree of impact that urbanization had on the CCD was approximately 8.4 times larger than that of the EEQ. Therefore, this explains that in 2013, the decisive factor that was affecting the CCD was the development of urbanization.

5.2. Limitations and Prospects

Although this study has made some progress compared to previous studies and has drawn some meaningful conclusions, this study also has some limitations. First, it should be noted that although this study and previous studies [69] show that using nighttime light data as an indicator of urbanization is theoretically reasonable, it cannot be ignored that nighttime light only has certain advantages in characterizing the intensity of regional night human activities and cannot provide direct evidence of daytime urbanization intensity. Because of this situation, people mainly work in urban areas during the day and return to the suburbs to live and rest at night, which will lead to a certain difference between the level of urbanization that is assessed by nighttime light data and the actual level of urbanization. Secondly, since NASA suspended DMSP night light data updates in 2013, the main limitation of this study is that it is temporarily unable to confirm our theoretical methods and conclusions using the latest data. In the future, we will work to correct DMSP and VIIRS nighttime light data to help us carry out longer-term serial studies on this topic.

In addition, there are great differences in the EEQ (ecological footprint) between urbans areas and the suburbs. The EEQ of urban areas will generally be lower than that of suburbs because the higher population density in urban areas will produce higher anthropogenic heat emissions and other pollutants. However, we believe that the level of "sustainable development" comprises the integration of EEQ and urbanization because the goal of "sustainable development" is benign social and economic development, i.e., we need to pay attention to the protection of the ecological environment while pursuing social and economic development. Therefore, as shown by the CCDM, the level of "sustainable development" depends on the coupling coordination level of EEQ and urbanization.

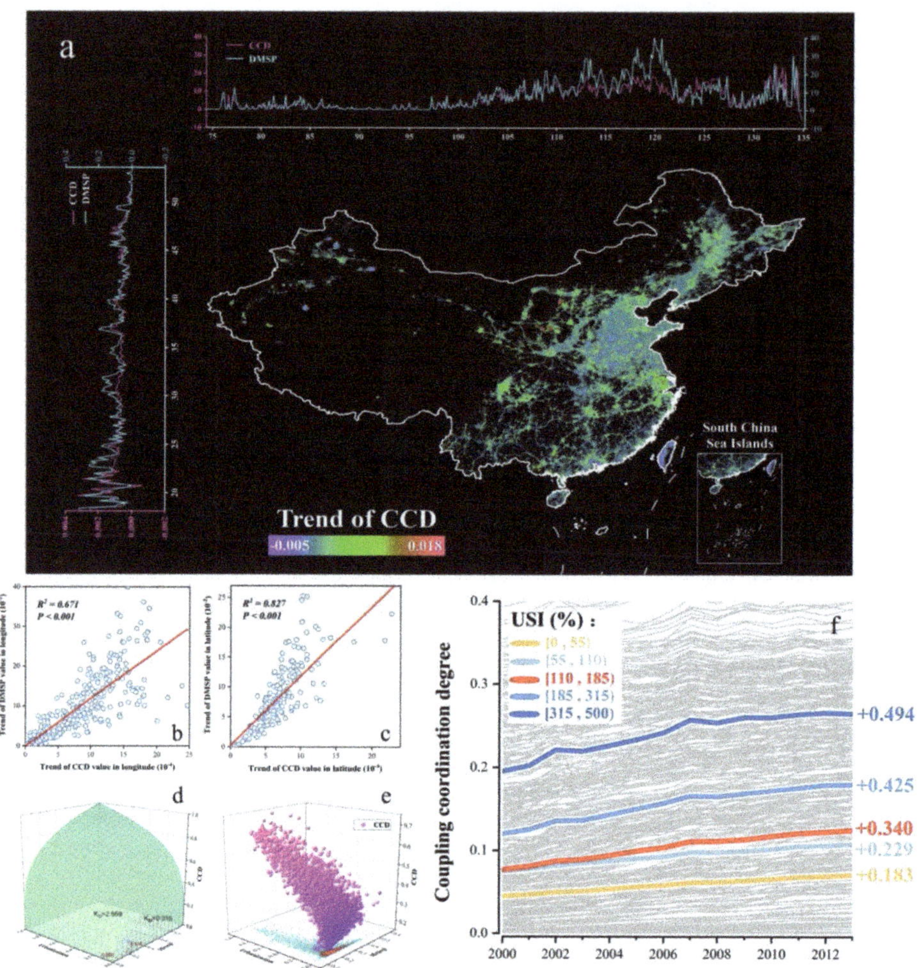

Figure 13. 3D scatter diagram of M-RSEQI, urbanization, and CCD in 2013 and 3D function diagram of the CCD model. (**a**) shows the spatial distribution of the CCD trends in China from 2001 to 2013; (**b,c**) show the correlation between the average trend value of CCD and nighttime light in latitude and longitude; (**d,e**) show that in 2013, the growth rate of the CCD in the urbanization component was 2.659; (**f**) the CCD value of most cities showed a continuous upward trend (gray broken line).

6. Conclusions

China's rapid urbanization has led to serious eco-environmental problems. With regard to the increasingly severe eco-environmental situation, it is important to understand the relationship between China's EEQ and urbanization for the realization of the 2030 agenda (Transforming Our World: The 2030 Agenda for Sustainable Development) [21]. Therefore, based on multisource remote sensing data and EM, we designed a spatiotemporal universal EEQ assessment system and used the CCDM in physics to construct the coupling relationship between the EEQ and urbanization in China from 2000 to 2013. Through this study, some meaningful conclusions are drawn:

1. The decisive factor affecting the CCD was urbanization development in 2013. The impact that the degree of urbanization had on the CCD was approximately 8.4 times higher than that of the EEQ.
2. From 2000 to 2013, the urbanization development of China showed the characteristics of "fast in the east and slow in the west" over the course of the past 14 years. The CCD between the EEQ and urbanization in China showed the characteristic of "strong in the east, weak in the west".
3. Most of China's cities were in an uncoordinated state and were concentrated in the central and western regions of China. The coupling pattern of EEQ and urbanization in China evolved from "uncoordinated cities into coordinated cities, with the characteristic of "urbanization lags behind EEQ" evolving into "EEQ lags behind urbanization".

Author Contributions: Conceptualization, D.X. and H.L.; methodology, D.X. and J.C.; software, S.X. and J.G.; validation, J.C. and H.F.; formal analysis, D.X. and H.F.; investigation, D.X.; resources, Y.W., J.X. and S.W.; data curation, D.X.; writing—original draft preparation, D.X.; writing—review and editing, H.L., F.Y. and R.Z.; visualization, S.X.; supervision, J.C.; project administration, M.L. and J.H.; funding acquisition, H.F. and J.G. All authors have read and agreed to the published version of the manuscript.

Funding: This research was funded by the Youth Program of National Natural Science Foundation of Zhejiang Province grant number LQ21D060001, Hainan Provincial Natural Science Foundation of China grant number 420RC673, 2021 Science and Technology Plan Project of Zhejiang Meteorological Bureau grant number 2021YB07, Fengyun Application Pioneering Project grant number FY-APP-2021.0105, and the National Natural Science Foundation of China grant number 42176180. And The APC was funded by the Youth Program of National Natural Science Foundation of Zhejiang Province grant number LQ21D060001.

Institutional Review Board Statement: Not applicable.

Informed Consent Statement: Not applicable.

Data Availability Statement: Not applicable.

Acknowledgments: This work is supported by the Youth Program of National Natural Science Foundation of Zhejiang Province (LQ21D060001), Hainan Provincial Natural Science Foundation of China grant number 420RC673, 2021 Science and Technology Plan Project of Zhejiang Meteorological Bureau (2021YB07), Fengyun Application Pioneering Project grant number FY-APP-2021.0105, and the National Natural Science Foundation of China grant number 42176180.

Conflicts of Interest: The authors declare no conflict of interest.

References

1. Shan, G.; Xianjin, H. Performance Evaluation of Eco-construction Based on PSR Model in China from 1953 to 2008. *J. Nat. Resour.* **2010**, *25*, 341–350.
2. Sun, D.; Zhang, J.; Zhu, C. An Assessment of China's Ecological Environment Quality Change and Its Spatial Variation. *Acta Geogr. Sin.* **2012**, *67*, 1599–1610.
3. McKinney, M.L. Effects of urbanization on species richness: A review of plants and animals. *Urban Ecosyst.* **2008**, *11*, 161–176. [CrossRef]
4. Xiuxing, Y. An Assessment of China's Ecological Environment Quality Change and the Spatial Variation. *Environ. Life* **2014**, *9X*, 249.
5. Wei, Y.L.; Bao, L.J.; Wu, C.C.; He, Z. Assessing the effects of urbanization on the environment with soil legacy and current-use insecticides: A case study in the Pearl River Delta, China. *Sci. Total Environ.* **2015**, *514*, 409–417. [CrossRef]
6. Liu, X.; Ma, L.; Li, X.; Ai, B.; Li, S.; He, Z. Simulating urban growth by integrating landscape expansion index (LEI) and cellular automata. *Int. J. Geogr. Inf. Sci.* **2014**, *28*, 148–163. [CrossRef]
7. Zhu, J. The 2030 Agenda for Sustainable Development and China's implementation. *Chin. J. Popul. Resour. Environ.* **2016**, *15*, 142–146. [CrossRef]

8. Wang, S.-X.; Yao, Y.; Zhou, Y. Analysis of Ecological Quality of the Environment and Influencing Factors in China during 2005–2010. *Int. J. Environ. Res. Public Health* **2014**, *11*, 1673–1693. [CrossRef]
9. Du, W. *Research on the Evaluation Index of Air Pollution Control Audit Based on PSR Model*; IOP Conference Series: Earth and Environmental Science; IOP Publishing: Bristol, UK, 2020; Volume 514.
10. McDonald Michael, E. *EMAP Overview: Objectives, Approaches, and Achievements*; Monitoring Ecological Condition in the Western United States; Springer: Dordrecht, The Netherlands, 2000; pp. 3–8.
11. Smith, E.R. An Overview of EPA's Regional Vulnerability Assessment (ReVA) Program. *Environ. Monit. Assess.* **2000**, *64*, 9–15 [CrossRef]
12. US Forest Service. Forest Service Resource Inventories: An Overview. 1992. Available online: https://www.fs.usda.gov/treesearch/pubs/123 (accessed on 23 December 2021).
13. Ahmed, Z.; Wang, Z.; Mahmood, F.; Hafeez, M.; Ali, N. Does globalization increase the ecological footprint? Empirical evidence from Malaysia. *Environ. Sci. Pollut. Res.* **2019**, *26*, 18565–18582. [CrossRef]
14. Shah, S.A.A.; Zhou, P.; Walasai, G.; Mohsin, M. Energy security and environmental sustainability index of South Asian countries: A composite index approach. *Ecol. Indic.* **2019**, *106*, 105507. [CrossRef]
15. Xu, C.; Wang, S. A comprehensive quantitative evaluation of new sustainable urbanization level in 20 Chinese urban agglomerations. *Sustainability* **2016**, *8*, 91. [CrossRef]
16. Xu, D.; Hou, G. The Spatiotemporal Coupling Characteristics of Regional Urbanization and Its Influencing Factors: Taking the Yangtze River Delta as an Example. *Sustainability* **2019**, *11*, 822. [CrossRef]
17. Zhang, W.; Wang, M. Spatial-temporal characteristics and determinants of land urbanization quality in China: Evidence from 285 prefecture-level cities. *Sustain. Cities Soc.* **2018**, *38*, 70–79. [CrossRef]
18. Elmqvist, T.; Fragkias, M.; Goodness, J.; Güneralp, B.; Marcotullio, P.J.; McDonald, R.I.; Parnell, S.; Schewenius, M.; Sendstad, M.; Seto, K.C.; et al. *Urbanization, Biodiversity and Ecosystem Services: Challenges and Opportunities: A Global Assessment*; Springer Nature: Dordrecht, The Netherlands, 2013.
19. Shao, Z.; Ding, L.; Li, D.; Altan, O.; Huq, E.; Li, C. Exploring the Relationship between Urbanization and Ecological Environment Using Remote Sensing Images and Statistical Data: A Case Study in the Yangtze River Delta, China. *Sustainability* **2020**, *12*, 5620. [CrossRef]
20. Chen, W.; Zhang, Y.; Pengwang, C.; Gao, W. Evaluation of Urbanization Dynamics and its Impacts on Surface Heat Islands: A Case Study of Beijing, China. *Remote Sens.* **2017**, *9*, 453. [CrossRef]
21. Estoque, R.C. A review of the sustainability concept and the state of SDG monitoring using remote sensing. *Remote Sens.* **2020**, *12*, 1770. [CrossRef]
22. Xu, H. Remote Sensing Evaluation Index of Regional Ecological Environment Changes. *China Environ. Sci.* **2013**, *33*, 889–897.
23. Shan, W.; Jin, X.; Ren, J.; Wang, Y.; Xu, Z.; Fan, Y.; Gu, Z.; Hong, C.; Lin, J.; Zhou, Y. Ecological environment quality assessment based on remote sensing data for land consolidation. *J. Clean. Prod.* **2019**, *239*, 118126. [CrossRef]
24. Guo, B.; Yelin, F.; Xiaobin, J. Monitoring the effects of land consolidation on the ecological environmental quality based on remote sensing: A case study of Chaohu Lake Basin, China. *Land Use Policy* **2020**, *95*, 104569. [CrossRef]
25. Xu, H.; Wang, M.; Shi, T.; Guan, H.; Fang, C.; Lin, Z. Prediction of ecological effects of potential population and impervious surface increases using a remote sensing based ecological index (RSEI). *Ecol. Indic.* **2018**, *93*, 730–740. [CrossRef]
26. Liao, W.; Jiang, W. Evaluation of the Spatiotemporal Variations in the EEQ in China Based on the Remote Sensing Ecological Index. *Remote Sens.* **2020**, *12*, 2462. [CrossRef]
27. Li, K.; Chen, Y. A Genetic Algorithm-based urban cluster automatic threshold method by combining VIIRS DNB, NDVI, and NDBI to monitor urbanization. *Remote Sens.* **2018**, *10*, 277. [CrossRef]
28. Zhang, Q.; Seto, K.C. Mapping urbanization dynamics at regional and global scales using multi-temporal DMSP/OLS nighttime light data. *Remote Sens. Environ.* **2011**, *115*, 2320–2329. [CrossRef]
29. Ma, T.; Zhou, Y.; Zhou, C.; Haynie, S.; Pei, T.; Xu, T. Night-time light derived estimation of spatio-temporal characteristics of urbanization dynamics using DMSP/OLS satellite data. *Remote Sens. Environ.* **2015**, *158*, 453–464. [CrossRef]
30. Zheng, Z.; Wu, Z.; Chen, Y.; Yang, Z.; Marinello, F. Exploration of eco-environment and urbanization changes in coastal zones: A case study in China over the past 20 years. *Ecol. Indic.* **2020**, *119*, 106847. [CrossRef]
31. Chen, J.; Zhuo, L.; Shi, P.J. The study on urbanization process in China based on DMSP/OLS data: Development of a light index for urbanization level estimation. *J. Remote Sens.* **2003**, *7*, 168–175.
32. Fang, C.; Wang, S.; Li, G. Changing urban forms and carbon dioxide emissions in China: A case study of 30 provincial capital cities. *Appl. Energy* **2015**, *158*, 519–531. [CrossRef]
33. Ozokcu, S.; Özdemir, O. Economic growth, energy, and environmental Kuznets curve. *Renew. Sustain. Energy Rev.* **2017**, *72*, 639–647. [CrossRef]
34. Fanning Andrew, L.; O'Neill, D.W.; Büchs, M. Provisioning systems for a good life within planetary boundaries. *Glob. Environ. Change* **2020**, *64*, 102135. [CrossRef]
35. O'Neill, D.W.; Fanning, A.L.; Lamb, W.F.; Steinberger, J.K. A good life for all within planetary boundaries. *Nat. Sustain.* **2018**, *1*, 88–95. [CrossRef]

36. Fang, C.; Ren, Y. Analysis of emergy-based metabolic efficiency and environmental pressure on the local coupling and telecoupling between urbanization and the eco-environment in the Beijing-Tianjin-Hebei urban agglomeration. *Sci. China Earth Sci.* **2017**, *60*, 1083–1097. [CrossRef]
37. Andrea, L.; Newig, J.; Challies, E. Globalization's limits to the environmental state? Integrating telecoupling into global environmental governance. *Environ. Politics* **2016**, *25*, 136–159.
38. Yi, Y.; Meng, G. A bibliometric analysis of comparative research on the evolution of international and Chinese ecological footprint research hotspots and frontiers since 2000. *Ecol. Indic.* **2019**, *102*, 650–665.
39. Rashid, A.; Irum, A.; Malik, I.A.; Ashraf, A.; Rongqiong, L.; Liu, G.; Ullah, H.; Ali, M.U.; Yousaf, B. Ecological footprint of Rawalpindi; Pakistan's first footprint analysis from urbanization perspective. *J. Clean. Prod.* **2018**, *170*, 362–368. [CrossRef]
40. Beloin-Saint-Pierre, D.; Rugani, B.; Lasvaux, S.; Mailhac, A.; Popovici, E.; Sibiude, G.; Benetto, E.; Schiopu, N. A review of urban metabolism studies to identify key methodological choices for future harmonization and implementation. *J. Clean. Prod.* **2017**, *163*, S223–S240. [CrossRef]
41. Wu, Y.; Que, W.; Liu, Y.-G.; Li, J.; Cao, L.; Liu, S.-B.; Zeng, G.-M.; Zhang, J. Efficiency estimation of urban metabolism via Emergy, DEA of time-series. *Ecol. Indic.* **2018**, *85*, 276–284. [CrossRef]
42. Yang, S.; Cao, D.; Lo, K. Analyzing and optimizing the impact of economic restructuring on Shanghai's carbon emissions using STIRPAT and NSGA-II. *Sustain. Cities Soc.* **2018**, *40*, 44–53. [CrossRef]
43. Song, Q.; Zhou, N.; Liu, T.; Siehr, S.A.; Qi, Y. Investigation of a "coupling model" of coordination between low-carbon development and urbanization in China. *Energy Policy* **2018**, *121*, 346–354. [CrossRef]
44. Lowe, R.; Wu, Y.; Tamar, A.; Harb, J.; Abbeel, P.; Mordatch, I. Multi-agent actor-critic for mixed cooperative-competitive environments. Advances in neural information processing systems. *arXiv* **2017**, arXiv:1706.02275.
45. Fang, C.; Liu, H.; Li, G. International progress and evaluation on interactive coupling effects between urbanization and the eco-environment. *J. Geogr. Sci.* **2016**, *26*, 1081–1116. [CrossRef]
46. Zhao, Y.; Wang, S.; Zhou, C. Understanding the relation between urbanization and the eco-environment in China's Yangtze River Delta using an improved EKC model and coupling analysis. *Sci. Total Environ.* **2016**, *571*, 862–875. [CrossRef] [PubMed]
47. Li, Y.; Li, Y.; Zhou, Y.; Shi, Y.; Zhu, X. Investigation of a coupling model of coordination between urbanization and the environment. *J. Environ. Manag.* **2012**, *98*, 127–133. [CrossRef] [PubMed]
48. Wang, S.J.; Ma, H.; Zhao, Y.B. Exploring the relationship between urbanization and the eco-environment—A case study of Beijing–Tianjin–Hebei region. *Ecol. Indic.* **2014**, *45*, 171–183. [CrossRef]
49. Li, W.; Yi, P. Assessment of city sustainability—Coupling coordinated development among economy, society and environment. *J. Clean. Prod.* **2020**, *256*, 120453. [CrossRef]
50. Qu, B.; Zhang, Y.; Kang, S.; Sillanpää, M. Water quality in the Tibetan Plateau: Major ions and trace elements in rivers of the "Water Tower of Asia". *Sci. Total Environ.* **2019**, *649*, 571–581. [CrossRef]
51. Fan, Y.; Fang, C.; Zhang, Q. Coupling coordinated development between social economy and ecological environment in Chinese provincial capital cities-assessment and policy implications. *J. Clean. Prod.* **2019**, *229*, 289–298. [CrossRef]
52. Peng, S.; Ding, Y.; Liu, W.; Li, Z. 1 km monthly temperature and precipitation dataset for China from 1901 to 2017. *Earth Syst. Sci. Data* **2019**, *11*, 1931–1946. [CrossRef]
53. Wang, S.; Hu, D.; Yu, C.; Chen, S.; Di, Y. Mapping China's time-series anthropogenic heat flux with inventory method and multi-source remotely sensed data. *Sci. Total Environ.* **2020**, *734*, 139457. [CrossRef]
54. Zhao, G.; Chen, L.; Mu, J. Discussion on Construction of Ecological Environment Quality Evaluation System. *Meteorol. Environ. Sci.* **2018**, *2018*, 1–11.
55. Lu, Y.; Xiang, P. Analysis of the Coupling Relationship between Ecological Environment and Urbanization:Chang-Zhu-Tan Urban Agglomeration in Hunan Province, China as a Case. *Urban Dev. Res.* **2020**, *1*, 19.
56. Cao, Z.; Wu, Z.; Kuang, Y.; Huang, N. Correction of DMSP/OLS Night-time Light Images and Its Application in China. *J. Earth Inf. Sci.* **2015**, *17*, 1092–1102.
57. Cao, Z.; Wu, Z.; Mi, S.; Yang, K. A Method for Classified Correction of Stable DMSP/OLS night-time Light Imagery across China. *J. Earth Inf. Sci.* **2020**, *22*, 246–257.
58. Liao, L.; Dai, W.; Huang, F.H.; Hu, Q. Coupling Coordination Analysis of Urbanization and Eco-environment System in Jinjiang Using Landsat Series Data and DMSP/OLS Nighttime Light Data. *J. Fujian Norm. Univ. Nat. Sci. Ed.* **2018**, *6*, 16.
59. Yajie, Z.; Huizhi, L. Spatial-Temporal Coupled Coordination between Urbanization and Ecological Environment in Yangtze River Economic Belt. *Bull. Soil Water Conserv.* **2018**, *37*, 334–340.
60. Li, S.; Yan, J.; Wan, J. The Spatial-temporal Changes of Vegetation Restoration on Loess Plateau in Shaanxi-Gansu-Ningxia Region. *Acta Geogr. Sin.* **2012**, *67*, 960–970.
61. Chen, J.; Jia, W.; Zhao, Z.; Zhang, Y.; Liu, Y. Research on Temporal and Spatial Variation Characteristics of Vegetation Cover of Qilian Mountains from 1982 to 2006. *Adv. Earth Sci.* **2015**, *30*, 834–845.
62. Fensholt, R.; Proud, S.R. Evaluation of earth observation based global long term vegetation trends—Comparing GIMMS and MODIS global NDVI time series. *Remote Sens. Environ.* **2012**, *119*, 131–147. [CrossRef]
63. Yuan, S.; Huichun, S.; Minhui, X.; Pengxia, Z. Spatiotemporal evolution pattern and influencing factors of EEQ in Gansu from 2000 to 2017. *J. Ecol.* **2019**, *38*, 3800–3808.

64. Zhou, Z.; Wang, X.; Ding, Z.; Chen, Y.; Wang, C. Remote sensing analysis of ecological quality change in Xinjiang. *J. Ecol.* **2020**, *40*, 2907–2919.
65. Xu, D.; Yang, F.; Yu, L.; Zhou, Y.; Li, H.; Ma, J.; Huang, J.; Wei, J.; Xu, Y.; Zhang, C.; et al. Quantization of the coupling mechanism between eco-environmental quality and urbanization from multisource remote sensing data. *J. Clean. Prod.* **2021**, *321*, 128948. [CrossRef]
66. Dewan, A.M.; Yamaguchi, Y. Land use and land cover change in Greater Dhaka, Bangladesh: Using remote sensing to promote sustainable urbanization. *Appl. Geogr.* **2009**, *29*, 390–401. [CrossRef]
67. Taubenböck, H.; Wegmann, M.; Roth, A.; Mehl, H.; Dech, S. Urbanization in India–Spatiotemporal analysis using remote sensing data [J]. *Comput. Environ. Urban Syst.* **2009**, *33*, 179–188. [CrossRef]
68. Xu, Y.; Yu, L.; Peng, D.; Zhao, J.; Cheng, Y.; Liu, X.; Li, W.; Meng, R.; Xu, X.; Gong, P. Annual 30-m land use/land cover maps of China for 1980–2015 from the integration of AVHRR, MODIS and Landsat data using the BFAST algorithm. *Sci. China Earth Sci.* **2020**, *63*, 1390–1407. [CrossRef]
69. Kyba, C.; Kuester, T.; de Miguel, A.S.; Baugh, K.; Jechow, A.; Hölker, F.; Bennie, J.; Elvidge, C.; Gaston, K.; Guanter, L. Artificially lit surface of Earth at night increasing in radiance and extent. *Sci. Adv.* **2017**, *3*, e1701528. [CrossRef]

 *remote sensing*

Article

Assessing Spatiotemporal Changes of SDG Indicators at the Neighborhood Level in Guilin, China: A Geospatial Big Data Approach

Liying Han [1,2,3], Linlin Lu [2,3,*], Junyu Lu [4], Xintong Liu [5], Shuangcheng Zhang [1], Ke Luo [2,3], Dan He [6], Penglong Wang [7], Huadong Guo [2,3] and Qingting Li [8]

1 College of Geological Engineering and Geomatics, Chang'an University, Xi'an 710054, China
2 International Research Center of Big Data for Sustainable Development Goals, Beijing 100094, China
3 Key Laboratory of Digital Earth Science, Aerospace Information Research Institute, Chinese Academy of Sciences, Beijing 100094, China
4 School of Community Resources and Development, Arizona State University, Phoenix, AZ 85004, USA
5 Faculty of Geographical Science, Beijing Normal University, Beijing 100875, China
6 Urban Science Department, College of Applied Arts and Science, Beijing Union University, Beijing 100191, China
7 Northwest Institute of Eco-Environment and Resources, Chinese Academy of Sciences, Lanzhou 730000, China
8 Airborne Remote Sensing Center, Aerospace Information Research Institute, Chinese Academy of Sciences, Beijing 100094, China
* Correspondence: lull@radi.ac.cn

Citation: Han, L.; Lu, L.; Lu, J.; Liu, X.; Zhang, S.; Luo, K.; He, D.; Wang, P.; Guo, H.; Li, Q. Assessing Spatiotemporal Changes of SDG Indicators at the Neighborhood Level in Guilin, China: A Geospatial Big Data Approach. *Remote Sens.* **2022**, *14*, 4985. https://doi.org/10.3390/rs14194985

Academic Editor: Danlin Yu

Received: 2 August 2022
Accepted: 5 October 2022
Published: 7 October 2022

Publisher's Note: MDPI stays neutral with regard to jurisdictional claims in published maps and institutional affiliations.

Abstract: Due to the challenges in data acquisition, especially for developing countries and at local levels, spatiotemporal evaluation for SDG11 indicators was still lacking. The availability of big data and earth observation technology can play an important role to facilitate the monitoring of urban sustainable development. Taking Guilin, a sustainable development agenda innovation demonstration area in China as a case study, we developed an assessment framework for SDG indicators 11.2.1, 11.3.1, and 11.7.1 at the neighborhood level using high-resolution (HR) satellite images, gridded population data, and other geospatial big data (e.g., road network and point of interest data). The findings showed that the proportion of the population with convenient access to public transport in the functional urban area gradually improved from 42% in 2013 to 52% in 2020. The increase in built-up land was much faster than the increase in population. The areal proportion of public open space decreased from 56% in 2013 to 24% in 2020, and the proportion of the population within the 400 m service areas of open public space decreased from 73% to 59%. The township-level results indicated that low-density land sprawling should be strictly managed, and open space and transportation facilities should be improved in the three fast-growing towns, Lingui, Lingchuan, and Dingjiang. The evaluation results of this study confirmed the applicability of SDG11 indicators to neighborhood-level assessment and local urban governance and planning practices. The evaluation framework of the SDG11 indicators based on HR satellite images and geospatial big data showed great promise to apply to other cities for targeted planning and assessment.

Keywords: SDG11; geospatial big data; sustainable development goals; earth observation; Guilin

1. Introduction

1.1. The SDG11 Indicators

The 2030 Agenda for Sustainable Development and 17 Sustainable Development Goals (SDGs) proposed by the United Nations in 2015 enable the international community to make a scientific understanding and accurate assessment on the sustainable development of global cities, thereby guiding their practical actions [1]. The SDGs seek to provide a comprehensive set of goals and indicators to measure progress towards sustainable development from 2015 to 2030 [2]. However, the holistic and complex nature of the SDGs has severely hampered progress towards these goals. With the date of achieving the goals of the 2030 Agenda approaching, a robust and unified assessment framework and reliable data are crucial for their accurate measurement and the fulfillment of the pledge—to ensure that "no one will be left behind" [1].

The sustainability in cities and urban settlements influences all aspects of sustainable development. The targets and indicators of the 11th Sustainable Development Goal (SDG11) provide a standardized indicator-based assessment framework to track the progress of sustainable urban development and inform policy implementation and practice. According to the Inter-agency and Expert Group on SDG Indicators (IAEG-SDGs), SDG11.2.1, 11.3.1 and 11.7.1 are Tier II indicators. These indicators are conceptually clear and have an internationally recognized methodology and standard, but data are not regularly reported. SDG11.2.1 refers to the proportion of the population that has convenient access to public transport disaggregated by age group, sex, and persons with disabilities. This indicator aims to monitor the use of and access to the public transportation system, alleviate the reliance on private means of transportation, and improve the traffic conditions in areas with a high proportion of transport disadvantaged people. SDG11.3.1 measures how efficiently cities utilize land and is measured as a ratio of the rate at which cities spatially consume land against the rate at which their populations grow. SDG11.7.1 refers to the average share of the built-up area of cities that is open space for public use for all, by sex, age, and persons with disabilities. It enables cities to collect accurate, timely, disaggregated data and information on open space by adopting a systemic approach.

1.2. The Role of Geospatial Big Data for SDG11 Indicators Monitoring

Geospatial big data played an important role in the monitoring of SDG11 indicators. The use of big data such as mobile phone data, transaction data, health records, and social media can complement traditional official statistical data and help fill data gaps in monitoring SDG indicators [3,4]. Earth observation data (EO) obtained from satellites and geospatial data collected by on-site sensors or citizens are recognized as an effective, timely, and continuous information source to support evidence-based decision-making for sustainable urban development [5,6]. Remotely-sensed EO data have the advantage of collecting extensive information on the Earth's surface at large spatial scales with repeat acquisition cycles, which can supplement or enhance the traditional data sources in urban areas [7–9]. The availability of open remote sensing data and high-performance cloud computing platforms makes it possible to map built-up urban areas or impervious surfaces over large areas with medium and high spatial resolution in recent decades [10,11]. Based on open satellite data, several global built-up area layers have been developed, including the Global Human Settlements Layer (GHSL) from the Joint Research Centre of the European Commission [12], the Global Urban Footprint (GUF) [13], and the World Settlements Footprint (WSF) jointly developed by ESA, the German Aerospace Center (DLR), and the Google Earth Engine team [14]. These products provide information on the global human settlement with spatial resolutions from 10 m to 30 m via processing millions of images from Landsat and Sentinel satellites [15,16]. The accessibility of open geospatial data such as WorldPop population grids and OpenStreetMap road networks facilitates accessibility measurements in cities around the world.

1.3. Research Questions, Motivation and Objectives

The process of sustainable development goals was mostly reported at the national level. Assessing the sustainable development goals locally can track the progress in local sustainable development and provide relevant strategies to guide sustainability practices [17]. The availability of big data (e.g., high-frequency satellite Earth observation data) can power the sustainability practitioners to better monitor and evaluate the progress of sustainable development. As an internationally agreed and reported assessment framework, knowledge and practice gaps still exist on how to link and integrate the indicator measurements into urban governance and planning in the local context [18]. In addition, the complexity of SDG indicator monitoring lies in the trade-offs and synergies between indicators [19–21]. To address the above research gaps and challenges, the following key research questions were raised: (1) whether an assessment framework to evaluate multiple SDG indicators conjunctively can be developed using geospatial big data, and (2) whether the evaluation results of SDG indicators at the neighborhood level can be linked with local urban governance and planning practices.

With a unique karst landform, Guilin city in China was added to the United Nations Educational, Scientific and Cultural Organization (UNESCO)'s world heritage list in 2014. In February 2018, with the theme of "sustainable utilization of landscape resources," Guilin was selected as the innovation demonstration zone of the National Sustainable Development Agenda in China. Taking Guilin city as a case study, this study aims to (1) develop a framework to monitor three SDG11 indicators (indicators 11.2.1, 11.3.1, and 11.7.1) at the neighborhood level using high-resolution satellite data, gridded population data, and other geospatial big data (e.g., road network and point of interest data), (2) to provide a holistic perspective of the progress of sustainable urban development in the study area, and to (3) evaluate the feasibility of integrating SDG11 indicators into urban governance practices in the local context.

2. Literature Review

2.1. Geospatial Datasets for SDG11 Indicators Monitoring

With the advantages of varying spatial and temporal resolution, large spatial coverage, and long temporal coverage, Earth observation data provide an optimal data source for the monitoring of SDG indicators both directly and indirectly [5]. Numerous studies used satellite images from different sensors to assist in monitoring the progress of SDG11. In these studies, the freely accessible global Landsat archive containing millions images was the main source of remote sensing data. Landsat 2/5/7/8 datasets have been used to analyze land use and landscape changes from local to global scales [22–24]. For the evaluation of SDG11.3.1 indicators, the combination of Landsat and satellite images with higher spatial resolution such as Sentinel and SPOT provided more accurate classification results [25]. The SPOT 2 panchromatic imagery has a spatial resolution of 10 m and multispectral imagery has a spatial resolution of 20 m. The SPOT 5 satellite imagery has a spatial resolution of 2.5 m in the panchromatic band and 10 m in multispectral bands. The fusion of remote sensing images and products was also used to obtain urban land cover classification results with higher accuracy [22].

UN-Habitat proposed to use free EO satellite data from Landsat and Sentinel-2 satellites to delineate potential public urban open spaces. Urban green areas comprise of many small-size green spaces, such as gardens, community parks, roadside trees, etc. Although open and free earth observations (10–30 m) with low and medium resolution can provide valuable insights for policymakers and urban managers [26], their relatively coarse spatial resolution tends to cause underestimation of small-sized open spaces [27,28] and leads to low accuracy of open space detection in complex urban areas [29,30]. Streets less than 10 m wide can hardly be detected from Sentinel-2 satellite imagery at 10 m resolution [31]. High resolution remote sensing images (spatial resolution higher than 10 m) are more suitable data sources for open space extraction in urban areas [28]. Most studies use remote sensing images with very high spatial resolution for SDG11.7.1 monitoring, such as PlanetScope [27], RapidEye [28], QuickBird-2 [28,32], WorldView-2 [32] images, etc. Although satellite images from these satellites can capture the details of land surface, they are very costly and their application over large areas is infeasible, especially in areas with cloudy landscapes [33].

Monitoring the progress of achieving the sustainable development goals through the global indicator framework increased the demand for data that are high in quality, broad in coverage, frequently available, and spatially disaggregated from countries around the world [34]. In addition to remote sensing data, geospatial data collected voluntarily using a wide range of technologies and methods provided supplementary datasets to insufficient official data and improved the monitoring of sustainable development goals [3]. Fried et al. [35] leveraged open data, including OpenStreetMap road network and World-Pop population data to derive the values of SDG 11.2.1 indicator and accessibility metrics and identified transport inequalities of low-income communities.

2.2. Methods for SDG11 Indicators Monitoring

Since the adoption of SDGs in 2015, the SDG11.2.1, SDG11.3.1, and SDG11.7.1 indicators have been used to monitor and assess the progress of sustainable urban development in numerous studies (Table 1). The popularity of SDG11.2.1 is attributed to its simple estimation methods and interpretation of results. However, researchers argued that it was not comprehensive to use a single SDG11.2.1 indicator to evaluate the traffic accessibility of cities or countries. Other indicators should also be measured in decision-making for transportation facility improvement. Tiznado-Aitken et al. [36] evaluated the accessibility of Santiago's pedestrian environment based on Lorenz curves, Gini coefficient, and Foster-Greer-Thorbecke (FGT) poverty measures. Brussel et al. [37] found that the SDG indicator 11.2 could not represent the traffic reality well and proposed accessibility indicators that could provide a more diversified, complete, and realistic picture of the transportation system's performance. Fried et al. [35] supplemented the analysis results of SDG11.2.1 through a more detailed location-based accessibility analysis and revealed traffic inequality in low-income communities.

Early studies used remote sensing images, machine learning classification methods, and GIS technology to analyze urban growth and sprawl processes [38]. More advanced techniques, such as deep learning and scenario modeling, were applied for SDG11.3.1 monitoring and prediction [39–41]. Kussul et al. [39] proposed a method for land cover classification and land productivity assessment using medium and high spatial resolution satellite data and deep learning methods. Wang et al. [40] used the spatio-temporal interaction method and Pearson's method to monitor the spatio-temporal changes of SDG 11.3.1. Lu et al. [41] monitored and predicted changes in urban land use efficiency indicators based on remote sensing and scenario modeling in a coastal megacity from 2000 to 2030. Remote sensing and geospatial big data can help understand the spatiotemporal dynamics of urban green space under the urbanization [42], accessibility [43], and walkability [44]. However, the detailed mapping of urban open public space and the measurement of SDG11.7.1 indicators still needs localized data and strong urban data collection capacity [45].

Table 1. Literature review of three SDG11 indicators using geospatial data.

SDG Indicator	Data Source	Spatial Resolution	Study Area	References
SDG11.2.1	An underlying road network,	100 m	Nairobi, Kenya	Fried et al. [35]
	a general transit feed specification package,			
	WorldPop population,			
	an opportunity dataset.			
SDG11.2.1	Public transport stops,		Santiago, Chile	Tiznado-Aitken et al. [36]
	road network,			
	georeferenced information.			
SDG11.3.1	Built-up areas,	30 m, 250 m, 1 km	10,000 urban centers	Melchiorri et al. [23]
	resident population,	250 m, 1 km		
	settlement typologies.	1 km		

Table 1. *Cont.*

SDG Indicator	Data Source	Spatial Resolution	Study Area	References
SDG11.3.1	A GIS raster dataset of built-up areas,	1 km	Global	Estoque et al. [24]
	a statistical dataset of population.	250 m, 1 km		
SDG11.3.1	Landsat-5/8 images,	30 m	Beijing–Tianjin–Hebei region, China	Zhou et al. [22]
	built-up area products,	30 m		
	WorldPop population,	100 m		
	ancillary datasets.	30 m		
SDG11.3.1	LULC,	30 m	Mainland China	Wang et al. [46]
	census data,			
	DMSP/OLS,	1 km		
	administrative boundary map.			
SDG11.3.1	Landsat-5/8 images,	30 m	Tianjin, China	Lu et al. [41]
	topographic data,	30 m		
	road network,			
	demographic data.	100 m		
SDG11.3.1	Built-up areas,	1 km	Global	Schiavina et al. [47]
	resident population,			
	settlement typologies,			
	functional urban area.			
SDG11.3.1	Landsat 5 TM images,	30 m	South Africa	Mudau et al. [25]
	SPOT 2/5 sensors images,	Panchromatic 10/2.5 m; multispectral 20/10 m		
	census data.			
SDG11.3.1	Landsat 2/5/7/8 images,	80/30 m	Southern Brazil	Moro et al. [21]
	Sentinel-3B OLCI-WFR satellite images.	300 m		
SDG11.3.1	Built-up area,	100 m	the Yangtze River Delta, the Middle Reaches of the Yangtze River, and Chengdu–Chongqing, China	Wang et al. [40]
	population data,	100 m		
	boundaries maps.			
SDG11.3.1	Resident population,	1 km	Poland and Lithuania	Calka et al. [48]
	CORINE land cover 2000/2018	12.5 m		
SDG11.7.1	PlanetScope images,	3.7–4.1 m	The Athens Metropolitan Area	Verde et al. [27]
	Sentinel-1 images,			
	ground range-detected products.			
SDG11.7.1	Sentinel-2A images,	10 m	Hangzhou, China	Deng et al. [49]
	SPOT-2/3/5 images,	XS 20 m/PAN 10 m/XS 10&20 m		
	reference and ancillary data.			

2.3. Research Challenges

A review of recent research shows that Earth observation data can effectively support the government in addressing sustainable development goals and monitoring the implementation of SDG indicators. The rapid development of Earth observation technologies and big data platforms will continue to play a role in expanding indicators and targets that can be effectively measured and monitored globally. However, the success of SDG11 implementation depends largely on the availability of high-quality assessment data [2,3] Although satellite data has been widely used in SDG indicator evaluation, the spatial resolution of remote sensing images used varies from 2.5 m to 1 km (Table 1). High-resolution satellite data has become increasingly available. Its potential for urban built-up areas and open space mapping and monitoring can be further evaluated. Data fusion methods and more advanced data processing techniques, such as deep learning, should be exploited to improve the accuracy and reliability of geospatial products and information derived from remote sensing images. Data sharing and openness should also be promoted to better support the monitoring of SDG11 indicators.

Analytical frameworks, tools, and analyses that enable interlinkages between targets and indicators can provide more insights on how various interacting forces led to specific outcomes, thereby helping establish connections between science and policy [50]. Researchers argued that it is not comprehensive to use a single indicator to evaluate the traffic accessibility of cities. However, the majority of previous studies using geospatial data focused on the measurement of one specific indicators. Multiple indicators can be assessed conjunctively to draw policy and practice implications for sustainable urban development. Comprehensive assessment framework involving multiple indicators should be developed and implemented to provide sound policy guide for city governors in future studies.

Despite the fact that the methodologies and approaches of SDG indicator monitoring with Earth observation data have been developed, the assessment was mainly performed at a city, regional and national level. Few studies conducted neighborhood level analysis and incorporated the assessment results with local urban planning and governance practices. Localized urban practices using open geospatial data and SDG indicators is beneficial for guiding cities and regions, especially in developing countries, to assess sustainable development progress and support policy-making processes.

3. Study Area

Guilin City (Figure 1) is located in the northeastern part of Guangxi Zhuang Autonomous Region, China. It is located between 109°36′ to 111°29′E and from 24°15′ to 26°23′N. The territory is 236 km long from north to south and 189 km wide from east to west. Guilin has a subtropical monsoon climate with an average annual precipitation of 1887.6 mm and an annual mean temperature of 18.9 °C. The core urban area of Guilin city includes six districts, Xiufeng, Qixing, Xiangshan, Diecai, Yanshan, and Lingui, covering an area of 2767 km². Guilin has a typical karst landform and is a world-famous scenic city. Considering its extraordinary natural beauty and aesthetic values, the World Heritage Committee added the Guilin Karst to UNESCO's world heritage list in 2014. Guilin is also the political, economic, cultural, and technological center in the northeastern part of Guangxi Province.

With the rapid economic development, anthropogenic activities have become increasingly intensive in Guilin in the last decades. The over-development and exploitation of natural landscapes have led to considerable pressure on the sensitive and fragile ecological environment [51]. The contradiction between the growth of natural resource utilization demand and the actual environmental carrying capacity has become increasingly prominent. The evaluation and monitoring of SDG indicators can provide valuable information for the construction of the innovation demonstration zone of the National Sustainable Development Agenda and for protecting the ecological environment in local areas.

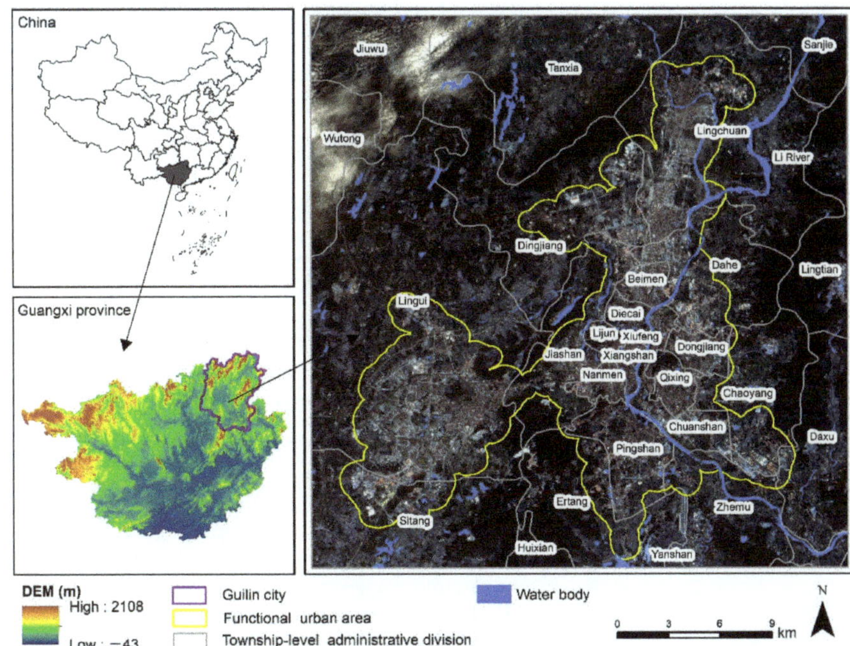

Figure 1. Geographic location of the study area.

4. Materials and Methods

4.1. Datasets

In this study, three SDG11 indicators were measured within the urban functional boundary of Guilin from 2013–2020. High-resolution remote sensing images of Chinese Gaofen-1/6 satellites and geospatial big data were mainly used (Table 2). Since the Gaofen-1 satellite was launched in 2013 and the high-resolution satellite images acquired by Gaofen-1 have been available since then, this study selected 2013–2020 as our study period. The Gaofen-1/6 satellites carry two 2 m panchromatic and 8 m multi-spectral high-resolution cameras with four bands (PMS). After preprocessing, image fusion was employed to fuse the multispectral images and panchromatic images, and 2 m resolution multi-band data was finally obtained for land use classification in the study area. Point of interest (POI) data including bus stations and train stations in 2015 and 2020 were collected from the AutoNavi electronic navigation map (https://ditu.amap.com, accessed on 10 April 2021). Due to the low quality of historical data, the road networks were only obtained for the year 2020. LandScan global population data, produced by the U.S. Department of Energy's Oak Ridge National Laboratory (ORNL) (https://landscan.ornl.gov/, accessed on 1 June 2021), was used to measure the number of populations. This dataset used spatial data, high-resolution imagery analysis techniques, and a dasymetric modeling approach to disaggregate census population numbers within administrative boundaries. LandScan is the finest resolution global population distribution data available representing an ambient population. The LandScan data covering the study area were retrieved in 2013, 2015, and 2020 for the SDG indicator measurement.

Table 2. Geospatial datasets used in this study.

Data Set	Acquisition Time	Spatial Resolution	Source
Road network	2020		Autonavi electronic navigation map
Point of interest	2015 2020		Autonavi electronic navigation map
Gaofen-1/6 satellite images	2013 2015 2020	2/8 m	Chinese Academy of Sciences
Population grid data	2013 2015 2020	1 km	LandScan
Urban park	2015 2020		Autonavi electronic navigation map

4.2. Methods

The workflow for SDG indicator assessment is shown in Figure 2. The road network data, point of interest data, and LandScan population data were used to perform a dynamic assessment of public transport accessibility and SDG11.2.1 indicators. Secondly, high-resolution satellite images were used for land cover classification to analyze changes in SDG11.3.1 indicators. Finally, high-resolution satellite images were used to extract green space and measure changes in SDG11.7.1 indicators. The LandScan population grid data was used to calculate the number of populations with convenient access to urban open spaces.

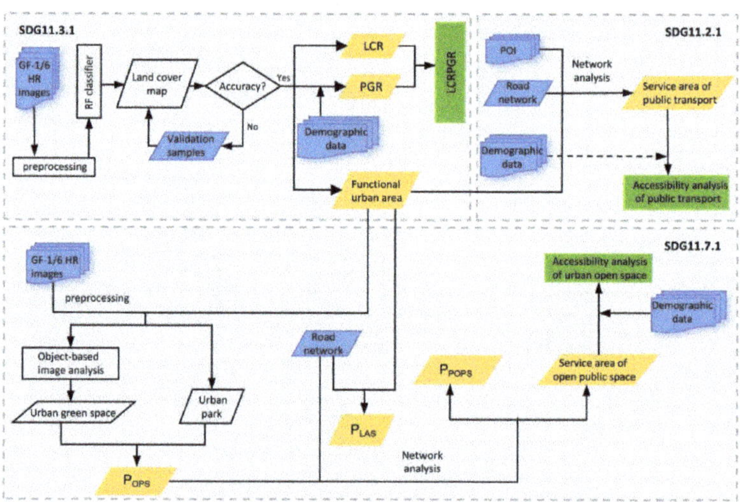

Figure 2. Workflow for SDG indicators assessment in this study.

The notations used for the calculation of three SDG11 indicators, SDG11.2.1, SDG11.3.1, and SDG11.7.1, are summarized in Table 3.

Table 3. Notations used in this study.

Terms	Definition	Unit
P_i	The total population served by public transport service area i.	-
P_{ij}	The number of the population of population zone j (j = 1 … n) that fully or partially intersect with a public transport service area i.	-
LCR	Land consumption rate.	-
Urb_t	The total area covered by the urban built-up area in the initial year t.	km^2
Urb_{t+n}	The total area covered by the urban built-up area in the final year t+n.	km^2
LCRPGR	The ratio of land consumption rate to population growth rate.	-
PGR	Population growth rate.	-
$LCPC_{t1}$	The land consumption per capita at time t1.	km^2
Urb_{t1}	The total built up area within the urban boundaries at time t1.	km^2
Pop_{t1}	The total population within the urban boundaries at time t1.	-
Change in $LCPC_{(t1-t2)}$	The percentage change in land consumption per capita between t1 and t2.	%
Change in Urban Infill	The percentage change rate of urban density.	%
$S_{streets}$	Total area occupied by streets in all locales.	km^2
S_{city}	Total area of all locales.	km^2
S_{OPS}	The total area occupied by open public spaces.	km^2
P_{LAS}	The share of city land occupied by streets.	%
P_{OPS}	The share of urban areas that is allocated to open public spaces.	%
P_{POPS}	The average share of built-up area of cities that is open public space and streets.	%
Subscripts		
i	The service area	
j	The population zone	
t	The initial year	
n	The number of years between the initial year and the final year	
t1	The initial year	
t2	The final year	
Streets	Urban streets	
City	Urban area	
OPS	Open public spaces	
POPS	Built-up area of cities that is open public space and streets	
Symbols		
Σ	The summation symbol	
%	Percentage means the percentage of one number that is the other number, expressed by "%"	
ln	The natural logarithm symbol is the logarithm with constant e as the base, which is recorded as lnN (N > 0)	

4.2.1. SDG11.2.1

SDG indicator 11.2.1 measures proportion of population that has convenient access to public transport. For SDG 11.2.1, public transport is considered "convenient" for people who live within 500 m walking distance from the nearest low-capacity station and 1 km from the nearest high-capacity station. According to the definition of low-capacity public transportation and high-capacity public transportation, the 500 m service area of the bus stations and the 1 km service area of the railway stations were created using network analysis [34]. A network service area was created along the road network at each public transport stop or around each public transport route per applicable walking distance thresholds. The use of network distance can reflect the configuration of the road network, and identify the existence of obstacles that hinder direct access to public transport facilities. All individual service areas are then merged to create continuous service area polygons. The service area and the population data are overlaid to calculate the population with access to each public transport stop. The number of populations served by the public transport service was calculated by:

$$P_i = \sum_{j=1}^{n} P_{ij},\tag{1}$$

where P_{ij} is the population served by the public transport service in buffer i of population zone j ($j = 1 \ldots n$) that completely or partially intersect with the service area i, and P_i refers to the total population served by public transport stations in service area i.

The SDG11.2.1 indicator was calculated as percentage of population with convenience assess to public transport. The higher the percentage of the population with convenient access to public transportation services, the better the accessibility, and vice versa. This indicator reflects the service status of the regional road network and transportation stations. The SDG 11.2.1 indicators in 2013, 2015, and 2020 were calculated to analyze the changes in public transportation accessibility in the study area.

4.2.2. SDG11.3.1

SDG indicator 11.3.1 measures the ratio of land consumption rate to population growth rate [40,41,48,52]. Using the remote sensing images of Gaofen-1 in 2013 and 2016 and Gaofen-6 in 2020, a random forest classifier was used to classify the land cover of the study area into five categories: built-up land, forest, cultivated land, water body, and bare land. Based on the classification results, changes in the urban functional boundaries and land use were analyzed for each period. The land consumption rate (LCR) was calculated using the following equation:

$$LCR = \frac{ln(\frac{Urb_{t+n}}{Urb_t})}{n},\tag{2}$$

where Urb_t is the total area covered by the urban built-up area in the initial year (km^2); Urb_{t+n} is the total area covered by the urban built-up area in the final year (km^2); n is the number of years between the two periods.

The global population grid data was used to calculate the population in the study area in the corresponding year, and the population growth rate (PGR) was calculated using similar method. Combining the changes in built-up land and population, the ratio of land consumption rate to population growth rate (LCRPGR) in the functional urban area was calculated using equation:

$$LCRPGR = \frac{LCR}{PGR},\tag{3}$$

To capture the urbanization process more comprehensively, two secondary indicators were also calculated. The per capita land consumption (LCPC) at t_1 was derived according to Equation (4), and the percentage change rate of per capita land consumption (Change in LCPC) and the percentage change rate of urban density (Change in Urban Infill) between t_1 and t_2 were calculated according to Equations (5) and (6).

$$LCPC_{t1} = \frac{Urb_{t1}}{Pop_{t1}},\tag{4}$$

$$ChangeinLCPC_{(t1-t2)} = \frac{LCPC_{t2} - LCPC_{t1}}{LCPC_{t1}},\tag{5}$$

$$ChangeinUrbanInfill = \frac{Urb_{t2} - Urb_{t1}}{Urb_{t1}} \times 100, \tag{6}$$

where Urb_{t1} refers to the total built-up area within the urban boundary at time t_1 (km^2); Pop_{t1} refers to the total population within the urban boundaries at time t_1; Urb_{t2} refers to the total built-up area at time t_2 within the same urban boundary (km^2).

4.2.3. SDG11.7.1

SDG Indicator 11.7.1 measures the average share of the built-up area of cities that is open space for public use. The road network data and the boundaries of the urban functional area were used to calculate the land allocated to streets (Equation (7)). Using an object-based image analysis method, green space was extracted using the high-resolution satellite images. The urban parks were extracted using the Autonavi electronic navigation map (AMAP). Then, the proportion of open public spaces was calculated using Equation (7). As for the core indicator, the proportion of public open space in the city was expressed as the proportion of the total open space area of streets and open public spaces in the urban area [49].

$$\begin{cases} P_{LAS} = \frac{S_{streets}}{S_{city}} \times 100\% \\ P_{OPS} = \frac{S_{OPS}}{S_{city}} \times 100\% \\ P_{POPS} = \frac{S_{streets} + S_{OPS}}{S_{city}} \times 100\% \end{cases}, \tag{7}$$

where P_{LAS} represents the share of city land occupied by streets (%), $S_{streets}$ represents the total area occupied by streets in all locales (km^2), and S_{city} represents the sum area of all locales (km^2); P_{OPS} represents the share of urban areas that is allocated to open public spaces (%), S_{OPS} represents the total area occupied by open public spaces (km^2); P_{POPS} represents the average share of built-up area of cities that is open space for public use for all (%).

A network analysis was performed to generate an urban open space service area with a road network distance of 400 m. First, a network dataset was create using road network data, and then a road network-based service area was created around each public open space using a 400 m threshold. All people living in the service area are deemed to have convenient use of open public space. Finally, combined with the grid population data, the population in the service area of the open space was calculated in each period in the study area.

5. Results

5.1. Spatiotemporal Variation of Population with Access to Public Transport Stops (SDG 11.2.1)

The public transportation service area of Guilin in 2015 and 2020 was shown in Figure 3. The service area of public transport stations was increasing and covers most of the population in the main functional areas of the city, but there were also some densely populated areas that were not covered by the service area.

Figure 4 shows the changes in population with convenient access to public transport services from 2013 to 2020. The level of public transport services continued to improve from 2013 to 2020. The total population with access to public transport stops increased from 458,861 in 2013 to 489,379 in 2015 and 573,957 in 2020. The accessibility indicator increased from 42.08% in 2013 to 52.31% in 2020.

The changes in SDG 11.2.1 indicators were further evaluated at the township level (Figure 5). The indicators of most towns showed a trend of improvement over time. Among them, the population in Xiangshan, Xiufeng, Qixing, and Lijun has been fully covered by public transportation services. However, in towns such as Dingjiang, Lingchuan, and Lingui, less than 40% of the population has access to convenient public transport. The construction of public transportation facilities in these areas was weak, and investment on public transportation infrastructures is highly needed in these areas.

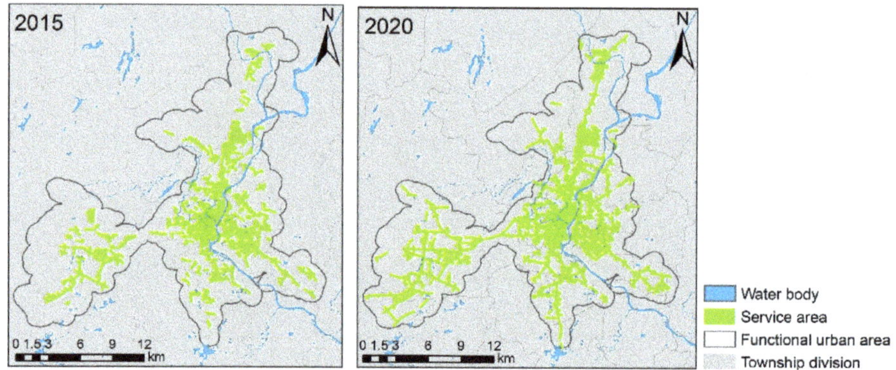

Figure 3. Service area of public transport stops in Guilin in 2015 and 2020.

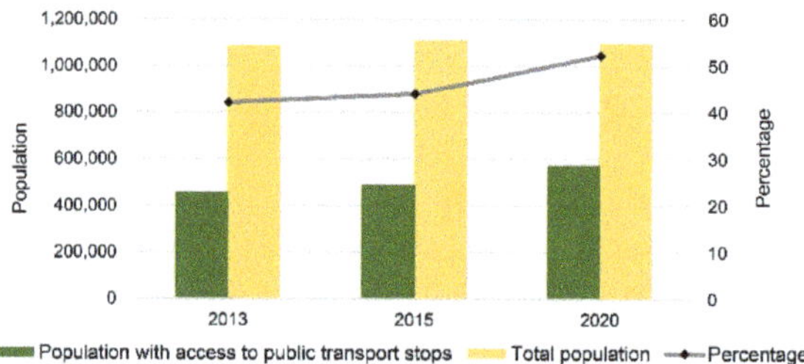

Figure 4. Number and percentage of the population with access to public transport stops from 2013 to 2020 in Guilin.

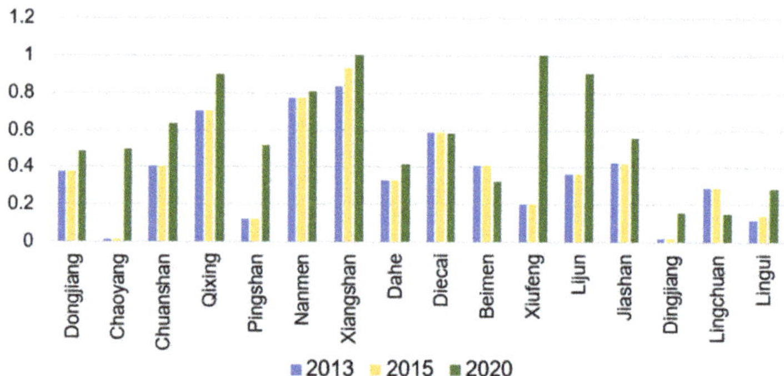

Figure 5. The proportion of populations with access to public transport stops from 2013 to 2020 in the townships of Guilin.

5.2. Spatiotemporal Variation of Land Consumption vs. Population Growth (SDG 11.3.1)

The accuracy of the land use classification results in the study area was evaluated using ground truth samples. The overall classification accuracy and kappa coefficient of the classification results of each period were higher than 90%. Based on the land use classification results, the temporal variations in LCR, PGR, and LCRPGR indicators and the corresponding secondary indicators (i.e., change in LCPC and change in urban infill) were measured in the study area (Table 4). The results showed that the urban expansion and population growth rates of Guilin were not well coordinated from 2013 to 2020. The urban expansion rate was faster than the population growth rate, and the per capita urban built-up area continued to increase at an accelerating rate. This indicates a sprawling urban growth pattern in the study area.

Table 4. SDG11.3.1 indicators from 2013 to 2015 in Guilin.

Time Span	LCR	PGR	Change in LCPC	Change in Urban Infill	LCRPGR
2013–2015	0.0525	0.0179	7.16%	4.77%	2.9343
2015–2020	0.0320	−0.0007	17.78%	23.30%	−45.7867

At the township level, the built-up area expanded more than 12 times faster than the population growth rate in Dingjiang, followed by Dahe, Pingshan, and Chuanshan from 2013 to 2015 (Figure 6). During the period 2013 to 2015, a large number of construction projects were started. However, those towns hardly attracted many people inflow. The towns of Diecai, Xiufeng, Lijun, Xiangshan, and Nanmen experienced a population decline. From 2015 to 2020, the expansion rates of built-up areas in Lingchuan, Lingui, and Dingjiang were still far greater than the population growth rate, which might be due to new settlement planning in these towns. The growth rate of built-up area in Lingui new area accelerated tremendously after 2015.

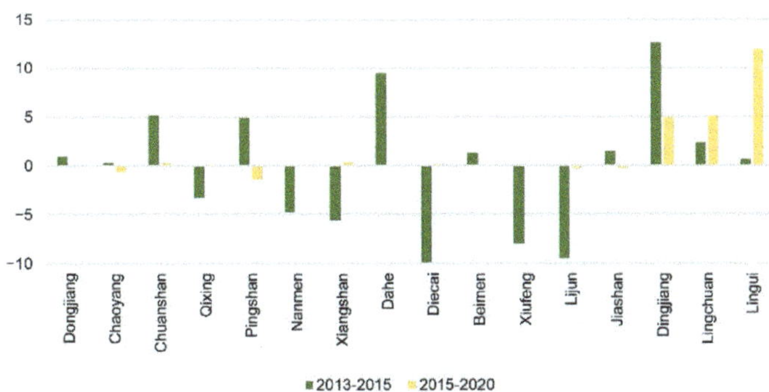

Figure 6. Ratio of land consumption rate to population growth rate(LCRPGR) from 2013 to 2020 in townships of Guilin.

From 2013 to 2015, the fastest growing area in per capita land consumption (LCPC) was in Dahe, up to 24.72% (Figure 7). The LCPC in Dingjiang, Lingchuan, Chuanshan, and Pingshan continued to grow, and the LCPC of other regions were decreasing. From 2015 to 2020, the LCPC of Jiashan, Lijun, Xiufeng, and Nanmen increased sharply due to the loss of population, while the per capita land consumption area of most of the remaining townships showed a downward trend. The changes in LCPC (Figure 7) indicates that land use efficiency continued to decrease, especially in Lingui, Lingchuan, and Dingjiang, where disorderly expansion occurred. In the townships where the expansion of built-up land was much faster than the population growth, the planning and management of land development should be strengthened to avoid low-density sprawl of urban land.

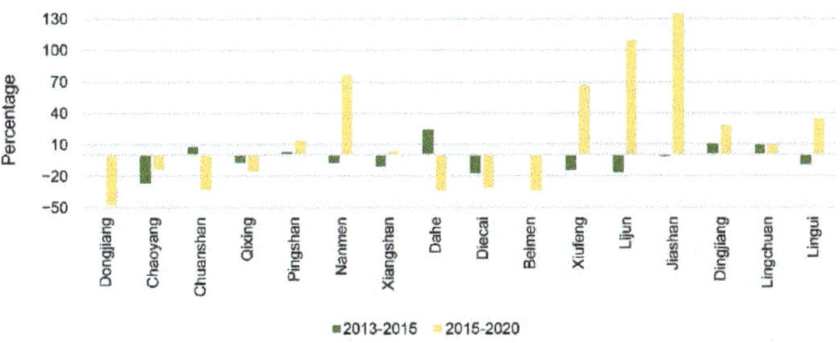

Figure 7. Changes in per capita land consumption (LCPC) from 2013 to 2020 in townships of Guilin.

5.3. Spatiotemporal Variation of Open Public Space (SDG 11.7.1)

For SDG11.7.1 indicator, the land of streets in the area was 16.18 km^2, which accounts for 5.13% of total urban area (Figure 8). The area of green space was 160.28 km^2, 123.10 km^2, and 59.32 km^2 in 2013, 2015 and 2020, respectively, showing a rapidly decreasing trend. The corresponding areal proportion of urban green space (POPS) accounted for 50.83%, 33.91%, and 18.81% of the urban area, respectively. The proportion of the overall open public space has gradually decreased from 55.97% in 2013 to 39.04% in 2015 and 23.95% in 2020.

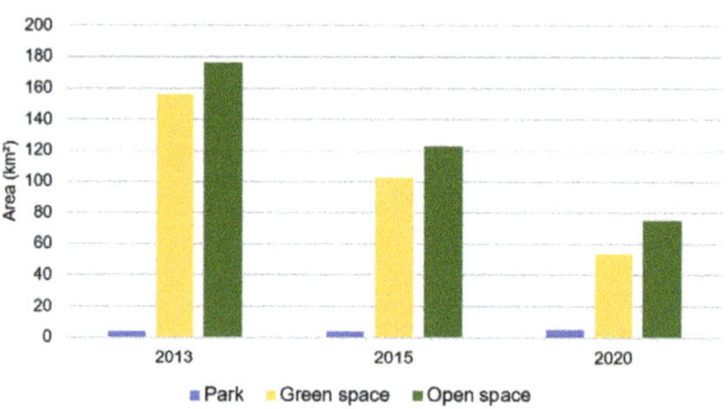

Figure 8. Area of open public space from 2013 to 2020 in Guilin.

The 400 m service area of parks and green spaces were shown in Figure 9. In 2013, 2015, and 2020, the service areas were 151.34 km^2, 96.09 km^2, and 81.76 km^2, respectively. The proportions of the population in the service area of open public space in Guilin were 73.2 %, 64.0%, and 59.3% in 2013, 2015, and 2020, respectively (Table 5). With the rapid decrease of green spaces, the service area has decreased over time. The total number of residents served by green open spaces also showed a rapid downward trend. Most of the green areas were converted into urban built-up land. From 2013 to 2020, the area of green space in the urban functional area dropped sharply, while the number of urban parks has only increased by 6 km^2 with a total area of 1.24 km^2. Most of them were distributed in the new urban areas in the west, however, the number and area cannot meet the needs of citizens due to rapid increase in population.

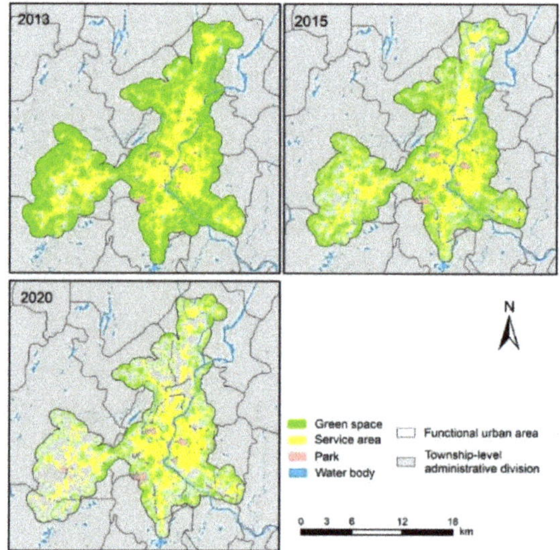

Figure 9. Open public spaces and service areas from 2013 to 2020 in Guilin.

Table 5. Population with access to open public spaces from 2013 to 2020 in Guilin.

	2013	**2015**	**2020**
Population with access to public open space	817,366	731,600	664,953
Total population	1,117,265	1,142,848	1,121,568
Proportion (%)	73.16	64.02	59.29

Figure 10 shows the changes in the proportion of the population in the green space service area of each town in the functional area of Guilin. At the township level, the service areas were relatively large for Xiufeng, Xiangshan and Qixing. The proportions of served population in most towns have decreased over time. The fastest decline was observed in Lingui, which has dropped from 50% to 24%. In 2020, the proportions of served population in Dingjiang, Lingui, Lingchuan, Dahe, Chaoyang, and Jiashan were lower than 40%. In addition, the entire area of Xiufeng was fully covered by the service area of open public space from 2013 to 2020, and all residents have convenient access to urban green space.

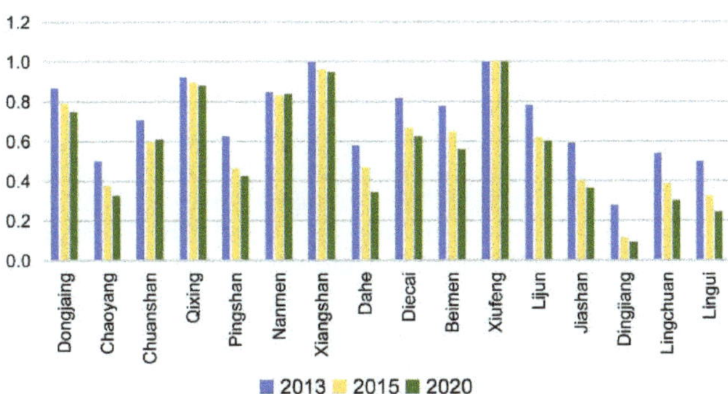

Figure 10. Proportion of population with access to open public space from 2013 to 2020 in townships of Guilin.

6. Discussion

The SDG11.2.1 evaluation results show that the public transport convenience of the regional central towns in Guilin is better than that of the peripheral towns. The research results of Tiznado-Aitken, Muñoz and Hurtubia [36] are consistent with our study. Building an integrated transportation network that meets the needs of social and economic development is very important for achieving sustainable development. The transportation links between the bordering towns and the central towns should be strengthened to promote the in-depth development of regional economics [53]. Meanwhile, intensive use of urban resources and livable environment should be maintained to build a modern town in Guilin with a good ecological environment during the road network construction. This study confirms that the application of high-resolution Earth observation data can improve the lack of information on SDG11.3.1 at a fine scale [48]. The growth rate of the built-up area exceeded the growth rate of the population in the same time period, increasing imbalance between rapid urban expansion and population growth. The uncoordinated population and land growth have been reported in small and medium-sized cities in China [22,46,54]. One possible reason is that low land acquisition costs and rapid industrial development have transformed a large number of agricultural land markets into non-agricultural land markets, which has resulted in the growth of LCR [54]. The proportion of land use types used for built-up areas in Guilin was increasing, while the land used to provide urban green spaces was decreasing. The benefits provided by the urban green space cannot meet the needs of the increasing population. The realization of the SDGs depends on the balance and cohesion of all the elements related to land [55]. It is necessary to balance the growth of urban built-up areas and green spaces to promote sustainable development. In the undergoing densification of urban areas, adequate and accessible green public spaces should be planned to solve the scarcity of green space and to sustain the quality of urban environments and social systems, and human wellbeing [56].

Actions taken at the neighborhood level can lead to positive urban changes and enhance the sustainability of communities [57]. The evaluation of SDG indicators 11.2.1, 11.3.1, and 11.7.1 suggested that the three towns of Lingui, Lingchuan, and Dingjiang should consider increasing their green area andpublic transportation facilities to prevent the low-density development of built-up land. Investments in urban infrastructure to enable a function can often create lock-in situations that last for decades or even centuries [58]. In urban planning, natural landscapes, trunk lines, and urban construction should be integrated, and ecological and livable towns should be built on the basis of existing landscapes. It is worth mentioning, however, that some scholars have found that the construction of extensive road networks across the landscape also destroys natural habitats and increases the pressure on sustainable development policies. For example, human activities have led to ecological degradation in the Qilian Mountains [59]. Monitoring of SDG 11.7.1 indicator aims to ensure that important ecological functional areas such as nature reserves, scenic spots, forest parks, wetlands, and vital water sources within the urban area of Guilin City are not damaged. It also aims to improve the construction of urban open space. According to the township-level analysis results of SDG11.3.1, policymakers in Lingui, Lingchuan, and Dingjiang should plan and consider building new districts based on their population density, resources, and environmental carrying capacity and should not pursue urbanization while ignoring the ecological environment. Due to the lack of conservation and preservation, deforestation in the past ten years was extensive, and the water storage capacity was weakened, which also impacted the water quality in Guilin. Since 2011, Guilin has implemented a series of policies to improve the ecological environment. These policies are encouraging and might help to formulate sustainable development strategies suitable for Guilin in the future.

A well-conceptualized and robust SDG monitoring framework can inform the strategic design of policies and interventions to address the challenges of growing urban areas and uncertainty in a variety of scenarios. SDG11 monitoring framework explicitly considers the linkages between policy, urban development, and the SDGs by integrating interdisciplinary knowledge. Implementing isolated targets without a comprehensive approach will undermine the unique dynamics of each city and endanger sustainable development. The comprehensive evaluation results in our study indicates that the evaluation results of multiple SDG indicators should be considered comprehensively under the local context to gain insights from local ecosystem and development conditions and guide sustainable policies and practices. Since the proposed approach is based on the United Nations' guidelines and uses open geospatial data, it can be easily adopted in cities in other countries or regions to support sustainable urban practices.

One limitation of our study is that the indicator assessment was performed in the years 2013, 2015, and 2020. The study periods highly depend on the availability of open geospatial data. With geospatial data becoming more readily available, an annual evaluation of the SDG indicators can be performed in future studies. Moreover, raw spatial and temporal resolution geospatial data such as gridded population data (1000 m) was used in this study. This may have caused uncertainties and bias in the evaluation results [60,61]. Geospatial data with higher spatiotemporal resolution and thematic accuracy should be used to obtain more reliable assessment results. The sharing of geospatial data should be endorsed, and multi-source data fusion technology should be developed to improve the resolution and accuracy of SDG indicator measurements.

7. Conclusions

Following the guidelines of SDG11 indicators, this study evaluated the spatiotemporal changes in transport accessibility, land consumption, and urban open space in the urban functional area of Guilin from 2013 to 2020. The evaluation results showed that the accessibility of public transport gradually improved from 2013 to 2020, and the SDG11.2.1 indicators increased from 42.08% in 2013 to 52.31% in 2020. However, the expansion of built-up land was faster than the increase in population, and the per capita land consumption continued to increase. The proportion of public open space area also decreased from 56.0% in 2013 to 24.0% in 2020, and the proportion of population with convenient access to open public space decreased from 73.2% to 59.3%. At the township level, the SDG11 indicators of Lingui, Lingchuan, and Dingjiang, which have been rapidly growing in recent years, ranked the lowest among the evaluated towns. Therefore, the construction of public transport facilities should be increased, the low-density sprawl of built-up land should be controlled, and the area of green space should be enlarged in these areas. This study proved the effectiveness of the United Nations Sustainable Development Goal SDG11 indicators in evaluating changes in urban transportation, urban public space, and urban land use efficiency that are closely related to urban sustainability at the neighborhood level. The evaluation framework for SDG11 indicators based on HR satellite images and open geospatial big data proposed in this study can be applied to other cities, thereby contributing to the achievement of the sustainable development goals. Geospatial data with enhanced spatial and temporal resolution should be produced and applied to improve the accuracy and reliability of SDG evaluation results in future studies.

Author Contributions: Conceptualization, L.L.; methodology, L.H. and L.L.; software, K.L.; validation, K.L.; formal analysis, L.H.; investigation, S.Z.; resources, L.L.; data curation, P.W.; writing—original draft preparation, L.H.; writing—review and editing, L.L. and J.L.; visualization, P.W. and X.L.; supervision, D.H.; project administration, Q.L.; funding acquisition, H.G. All authors have read and agreed to the published version of the manuscript.

Funding: This research was funded by the Director Fund of the International Research Center of Big Data for Sustainable Development Goals (grant number CBAS2022DF016); and the National Natural Science Foundation of China (grant number 42071321).

Data Availability Statement: The data presented in this study are available on request from the corresponding author. The data are not publicly available due to the data management policies of high-resolution satellite images.

Conflicts of Interest: The authors declare no conflict of interest.

References

1. UN. Transforming our World: The 2030 Agenda for Sustainable Development. 2015. Available online: http://www.un.org/ga/search/view_doc.asp?symbol=A/RES/.70/1&Lang=E (accessed on 30 June 2022).
2. Patel, Z.; Greyling, S.; Simon, D.; Arfvidsson, H.; Moodley, N.; Primo, N.; Wright, C. Local responses to global sustainability agendas: Learning from experimenting with the urban sustainable development goal in Cape Town. *Sustain. Sci.* **2017**, *12*, 785–797. [CrossRef]
3. Ballerini, L.; Bergh, S.I. Using citizen science data to monitor the Sustainable Development Goals: A bottom-up analysis. *Sustain. Sci.* **2021**, *16*, 1945–1962. [CrossRef] [PubMed]
4. Yin, J.; Dong, J.; Hamm, N.A.S.; Li, Z.; Wang, J.; Xing, H.; Fu, P. Integrating remote sensing and geospatial big data for urban land use mapping: A review. *Int. J. Appl. Earth Obs. Geoinf.* **2021**, *103*, 102514. [CrossRef]
5. Estoque, R.C. A Review of the Sustainability Concept and the State of SDG Monitoring Using Remote Sensing. *Remote Sens.* **2020**, *12*, 1770. [CrossRef]
6. Jia, Y.; Ge, Y.; Ling, F.; Guo, X.; Wang, J.; Wang, L.; Chen, Y.; Li, X. Urban Land Use Mapping by Combining Remote Sensing Imagery and Mobile Phone Positioning Data. *Remote Sens.* **2018**, *10*, 446. [CrossRef]
7. Mariathasan, V.; Bezuidenhoudt, E.; Olympio, K.R. Evaluation of earth observation solutions for Namibia's SDG monitoring system. *Remote Sens.* **2019**, *11*, 1612. [CrossRef]
8. Anderson, K.; Ryan, B.; Sonntag, W.; Kavvada, A.; Friedl, L. Earth observation in service of the 2030 Agenda for Sustainable Development. *Geo Spat. Inf. Sci.* **2017**, *20*, 77–96. [CrossRef]
9. Kavvada, A.; Metternicht, G.; Kerblat, F.; Mudau, N.; Haldorson, M.; Laldaparsad, S.; Friedl, L.; Held, A.; Chuvieco, E. Towards delivering on the Sustainable Development Goals using Earth observations. *Remote Sens. Environ.* **2020**, *247*, 111930. [CrossRef]
10. Molina-Azorín, J.F.; Font, X. Mixed methods in sustainable tourism research: An analysis of prevalence, designs and application in JOST (2005–2014). *J. Sustain. Tour.* **2016**, *24*, 549–573. [CrossRef]

11. Lu, L.; Guo, H.; Corbane, C.; Li, Q. Urban sprawl in provincial capital cities in China: Evidence from multi-temporal urban land products using Landsat data. *Sci. Bull.* **2019**, *64*, 955–957. [CrossRef]

12. Pesaresi, M.; Huadong, G.; Blaes, X.; Ehrlich, D.; Ferri, S.; Gueguen, L.; Halkia, M.; Kauffmann, M.; Kemper, T.; Lu, L. A global human settlement layer from optical HR/VHR RS data: Concept and first results. *IEEE J. Sel. Top. Appl. Earth Obs. Remote Sens.* **2013**, *6*, 2102–2131. [CrossRef]

13. Esch, T.; Heldens, W.; Hirner, A.; Keil, M.; Marconcini, M.; Roth, A.; Zeidler, J.; Dech, S.; Strano, E. Breaking new ground in mapping human settlements from space–The Global Urban Footprint. *ISPRS J. Photogramm. Remote Sens.* **2017**, *134*, 30–42. [CrossRef]

14. Marconcini, M.; Metz-Marconcini, A.; Üreyen, S.; Palacios-Lopez, D.; Hanke, W.; Bachofer, F.; Zeidler, J.; Esch, T.; Gorelick, N.; Kakarla, A. Outlining where humans live, the World Settlement Footprint 2015. *Sci. Data* **2020**, *7*, 242. [CrossRef]

15. Ghazaryan, G.; Rienow, A.; Oldenburg, C.; Thonfeld, F.; Trampnau, B.; Sticksel, S.; Jürgens, C. Monitoring of urban sprawl and densification processes in western Germany in the light of SDG indicator 11.3.1 based on an automated retrospective classification approach. *Remote Sens.* **2021**, *13*, 1694. [CrossRef]

16. Xiao, D.; Lu, L.; Wang, X.; Nitivattananon, V.; Guo, H.; Hui, W. An urbanization monitoring dataset for world cultural heritage in the Belt and Road region. *Big Earth Data* **2022**, *6*, 127–140. [CrossRef]

17. Neumann, B.; Ott, K.; Kenchington, R. Strong sustainability in coastal areas: A conceptual interpretation of SDG 14. *Sustain. Sci.* **2017**, *12*, 1019–1035. [CrossRef] [PubMed]

18. Yamasaki, K.; Yamada, T. A framework to assess the local implementation of Sustainable Development Goal 11. *Sustain. Cities Soc.* **2022**, *84*, 104002. [CrossRef]

19. Biggeri, M.; Clark, D.A.; Ferrannini, A.; Mauro, V. Tracking the SDGs in an 'integrated' manner: A proposal for a new index to capture synergies and trade-offs between and within goals. *World Dev.* **2019**, *122*, 628–647. [CrossRef]

20. Xu, Z.; Chau, S.N.; Chen, X.; Zhang, J.; Li, Y.; Dietz, T.; Wang, J.; Winkler, J.A.; Fan, F.; Huang, B.; et al. Assessing progress towards sustainable development over space and time. *Nature* **2020**, *577*, 74–78. [CrossRef] [PubMed]

21. Moro, L.D.; Maculan, L.S.; Pivoto, D.; Cardoso, G.T.; Pinto, D.; Adelodun, B.; Bodah, B.W.; Santosh, M.; Bortoluzzi, M.G.; Branco, E.; et al. Geospatial Analysis with Landsat Series and Sentinel-3B OLCI Satellites to Assess Changes in Land Use and Water Quality over Time in Brazil. *Sustainability* **2022**, *14*, 9733. [CrossRef]

22. Zhou, M.; Lu, L.; Guo, H.; Weng, Q.; Cao, S.; Zhang, S.; Li, Q. Urban Sprawl and Changes in Land-Use Efficiency in the Beijing–Tianjin–Hebei Region, China from 2000 to 2020: A Spatiotemporal Analysis Using Earth Observation Data. *Remote Sens.* **2021**, *13*, 2850. [CrossRef]

23. Melchiorri, M.; Pesaresi, M.; Florczyk, A.J.; Corbane, C.; Kemper, T. Principles and applications of the global human settlement layer as baseline for the land use efficiency indicator—SDG 11.3.1. *ISPRS Int. J. Geo-Inf.* **2019**, *8*, 96. [CrossRef]

24. Estoque, R.C.; Ooba, M.; Togawa, T.; Hijioka, Y.; Murayama, Y. Monitoring global land-use efficiency in the context of the UN 2030 Agenda for Sustainable Development. *Habitat Int.* **2021**, *115*, 102403. [CrossRef]

25. Mudau, N.; Mwaniki, D.; Tsoeleng, L.; Mashalane, M.; Beguy, D.; Ndugwa, R. Assessment of SDG indicator 11.3. 1 and urban growth trends of major and small cities in South Africa. *Sustainability* **2020**, *12*, 7063. [CrossRef]

26. Prakash, M.; Ramage, S.; Kavvada, A.; Goodman, S. Open Earth observations for sustainable urban development. *Remote Sens.* **2020**, *12*, 1646. [CrossRef]

27. Verde, N.; Patias, P.; Mallinis, G. A Cloud-Based Mapping Approach Using Deep Learning and Very-High Spatial Resolution Earth Observation Data to Facilitate the SDG 11.7.1 Indicator Computation. *Remote Sens.* **2022**, *14*, 1011. [CrossRef]

28. Aguilar, R.; Kuffer, M. Cloud computation using high-resolution images for improving the SDG indicator on open spaces. *Remote Sens.* **2020**, *12*, 1144. [CrossRef]

29. Kearney, S.P.; Coops, N.C.; Sethi, S.; Stenhouse, G.B. Maintaining accurate, current, rural road network data: An extraction and updating routine using RapidEye, participatory GIS and deep learning. *Int. J. Appl. Earth Obs. Geoinf.* **2020**, *87*, 102031. [CrossRef]

30. Zhang, C.; Sargent, I.; Pan, X.; Li, H.; Gardiner, A.; Hare, J.; Atkinson, P.M. An object-based convolutional neural network (OCNN) for urban land use classification. *Remote Sens. Environ.* **2018**, *216*, 57–70. [CrossRef]

31. Chen, B.; Tu, Y.; Song, Y.; Theobald, D.M.; Zhang, T.; Ren, Z.; Li, X.; Yang, J.; Wang, J.; Wang, X. Mapping essential urban land use categories with open big data: Results for five metropolitan areas in the United States of America. *ISPRS J. Photogramm. Remote Sens.* **2021**, *178*, 203–218. [CrossRef]

32. Furberg, D.; Ban, Y.; Mörtberg, U. Monitoring Urban Green Infrastructure Changes and Impact on Habitat Connectivity Using High-Resolution Satellite Data. *Remote Sens.* **2020**, *12*, 3072. [CrossRef]

33. Deilami, K.; Hashim, M. Very high resolution optical satellites for DEM generation: A review. *Eur. J. Sci. Res.* **2011**, *49*, 542–554.

34. Fritz, S.; See, L.; Carlson, T.; Haklay, M.; Oliver, J.L.; Fraisl, D.; Mondardini, R.; Brocklehurst, M.; Shanley, L.A.; Schade, S.; et al. Citizen science and the United Nations Sustainable Development Goals. *Nat. Sustain.* **2019**, *2*, 922–930. [CrossRef]

35. Fried, T.; Tun, T.H.; Klopp, J.M.; Welle, B. Measuring the Sustainable Development Goal (SDG) transport target and accessibility of Nairobi's matatus. *Transp. Res. Rec.* **2020**, *2674*, 196–207. [CrossRef]

36. Tiznado-Aitken, I.; Muñoz, J.C.; Hurtubia, R. The Role of Accessibility to Public Transport and Quality of Walking Environment on Urban Equity: The Case of Santiago de Chile. *Transp. Res. Rec.* **2018**, *2672*, 129–138. [CrossRef]

37. Brussel, M.; Zuidgeest, M.; Pfeffer, K.; van Maarseveen, M. Access or Accessibility? A Critique of the Urban Transport SDG Indicator. *ISPRS Int. J. Geo. Inf.* **2019**, *8*, 67. [CrossRef]

38. El Garouani, A.; Mulla, D.J.; El Garouani, S.; Knight, J. Analysis of urban growth and sprawl from remote sensing data: Case of Fez, Morocco. *Int. J. Sustain. Built. Environ.* **2017**, *6*, 160–169. [CrossRef]
39. Kussul, N.; Lavreniuk, M.; Kolotii, A.; Skakun, S.; Rakoid, O.; Shumilo, L. A workflow for Sustainable Development Goals indicators assessment based on high-resolution satellite data. *Int. J. Digit. Earth* **2020**, *13*, 309–321. [CrossRef]
40. Wang, Y.; Li, B.; Xu, L. Monitoring Land-Use Efficiency in China's Yangtze River Economic Belt from 2000 to 2018. *Land* **2022**, *11*, 1009. [CrossRef]
41. Lu, L.; Qureshi, S.; Li, Q.; Chen, F.; Shu, L. Monitoring and projecting sustainable transitions in urban land use using remote sensing and scenario-based modelling in a coastal megacity. *Ocean. Coast. Manag.* **2022**, *224*, 106201. [CrossRef]
42. Zhou, X.; Wang, Y.C. Spatial–temporal dynamics of urban green space in response to rapid urbanization and greening policies. *Landsc. Urban Plan.* **2011**, *100*, 268–277. [CrossRef]
43. Cetin, M. Using GIS analysis to assess urban green space in terms of accessibility: Case study in Kutahya. *Int. J. Sustain. Dev. World Ecol.* **2015**, *22*, 420–424. [CrossRef]
44. Lwin, K.K.; Murayama, Y. Modelling of urban green space walkability: Eco-friendly walk score calculator. *Comput. Environ. Urban Syst.* **2011**, *35*, 408–420. [CrossRef]
45. Klopp, J.M.; Petretta, D.L. The urban sustainable development goal: Indicators, complexity and the politics of measuring cities. *Cities* **2017**, *63*, 92–97. [CrossRef]
46. Wang, Y.; Huang, C.; Feng, Y.; Zhao, M.; Gu, J. Using Earth Observation for Monitoring SDG 11.3.1-Ratio of Land Consumption Rate to Population Growth Rate in Mainland China. *Remote Sens.* **2020**, *12*, 357. [CrossRef]
47. Schiavina, M.; Melchiorri, M.; Freire, S.; Florio, P.; Ehrlich, D.; Tommasi, P.; Pesaresi, M.; Kemper, T. Land use efficiency of functional urban areas: Global pattern and evolution of development trajectories. *Habitat Int.* **2022**, *123*, 102543. [CrossRef] [PubMed]
48. Calka, B.; Orych, A.; Bielecka, E.; Mozuriunaite, S. The Ratio of the Land Consumption Rate to the Population Growth Rate: A Framework for the Achievement of the Spatiotemporal Pattern in Poland and Lithuania. *Remote Sens.* **2022**, *14*, 1074. [CrossRef]
49. Deng, J.; Huang, Y.; Chen, B.; Tong, C.; Liu, P.; Wang, H.; Hong, Y. A Methodology to Monitor Urban Expansion and Green Space Change Using a Time Series of Multi-Sensor SPOT and Sentinel-2A Images. *Remote Sens.* **2019**, *11*, 1230. [CrossRef]
50. Allen, C.; Metternicht, G.; Wiedmann, T. Initial progress in implementing the Sustainable Development Goals (SDGs): A review of evidence from countries. *Sustain. Sci.* **2018**, *13*, 1453–1467. [CrossRef]
51. He, G.; Zhao, X.; Yu, M. Exploring the multiple disturbances of karst landscape in Guilin World Heritage Site, China. *CATENA* **2021**, *203*, 105349. [CrossRef]
52. Gao, K.; Yang, X.; Wang, Z.; Zhang, H.; Huang, C.; Zeng, X. Spatial Sustainable Development Assessment Using Fusing Multisource Data from the Perspective of Production-Living-Ecological Space Division: A Case of Greater Bay Area, China. *Remote Sens.* **2022**, *14*, 2772. [CrossRef]
53. Su, X.; Zheng, C.; Yang, Y.; Yang, Y.; Zhao, W.; Yu, Y. Spatial Structure and Development Patterns of Urban Traffic Flow Network in Less Developed Areas: A Sustainable Development Perspective. *Sustainability* **2022**, *14*, 8095. [CrossRef]
54. Wang, Y.; Li, B. The Spatial Disparities of Land-Use Efficiency in Mainland China from 2000 to 2015. *Int. J. Environ. Res. Public Health* **2022**, *19*, 9982. [CrossRef]
55. Tirumala, R.D.; Tiwari, P. Importance of Land in SDG Policy Instruments: A Study of ASEAN Developing Countries. *Land* **2022**, *11*, 218. [CrossRef]
56. Zhang, X. Incremental Production of Urban Public Green Space: A 'Spiral Space' Building Typology. *Buildings* **2022**, *12*, 1330. [CrossRef]
57. Arslan, T.V.; Durak, S.; Aytac, D.O. Attaining SDG11: Can sustainability assessment tools be used for improved transformation of neighbourhoods in historic city centers? *Nat. Resour. Forum* **2016**, *40*, 180–202. [CrossRef]
58. Elmqvist, T.; Siri, J.; Andersson, E.; Anderson, P.; Bai, X.; Das, P.K.; Gatere, T.; Gonzalez, A.; Goodness, J.; Handel, S.N. Urban tinkering. *Sustain. Sci.* **2018**, *13*, 1549–1564. [CrossRef] [PubMed]
59. Liu, L.; Liang, Y.; Hashimoto, S. Integrated assessment of land-use/coverage changes and their impacts on ecosystem services in Gansu Province, northwest China: Implications for sustainable development goals. *Sustain. Sci.* **2020**, *15*, 297–314. [CrossRef]
60. Leyk, S.; Gaughan, A.E.; Adamo, S.B.; de Sherbinin, A.; Balk, D.; Freire, S.; Rose, A.; Stevens, F.R.; Blankespoor, B.; Frye, C.; et al. The spatial allocation of population: A review of large-scale gridded population data products and their fitness for use. *Earth Syst. Sci. Data* **2019**, *11*, 1385–1409. [CrossRef]
61. Tuholske, C.; Gaughan, A.E.; Sorichetta, A.; de Sherbinin, A.; Bucherie, A.; Hultquist, C.; Stevens, F.; Kruczkiewicz, A.; Huyck, C.; Yetman, G. Implications for Tracking SDG Indicator Metrics with Gridded Population Data. *Sustainability* **2021**, *13*, 7329. [CrossRef]

Article

Assessment of Intra-Urban Heat Island in a Densely Populated City Using Remote Sensing: A Case Study for Manila City

Mark Angelo Purio [1,2,*], Tetsunobu Yoshitake [3] and Mengu Cho [1]

1 Laboratory of Spacecraft Environmental Interaction Engineering, Kyushu Institute of Technology, Kitakyushu 804-8550, Japan
2 Electronics Engineering Department, Adamson University, Manila 1000, Philippines
3 Department of Civil and Architectural Engineering, Kyushu Institute of Technology, Kitakyushu 804-8550, Japan
* Correspondence: purio.mark-angelo894@mail.kyutech.jp or markangelo.purio@adamson.edu.ph; Tel.: +63-8-524-2011

Abstract: Changes in the environment occur in cities due to increased urbanization and population growth. Sustainable Development Goal (SDG) 11 is intrinsically linked to the environment, one facet of which is the need for universal access to secure, inclusive, and accessible green and public places. As urban heat islands (UHI) have the potential to negatively influence cities and their residents, existing resources and data must be used to identify and quantify these effects. To address this, we present the use of satellite-derived (2013–2022) and meteorological data (2014–2020) to assess intra-urban heat islands in Manila City, Philippines. The assessment includes (a) understanding the temporal variability of air temperature measurements and outdoor thermal comfort based on meteorological data, (b) comparative and correlative analysis between common Land-Use Land Cover indicators (Normalized Difference Vegetation Index (NDVI), Normalized Difference Water Index (NDWI) and Normalized Difference Built-up Index (NDBI)) and Land Surface Temperature (LST), (c) spatial and temporal analysis of LST using spatial statistics techniques, and (d) generation of an intra-urban heat island (IUHI) map with a recommended class of action using a suitability analysis model. Finally, the areas that need intervention are compared to the affected population, and suggestions to enhance the thermal characteristics of the city and mitigate the effects of UHI are established.

Keywords: intra-urban heat island; remote sensing; space-time; GIS; SDGs

Citation: Purio, M.A.; Yoshitake, T.; Cho, M. Assessment of Intra-Urban Heat Island in a Densely Populated City Using Remote Sensing: A Case Study for Manila City. *Remote Sens.* 2022, 14, 5573. https://doi.org/10.3390/rs14215573

Academic Editor: Ronald C. Estoque

Received: 19 September 2022
Accepted: 24 October 2022
Published: 4 November 2022

Publisher's Note: MDPI stays neutral with regard to jurisdictional claims in published maps and institutional affiliations.

1. Introduction

More focus has been placed on global urbanization recently as more people around the globe move to urban areas every year. Today, more than half of the world's population resides in urban areas, and forecasts indicate that an increasing share of urban residents will be responsible for almost all future population increases. The complicated socioeconomic process of urbanization affects the built environment, relocating the population's spatial distribution from rural to urban regions, and converting once rural areas into urban ones. It has an impact on dominant occupations, lifestyles, cultures, and behaviors in both urban and rural regions, altering both the demographic and social structure. The key effects of urbanization include the quantity, size, and density of urban settlements as well as the population share between urban and rural inhabitants [1,2]. By ensuring that cities and human settlements are inclusive, safe, and resilient, SDG 11—one of the United Nations' "17 Sustainable Development Goals"—highlights the importance that cities play in the world's political agenda [3]. In the review of Estoque [4], despite initial efforts, the UN Global Sustainable Development Report 2019 [5] found that the world is not on track to achieve most SDG objectives including indicators related to SDG 11.

Due to urban regions developing more quickly with population expansion, environmental changes ensue [6]. Loss of open space and animal habitats, water and air pollution, transportation, health concerns, and agricultural capacity are a few implications, while changing thermal properties is another result of urbanization and city growth. In terms of the increase in the Land Surface Temperature (LST) of the landscape, ongoing urbanization and the growth of impermeable surfaces are both factors [7,8]. As urban regions expand, the topography changes. Buildings, roads, and other forms of infrastructure take the place of open space and plants, for example, and permeable and moist surfaces eventually become impermeable and dry [9].

As a result of this development, urban heat islands (UHI) occur—a phenomenon in which urban areas experience warmer temperatures than their rural surroundings [10]. In particular, densely packed structures with little greenery develop "islands" with greater temperatures than their surroundings [9,11–13]. UHI may influence the increased risk of health-related conditions, increase in energy consumption, elevated pollutants, and water quality [14]. Urban heat islands (UHI) have the potential to have a detrimental impact on cities and their inhabitants, and as such, available resources and data must be used to detect and quantify these consequences. SDG 11 works toward making societies more sustainable and resilient by giving us a unique chance to make sure that the infrastructure we build today will still be useful in the future. This can be done by investing in parks and green spaces in cities, which will help reduce the "urban heat island effect" [3].

Aside from this, according to a growing body of research [15–17], "intra-urban" heat islands (IUHI), or regions within a city that are hotter than others due to an unequal distribution of heat-absorbing buildings and pavements, as well as cooler zones with trees and greenery, are becoming more prevalent [18]. Intra-Urban Heat Islands (IUHI) detection is of major interest to city planners since high temperatures influence energy usage and human health [16]. In 2015, Martin et al. [19] referred to surface intra-UHI as the detection of hotspots in a metropolis which is made possible by determining temperature thresholds by spatial reference. Consequently, the data can then be used to identify regions of interest in a city and potentially trigger alarms at a finer spatial scale. An example is a study conducted by Igergård et al. [20] in the Stockholm municipality.

In the literature, remote sensing is a good resource to understand the link between urban expansion and the characteristics describing the thermal changes in both geographical and temporal contexts [7,14,21]. Among remote sensing data, satellites are used more to estimate LST due to the thermal and passive microwave sensors aboard them. Although satellite data are very useful, Zhou et al. [14] stressed in their systematic review that retrieved satellite LST and air temperature differences, the effect of clouds, spatial and temporal resolution trade-off, SUHI quantification methods, varying land use land cover methods, and SUHI accuracy assessment are among the current challenges faced by UHI researchers. Worse, the limited availability of datasets for SUHI studies and applications exacerbates the challenges.

The increasing number of publications on the effect of UHI, particularly after 2016, reflects the scientific community's interest in disseminating information about this subject, which investigates its causes and ramifications from several viewpoints, including environmental, social, and economic [22]. The Philippines, like the rest of the world, is experiencing fast urbanization and a population density increase. Furthermore, these densely populated cities are largely clustered in Metro Manila [23,24]. In this context, statistically analyzing satellite data geographically and temporally, Landicho and Blanco [25] confirmed that intra-urban heat islands (IUHI) in Metro Manila are prevalent in 2019 while Alcantara et al. [26,27] conducted UHI studies in Quezon City. Estoque et al. [28], moreover, used satellite-derived surface temperature data and socio-ecological factors to analyze the present health risk in 139 Philippines cities. In addition, cities outside of Metro Manila were part of the Project GUHeat [24], which conducted urban heat island studies in cities such as Baguio [29], Cebu [30], Davao [31], Iloilo [32], Mandaue [33], and Zamboanga [34].

Given prior geographic biases in the literature, greater attention should be placed on understudied areas or cities, as proposed by Zhou et al. [14] and Almeida et al. [22] in their reviews. Furthermore, little published research explores how UHI affects the population because of a lack of fine-scale geographic population data [35]. Consequently, as there is inadequate research about UHI conducted in the country, area-specific assessment in cities like Manila would provide further details on how changes in the landscape impact the city's heat situation and will serve as a basis for urban planners and policymakers for mitigation and improvement. This also supports the goals of SDG 11 to aid the futureproofing of infrastructures for cleaner and greener cities.

The novelty of the present work is the use of space-time pattern mining to assess the presence of intra-urban heat islands using remote sensing data. Although this type of methodology is well established for space-time analysis applications, its usage on remote sensing data such as land surface temperature has not been extensively studied. Moreover, according to the author's knowledge, no work was dedicated to including the population and settlement data in such an assessment method for Manila City or any highly urbanized cities in the Philippines.

Its main purpose is to use satellite-derived and in situ meteorological remote sensing data to assess the presence of intra-urban heat islands in Manila City. Moreover, demographic data such as population and settlement data were used to enhance the assessment. Data represented in a space-time cube were used to carry out a space-time pattern mining approach in generating an Intra-Urban Heat Island (IUHI) map for Manila City. Finally, city-specific strategies to promote outdoor thermal comfort and hotspot interventions were also suggested. This paper is divided into five sections:

- Section 1 introduces the research, the state-of-the-art review, research gaps, and a statement of purpose.
- Section 2 presents the data and a detailed discussion of the methods employed.
- Section 3 shows the description of the results and output of the analysis.
- Section 4 discusses the results in detail, interprets the findings concerning previous studies, and examines the context of the outcomes of the study.
- Section 5 summarizes what was done in the study, the findings, and future work.

2. Materials and Methods

2.1. Study Area

As shown in Figure 1, Manila City is located in the northern Philippines archipelago, on the island of Luzon, on the eastern side of the old Manila Bay, with the Pasig River running through it [36,37]. As the Philippines' capital, Manila is considered to have the highest population density among the country's highly urbanized cities, and even among the world's densest cities. In 2020, the Philippine Statistics Authority [38] recorded that 1.84 million population reside in its 24.98 square kilometer land area, which translates to about 74,000 inhabitants per square km. The spatial attributes of the city are shown in Table 1.

Table 1. Spatial Attributes of Manila City.

Data Attributes	Description
Spatial Reference	GCS WGS 1984
Spatial Resolution	Approximately 30 m
Number of Pixels Covered	48,667
Data Format	Geo tiff

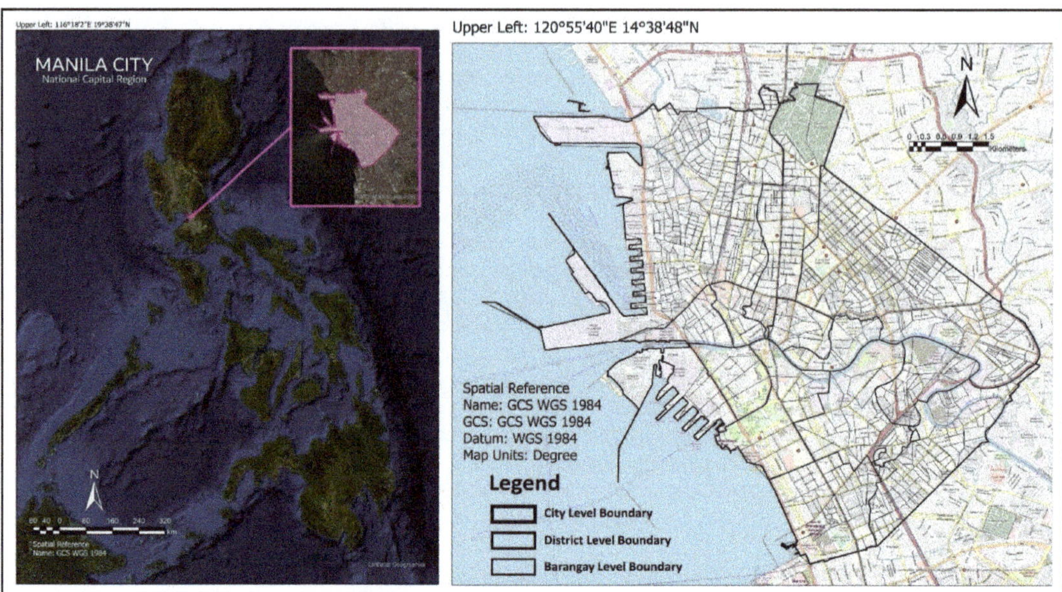

Figure 1. Manila City's geographical location (**left**) and administrative boundary (**right**).

According to the Koppen Climate Classification [39] Manila has a tropical rainforest climate (Af). There is no dry season in a tropical rainforest environment, and it rains at least 60 mm per month throughout the year (2.36 in). Tropical rainforest climates do not have distinct seasons; it is hot and humid year-round, with frequent and heavy rains. Manila has an annual average temperature of 27.8 degrees Celsius, or 82.0 degrees Fahrenheit. With an average temperature of 85.0 °F (29.4 °C), April is the hottest month of the year, while the lowest month is January at 79.0 °F (26.1 °C) [39].

2.2. Data and Data Sources

This section enumerates the data and their sources including their descriptions, attributes, and the methods employed to obtain and prepare the data.

2.2.1. Manila City Administrative Boundary

An administrative boundary represents subdivisions of areas, territories, or jurisdictions recognized by governments for administrative purposes [40]. The Philippines follows the Philippine Standard Geographic Code (PSGC) with different geographic levels such as region, province, city/municipality, and the smallest unit, barangay [41]. For the research, we need the shapefiles for Manila City at the city, district, and barangay levels. A published GitHub repository [42] was used since it is complete with all the needed geographic levels projected using the WGS 1984, latitude/longitude projection. These shapefiles were sourced from reliable webpages such as the OCHA Services Website [43] and GADM.org [44].

2.2.2. In-Situ Meteorological Data

Meteorological raw data taken daily from 2014 to 2018 were provided by the weather bureau of the Philippines. The meteorological parameters include rainfall amount, mean temperature, maximum temperature, minimum temperature, wind speed, wind direction, and relative humidity. Since just one synoptic station is in Port Area, Manila (14.5878°N latitude and 120.9690°E longitude), only point data are available for Manila City.

2.2.3. Population Data

Population density is a key metric for assessing domestic living circumstances. Due to the statistical approach used, traditional census statistics cannot represent the population's geographical distribution with a high degree of precision [35]. A high-resolution map estimate of the population density inside 30-m grid tiles was supplied by Data for Good Meta, which we used in this research. In this study, the population density demographic data for the year 2018 was used to give an insight into the distribution of people affected by the intra-urban heat island in Manila City. Since the downloaded data represent the whole country, we used ArcGIS Pro software to clip the region of interest based on the administrative boundary of Manila City. Aside from population density, those pixel grids with data are considered settlement areas while empty grids denote non-settlements areas in the city. Each cell's value represents the population density of that pixel/grid. This density may be expressed as a grid's area.

2.2.4. Satellite Data

Satellite-derived remote sensing data in the study were taken from MODIS and Landsat 8 satellite data products. Daily land surface temperatures (day and night) were obtained from MODIS between 2014 to 2018 as complementary data for the meteorological data mentioned above. Consequently, spatial yearly data raster for land surface temperature and spectral indices were downloaded from Landsat 8.

- MODIS Land Surface Temperature Product

Land Surface Temperature data were derived from the Collection-6 MODIS Land Surface Temperature product to complement the available meteorological data at hand. Details of their retrieval were reported in [45]. In this study, data were retrieved to complement the meteorological data since global daily LST data can be obtained with this. Although the spatial resolution is low, the temporal resolution of the MODIS dataset is good and was deemed applicable for the correlation analysis presented in Ref [45].

- Landsat 8 Data Product

The Climate Engine web app (https://app.climateengine.com/climateEngine# (accessed on 9 February 2020)) [46] was used to download analysis-ready Landsat 8 data which were preprocessed using the Google Earth Engine [47] platform. The web application allows easy download of Landsat Bands, Spectral Indices, and Land Surface Temperature aggregated per year of study. In ecological studies, digital numbers and reflectance are the most used while studies involving thermal bands often use digital numbers and temperatures. For this study, we used top-of-atmosphere (TOA) reflectance products to obtain the land surface temperature (LST) and surface reflectance (SR) products for spectral indices such as Normalized Difference Vegetation Index (NDVI), Normalized Difference Water Index (NDWI), and Normalized Difference Built-up Index (NDBI).

- Land Surface Temperature

According to ESA [48], "Land Surface Temperature (LST) is the radiative skin temperature of the land derived from solar radiation. A simplified definition would be how hot the "surface" of the Earth would feel to the touch in a particular location. From a satellite's point of view, the "surface" is whatever it sees when it looks through the atmosphere to the ground. It could be snow and ice, the grass on a lawn, the roof of a building, or the leaves in the canopy of a forest. Land surface temperature is not the same as the air temperature that is included in the daily weather report."

Landsat 8 passes the equator at 10:00 am +/− 15 min (mean local time) [49] so the maps that will be generated are only based on measurements from this specific time of the day. While IUHI can be measured better in Manila City in the afternoon than in the morning, the limitation of satellite data to provide this led the researchers to use such Landsat data for the investigation. Deilami et al. [50] stressed in their review that the popularity of Landsat images for UHI studies can be attributed to factors such as being freely available to researchers, their worldwide coverage with a reasonable spatial resolution of 30×30 m, and the long-term temporal coverage which enables researchers to extract the required information over a long period to monitor changes. Moreover, in their review article, about 22% of the papers reviewed use Landsat 8 data for UHI investigation with data available from 2013 to the present [50].

Raster data of land surface temperature data were taken from 2013 to 2022 on a yearly interval. Because of constraints in maximum cloud cover, LST within the year was obtained to depict maximum temperatures occurrence for that year. The top-of-atmosphere (TOA) product was used to illustrate the presence of cold and hotspots in the yearly intra-urban heat island map generated. Although the actual resolution Landsat 8 LST is 100 m, the analysis product downloaded from the climate engine is provided at 30 m.

- Normalized Difference Vegetation Index (NDVI)

The NDVI is a dimensionless index that describes the difference between visible and near-infrared reflectance of vegetation cover and can be used to estimate the density of green on an area of land. No green leaves produce a value near zero, yet calculations of NDVI for a particular pixel always yield a figure that falls between a negative one (-1) and a positive one $(+1)$. A value of zero denotes no vegetation, whereas a value of close to one (0.8–0.9) represents the greatest potential density of green leaves [51]. The following formula is used to calculate NDVI:

$$\text{NDVI} = \frac{(\text{NIR} - \text{Red})}{(\text{NIR} + \text{Red})} \tag{1}$$

For Landsat data, NDVI $=$ (Band 5 $-$ Band 4/(Band 5 $+$ Band 4). This can be directly downloaded from the climate engine. Table 2 shows the ranges of NDVI and their corresponding land use land cover (LULC) classification.

Table 2. NDVI ranges for LULC Classification.

NDVI Ranges	Land Use Land Cover (LULC) Classification	Class
−1.0 to 0.0	Water Body	1
0.0 to +0.2	Urban Built-up	2
+0.2 to +1.0	Vegetation	3

- Normalized Difference Water Index (NDWI)

NDWI is a measure of liquid water molecules in vegetation canopies that interacted with the incoming solar radiation. It is less sensitive to atmospheric scattering effects than NDVI [52]. This index uses NIR and SWIR bands where the resulting value ranges from minus one (-1) to plus one $(+1)$. Positive values of NDWI correspond to high vegetation water content and high vegetation fraction cover. Negative NDWI values correspond to low vegetation water content and low vegetation fraction cover. In a period of water stress, NDWI will decrease. The following formula gives the NDWI value.

$$\text{NDWI} = \frac{(\text{NIR} - \text{SWIR1})}{(\text{NIR} + \text{SWIR1})} \tag{2}$$

For Landsat data, NDWI $=$ (Band 5 $-$ Band 6)(Band 5 $+$ Band 6). This can be directly downloaded from the climate engine. Table 3 shows the ranges of NDWI values and the corresponding water content classification.

Table 3. NDWI ranges for Water Content Classification.

NDWI Ranges	Water Content Classification	Class
−1.0 to 0.0	Low Water Content	1
0.0 to +0.1	High Water Content	2

- Normalized Difference Built-up Index (NDBI)

The Normalized Difference Built-up Index (NDBI) uses the NIR and SWIR bands to emphasize constructed built-up areas. It is a ratio based on mitigating the effects of terrain illumination differences as well as atmospheric effects [53,54]. A negative value of NDBI represents water bodies whereas a higher value represents build-up areas. NDBI value for vegetation is low. The following formula gives the NDBI value.

$$NDBI = \frac{(SWIR1 - NIR)}{(SWIR1 + NIR)} \tag{3}$$

For Landsat 8 data, NDBI $= ($Band $6 -$ Band $5)/($Band $6 +$ Band $5)$. This cannot be directly downloaded from the climate engine, so the individual NIR and SWIR1 bands were downloaded, then NDBI was calculated using the raster calculator tool in ArcGIS Pro. Table 4 shows the ranges of NDBI values and the corresponding build-up area classification.

Table 4. NDBI ranges for Build-up Area Classification.

NDBI Ranges	Build-Up Area Classification	Class
−1.0 to 0.0	Non-Built-up areas	1
0.0 to +0.1	Built-up areas	2

2.3. Methodology

The workflow is divided into the following parts: (a) Meteorological Data and Land Surface Temperature Evaluation Methods, (b) LULC and LST Comparative and Correlation Analysis, (c) LST Spatiotemporal Pattern Analysis and Hotspots/Cold spots Identification, and (d) Intra-Urban Heat Island Map Generation.

The overall workflow of this methodology is shown in Figure 2. Finally, using the information obtained, data assessment and suggested area-specific mitigation strategies are provided.

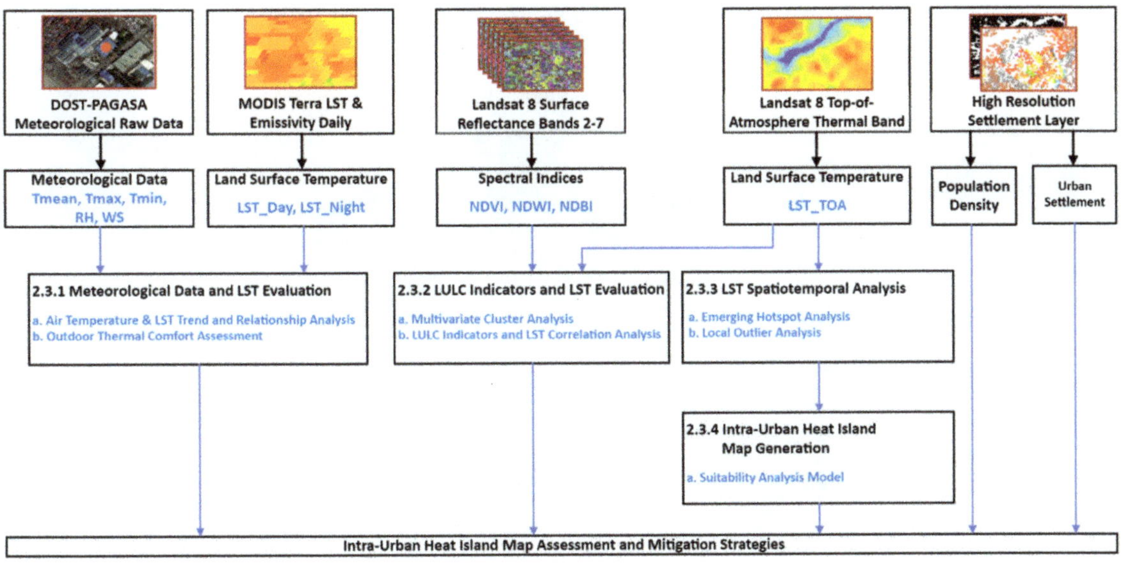

Figure 2. Overview of the overall workflow of the study to assess the IUHI map and provide mitigation strategies.

2.3.1. Meteorological Data and Land Surface Temperature Evaluation Methods

This section focuses on the use of meteorological data collected at Port Area, Manila City, and how they are used to understand the temporal variability of air temperature, the relationship of meteorological parameters to land surface temperature during the day and night, and outdoor thermal comfort assessment.

a.   Air Temperature and LST Trend and Relationship Analysis

This analysis's methodology and findings were already published by the authors in ref. [45]. There was no gap-filling technique used for missing information related to the in-situ measurements nor with the derived MODIS data specific to the meteorological data point. The in-situ data were directly taken from the weather agency which processed and prepared the data, while the MODIS data are directly downloaded from the Google earth engine. All data used were analysis-ready while any data point with a missing parameter entry was discarded and not used.

b.   Outdoor Thermal Comfort Assessment

The RayMan Model was proposed by Matzarakis, a micro-scale model developed to calculate radiation fluxes in simple and complex environments [55,56]. This research used this model to assess the thermal comfort in Port Area. The scientific basis for the computations is thoroughly detailed in the Rayman Pro tool handbook [55].

Thermal indices have been developed to approximate human thermal perception [55]. In particular, Physiological Equivalent Temperature (PET) is "the air temperature at which, in a typical indoor setting (without wind and solar radiation), the energy budget of the human body is balanced with the same core and skin temperature as under the complex outdoor conditions to be assessed" [57,58].

The Thermal Comfort Assessment workflow is as follows:

1.   Preparation of input parameters (Air Temperature, Relative Humidity, and Wind Velocity) in a .csv file as input to the RayMan Model.
2.   Calculate the Tmrt and Thermal Index (PET) using the RayMan Pro Software. The Graphical User Interface which contains the geographic data, personal data, and clothing & activity information used is shown in Figure 3.
3.   Graph the calculated values for comparison.
4.   Assess the thermal comfort by getting the equivalent physiological stress associated with the derived thermal index values as shown in Table 5.

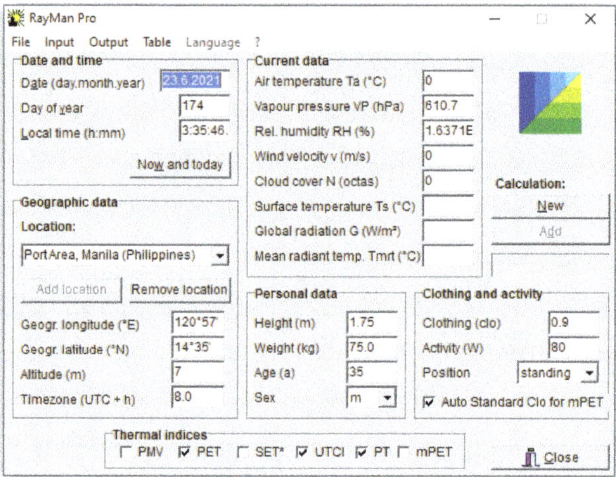

Figure 3. RayMan Pro Graphical User Interface. Geographic data, personal data, and clothing and activity information are shown.

Table 5. PET Thermal Index, corresponding classes, thermal sensation, and physiological stress.

Thermal Sensation	PET Range for Taiwan (°C PET) [59]	PET Range for Western/Middle Europe (°C PET) [59]	Physiological Stress
Very Cold	<+14	<+4	Extreme cold stress
–	–	–	Very strong cold stress
Cold	+14–+18	+4–+8	Strong cold stress
Cool	+18–+22	+8–+13	Moderate cold stress
Slightly Cool	+22–+26	+13–+18	Light cold stress
Neutral	+26–+30	+18–+23	No thermal stress (Thermal Comfort Zone)
Slightly Warm	+30–+34	+23–+29	Light heat stress
Warm	+34–+38	+29–+35	Moderate heat stress
Hot	+38–+42	+35–+41	Strong heat stress
–	–	–	Very strong heat stress
Very Warm	>+42	>+41	Extreme heat stress

It should be emphasized that the data being used in this analysis are solely temporal point data from Manila City's Port Area. It is deemed that these values do not represent the entire city; therefore, meteorological data-point locations should be explored to offer a better understanding of the thermal comfort in Manila City.

2.3.2. LULC Indicators and LST Evaluation Methods

This section discusses methods to evaluate satellite-derived data such as spectral indices (NDVI, NDWI, and NDBI, which are used as LULC indicators) and land surface temperature in Manila City. These methods include multivariate cluster analysis and correlation analysis.

a. Multivariate Cluster Analysis

Cluster analysis is a statistical method to use the values of the variables in devising a scheme for grouping the objects into classes so that similar objects are in the same class [60]. It is a multivariate method for classifying a sample of subjects (or objects) into several groups based on a set of measured characteristics, with related subjects placed in the same group.

Given that the group of values for each parameter is not known, we used the satellite-derived data to group the values in each parameter (NDVI, NDWI, NDBI) together with land surface temperature (LST) and observed how each of these LULC indicators relate to LST. Specifically, since the indicator values can be used to classify land use and land cover, this is an initial step to see how the land use and land cover of different areas in Manila City relate to their thermal characteristic.

For this, the k-means algorithm as shown in Algorithm 1 was used to identify the clusters within the dataset. It is an iterative algorithm that divides the unlabeled dataset into k different clusters in such a way that each dataset belongs to only one group that has similar properties [61]. The k-means clustering algorithm mainly performs two tasks: (1) determine the best value for k-center points or centroids by an iterative process and (2) assign each data point to its closest k-center. Those data points which are near a particular k-center create a cluster. Hence, each cluster has data points with some commonalities, and it is away from other clusters. Shown below is the k-means clustering algorithm flow.

Algorithm 1: k-means algorithm

1:	Specify the number k of clusters to assign.
2:	Randomly initialize k centroids.
3:	Repeat
4:	expectation: Assign each point to its closest centroid
5:	maximization: Compute the new centroid (mean) of each cluster.
6:	until the centroid positions do not change.

In this study, we used the multivariate clustering tool in ArcGIS Pro [62] to find these natural clusters of features based solely on the feature attribute values. Given the number of clusters to create, it will look for a solution where all the features within each cluster are as similar as possible, and all the clusters themselves are as different as possible. This tool utilizes unsupervised machine learning methods to determine natural clusters in the data. The classification method is considered unsupervised as they do not require a set of reclassified features to guide or train the method to find the clusters in the data. Since the tool is used to run the clustering algorithm, the following workflow was employed:

1. Extract the values from the raster map at different years to create a feature layer. The spectral indices (NDVI, NDWI, & NDBI) are in values between -1 and 1 while land surface temperature is in degrees Celsius (°C). All the raster data are taken from Landsat 8 as explained in Section 2.2.4.
2. Import the data into the ArcGIS Pro software and use the generated feature layer as input.
3. Execute the k-means clustering algorithm with the following:
4. Clustering method: k-means
5. Initialization Method: Optimized seed locations
6. Number of clusters: 4
7. Generate the cluster chart and interpret the results according to each of the input variables.

It should be noted that cluster analysis has no mechanism for differentiating between relevant and irrelevant variables. Therefore, the choice of variables included in a cluster analysis must be underpinned by conceptual considerations. This is very important because the clusters formed can be very dependent on the variables included. To see the relationship and extent of the values used in clustering, we also employed correlation analysis with the data.

b. LULC Indicators and LST Correlation Analysis

We use correlation analysis in addition to multivariate clustering analysis to evaluate the relationship of NDVI, NDWI, and NDBI with LST. The same method as explained in Section 2.3.1-a was used to analyze the extent and nature of the relationship between the abovementioned parameters. On the contrary, Pearson product correlation in GeoDa software was used.

2.3.3. LST Spatiotemporal Pattern Analysis

In this section, we focus on analyzing the spatial and temporal pattern of Land Surface Temperature in Manila City Philippines. Since data have both spatial and temporal context, several analytical tools in the Space-Time Pattern Analysis toolset in ArcGIS Pro software [62] were used. Before doing the analysis, a space-time cube was created based on the downloaded LST raster over the period (2014 to 2021) as shown in Figure 4.

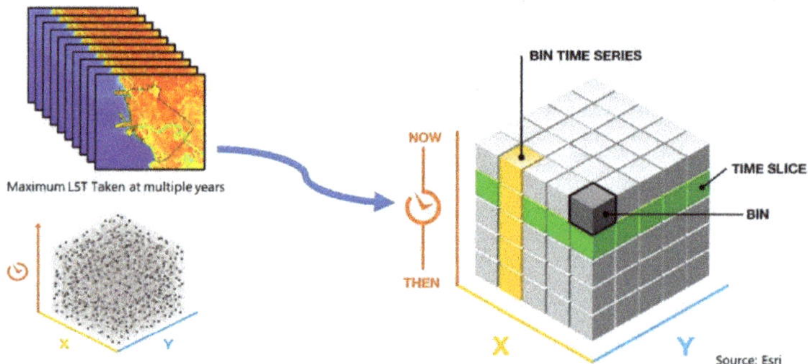

Figure 4. Creating a space-time cube based on yearly maximum LST from 2013 to 2022.

A time series analysis or an integrated spatial and temporal pattern analysis may be used to view and analyze spatial-temporal data using this approach. Using the prepared space-time cube as input, we perform emerging hotspot analysis and local outlier analysis to better understand the thermal situation in Manila City.

a.   Emerging Hotspot Analysis

The Emerging Hot Spot Analysis tool detects statistically significant hot and cold spot patterns over time. It is used to examine land surface temperature (LST) data in Manila City to identify new, intensifying, persistent, or sporadic hot spot trends at various time-step intervals. The workflow for this is as follows:

1.   Taking the space-time NetCDF cube created for LST as input.
2.   Conceptualize the spatial relationships of LST values using the k-nearest neighbor method with k = 8, where the eight closest neighbors to the target feature will be included in computations for that feature.
3.   Calculate the Getis-Ord Gi^* statistic [63] for each bin (pixel), represented in Table 6. The Getis-Ord local statistic is given as:

$$G_i^* = \frac{\sum_{j=1}^{n} w_{i,j} x_j - \overline{X} \sum_{j=1}^{n} w_{i,j}}{S\sqrt{\frac{\left[n \sum_{j=1}^{n} w_{i,j}^2 - \left(\sum_{j=1}^{n} w_{i,j}\right)^2\right]}{n-1}}} \tag{4}$$

where x_j is the attribute value for feature j, $w_{i,j}$ is the subscript weight between feature i and j, n is equal to the number of features; also:

$$\overline{X} = \frac{\sum_{j=1}^{n} x_j}{n} \tag{5}$$

$$S = \sqrt{\frac{\sum_{j=1}^{n} x_j^2}{n} - \left(\overline{X}\right)^2} \tag{6}$$

The G_i^* is a z-score so no further calculations are required.

Table 6. G_i^* statistic values for cold spot and hotspot classes at different significance levels.

Statistical Significance Level	G_i^* Statistic Pixel Representation	
	Cold Spot	**Hotspot**
99% confidence	−3	+3
95% confidence	−2	+2
90% confidence	−1	+1
Statistically not significant	0	

The G_i^* statistic returned for each point is a z-score. The more concentrated the clustering of high values (hot spots) of LST, the bigger the z-score for statistically significant positive z-scores. The clustering of low values (called a "cold spot") of LST is stronger, the smaller the z-score is for statistically significant negative z-scores.

4.   Once the space-time hot spot analysis completes, each bin (pixel) in the input NetCDF cube has an associated z-score, p-value, and hot spot bin classification added to it.
5.   Next, these hot and cold spot trends are evaluated using the Mann–Kendall trend test. As an independent bin time-series test, the Mann–Kendall trend test [64] is done for every location/point with LST data. For the point value and their time sequence, the Mann–Kendall statistic is a rank correlation analysis. The first time's point value is compared to the second time's point value. The outcome is +1 if the first is smaller than the second. The outcome is −1 if the first is greater than the second. The outcome is 0 if the two numbers are equal. The results are added together for each pair of periods compared. The predicted sum is 0, indicating that the numbers do not show

any trend over time. Based on the variance for the values in the point time series, the number of ties, and the number of periods, the observed sum is compared to the expected sum (zero) to determine if the difference is statistically significant. A z-score and a *p*-value are used to represent the trend for each point time series. A small *p*-value indicates that the trend is statistically significant. The sign associated with the z-score determines if the trend is an increase in point values (positive z-score) or a decrease in bin values (negative z-score).

With the resultant trend z-score and *p*-value for each location with data, and with the hot spot z-score and *p*-value for each bin, the Emerging Hot Spot Analysis tool in ArcGIS Pro categorizes each study area location as shown in Table 7 and is then reclassified as "monitor", "intervene", and "preserve". With the new classification, those categorized as diminishing, oscillating, and historical for both hot and cold spots will be reclassified as "monitor". Those with no pattern detected will be classified as "monitor" as well. On the other hand, categories such as new, consecutive, intensifying, and sporadic will have "preserve" as their new class for a cold spot and "intervene" for a hotspot.

6. An Emerging Hotspot Analysis (EHSA) Map showing areas to preserve, monitor, and intervene is generated based on the reclassification shown in Table 7.

Table 7. Emerging hot spot analysis trend categories, their definition, and equivalent new class.

Category		Definition	New Class
No Pattern Detected		Does not fall into any of the hot or cold spot patterns defined below	Monitor
Hot Spot	New	the most recent time step interval is hot for the first time	Intervene
	Consecutive	a single uninterrupted run of hot time step intervals, with of less than 90% of all intervals	Intervene
	Intensifying	at least 90% of the time step intervals are hot and become hotter over time	Intervene
	Persistent	at least 90% of the time step intervals are hot, with no trend up or down	Intervene
	Sporadic	some of the time step intervals are hot	Intervene
	Diminishing	at least 90% of the time step intervals are hot and become less hot over time	Monitor
	Oscillating	some of the time step intervals are hot, some are cold	Monitor
	Historical	at least 90% of the time step intervals are hot, but the most recent time step interval is not	Monitor
Cold Spot	New	the most recent time step interval is cold for the first time	Preserve
	Consecutive	a single uninterrupted run of cold time step intervals, withof less than 90% of all	Preserve
	Intensifying	at least 90% of the time step intervals are cold and become colder over time	Preserve
	Persistent	at least 90% of the time step intervals are cold, with no trend up or down	Preserve
	Sporadic	some of the time step intervals are cold	Preserve
	Diminishing	at least 90% of the time step intervals are cold and become less cold over time intervals	Monitor
	Oscillating	some of the time step intervals are cold, some are hot	Monitor
	Historical	at least 90% of the time step intervals are cold, but the most recent time step interval is not	Monitor

b. Local Outlier Analysis

The Local Outlier Analysis tool identifies statistically significant clusters of high or low land surface temperature LST values as well as outliers that have values that are statistically different from their neighbors in space and time.

The workflow for this is as follows:

1. Use the space-time NetCDF cube created for LST as input.
2. Conceptualize the spatial relationships of LST values using the k-nearest neighbor method with k = 8, where the eight closest neighbors to the target feature will be included in computations for that feature.

3. Calculate the Anselin Local Moran's *I* statistic of special association for each bin which includes a pseudo *p*-value and a CO_Type code. The Local Moran's *I* statistic of spatial association is given as

$$I_i = \frac{x_i - \overline{X}}{S_i^2} \sum_{j=1,\, j\neq i}^{n} w_{i,j}\left(x_i - \overline{X}\right) \tag{7}$$

where x_i is an attribute for feature i, $\overline{X}$ is the mean corresponding attribute, $w_{i,j}$ is the spatial weight between features i and j, and:

$$S_i^2 = \frac{\sum_{j=1,\, j\neq i}^{n}\left(x_j - \overline{X}\right)^2}{n - 1} \tag{8}$$

with n equating to the total number of features. The z_{I_i} score for the statistics is computed as

$$z_{I_i} = \frac{I_i - E[I_i]}{\sqrt{V[I_i]}} \tag{9}$$

$$V[I_i] = E\left[I_i^2\right] - E[I_i]^2 \tag{10}$$

A positive value for *I* indicates that a feature has neighboring features with similarly high or low attribute values; this feature is part of a cluster. A negative value for I indicates that a feature has neighboring features with dissimilar values; this feature is an outlier. In either instance, the *p*-value for the feature must be small enough for the cluster or outlier to be considered statistically significant.

In Table 8, the cluster/outlier type (CO Type) field distinguishes between a statistically significant cluster of high values (HH), a cluster of low values (LL), an outlier in which a high value is surrounded primarily by low values (HL), and an outlier in which a low value is surrounded primarily by high values (LH). Statistical significance is set at the 95 percent confidence level. This significance represents an FDR correction, which adjusts the *p*-value threshold from 0.05 to a value that better reflects the 95 percent confidence level taking into consideration multiple testing.

4. A two-dimensional map summarizing each location over time is created with the following categories shown in Table 9. Then, a new class is created based on these categories wherein pixels categorized as never significant, multiple types and outliers will be reclassified as "monitor" while only the high-high cluster and the low-low cluster will be reclassified as intervene and preserve, respectively.

5. Finally, a Local Outlier Analysis (LOA) Map showing areas to preserve, monitor, and intervene will be generated.

Table 8. Pixel representation of cluster and outliers based on the Anselin Local Moran's *I* statistic.

Cluster/Outlier Type	Definition
Never Significant	A location that is not statistically significant.
High-High Cluster (HH)	Locations that are part of a cluster of high LST_TOA values.
High-Low Outlier (HL)	Locations that represent high outliers within a cluster of low LST_TOA values.
Low-High Outlier (LH)	Locations that represent low outliers within a cluster of high LST_TOA values.
Low-Low Cluster (LL)	Locations that are part of a cluster of low LST_TOA values.

Table 9. Local outlier analysis trend categories, their definition, and equivalent new class.

Category	Definition	New Class
Never Significant	A location where there has never been a statistically significant CO_TYPE.	Monitor
Only High-High Cluster	A location where the only statistically significant type throughout time has been High-High Clusters.	Intervene
Only High-Low Outlier	A location where the only statistically significant type throughout time has been High-Low Outliers.	Monitor
Only Low-High Outlier	A location where the only statistically significant type throughout time has been Low-High Outliers.	Monitor
Only Low-Low Cluster	A location where the only statistically significant type throughout time has been Low-Low Clusters.	Preserve
Multiple Types	A location where there have been multiple types of statistically significant clusters and outlier types throughout time.	Monitor

2.3.4. Intra-Urban Heat Island Map Generation

This section discusses the method of generating the intra-urban island map for Manila City, Philippines, using results from EHSA and LOA through a Suitability Analysis Model.

Figure 5 shows the overall process to produce the needed map for further assessment. The Emerging Hot Spot Analysis identifies trends in the data, such as new, intensifying, diminishing, and sporadic hot and cold spots, while the Local Outlier Analysis identifies significant clusters and outliers in the data. Through the suitability analysis, the combination of both methods ensures that locations of hot and cold spots in the city are precisely identified by eliminating outlier clusters in the final map produced. The suitability analysis model was used to combine the resulting raster map from the emerging hotspot analysis and local outlier analysis.

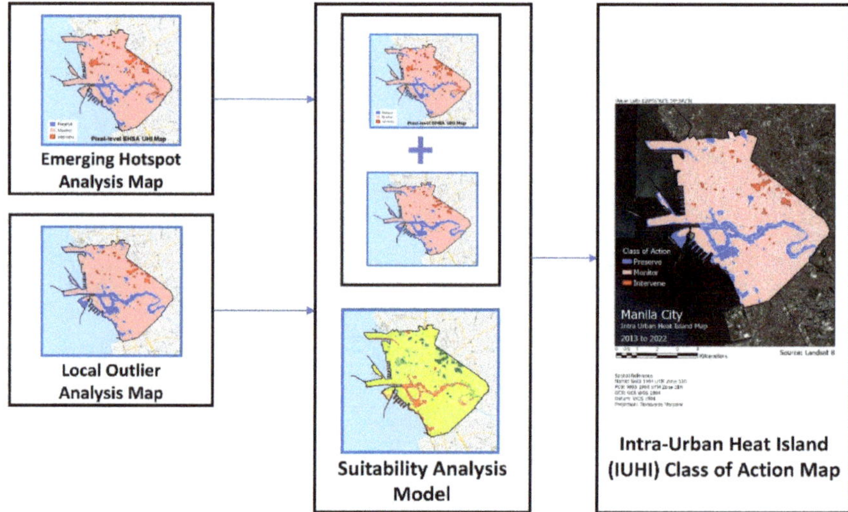

Figure 5. Overview of the Intra-Urban Heat Island (IUHI) Class of Action map generation based on EHSA and LOA maps using the suitability analysis model.

To carry out the suitability analysis, the classification classes of emerging hotspot analysis and local outlier analysis were given numerical equivalents to provide a common suitability scale.

Specifically, the following workflow was followed:

1. Preparation of criteria data. The resulting maps from the emerging hotspot analysis and local outlier analysis were prepared with their corresponding classes.
2. Transforming the classes of each criterion to a common suitability scale is shown in Table 10.
3. Assigning weight relative to each of the criteria and combining them to create a suitability map. In this application, we treat each criterion as equally important, so weight is assigned as a percentage: 50% for EHSA Classification and 50% for LOA Classification.
4. Finally, the pixel values were reclassified according to Table 11, shown to give an Intra-Urban Heat Island (IUHI) Class of Action Map.

Table 10. Common suitability scale used to transform EHSA and LOA Classification maps.

Emerging Hotspot Analysis (EHSA) Classification	Local Outlier Analysis (LOA) Classification	Suitability Scale
Preserve	Preserve	1
Monitor	Monitor	2
Intervene	Intervene	3

Table 11. Suitability values and their equivalent IUHI Class of Action.

Emerging Hotspot Analysis (EHSA) Classification	Local Outlier Analysis (LOA) Classification	Suitability Model Suitability Value	IUHI Class of Action
1	1	1.0	Preserve
1	2	1.5	Preserve
2	1	1.5	Preserve
1	3	2.0	Monitor
2	2	2.0	Monitor
3	1	2.0	Monitor
2	3	2.5	Monitor
3	2	2.5	Monitor
3	3	3.0	Intervene

2.3.5. Intra-Urban Heat Island Map Assessment and Mitigation Strategies

The results in Sections 2.3.1–2.3.4 are then used to evaluate the Intra-Urban Heat Island map with the population data and urban settlement raster from the high-resolution settlement layer. Moreover, area-specific mitigation strategies will be suggested based on the visual inspection of the areas that need intervention. Possible strategies may also be taken from the identified areas to be preserved in the city. Assessment and mitigation strategies are simplified so that they serve as a basis for urban planners and policymakers for mitigation and improvement.

3. Results

3.1. Satellite Data Retrieved from Landsat 8

Ten distribution maps from 2013 to 2022 were obtained from Landsat 8 data through the climate engine web application. These data were further processed in ArcGIS Pro by providing an equalized histogram stretch and a specific color scheme in its symbology.

3.2. Meteorological Data and Land Surface Temperature Evaluation

3.2.1. Air Temperature and LST Trend and Relationship Analysis

Figure 6 shows the monthly maximum (Tmax), mean (Tmean), and minimum (Tmin) air temperature trends from 2014 to 2018. The values were taken from the diurnal data and were averaged per month to clearly show the monthly trend. This observation was discussed in [45] showing an upward trend in the values starting from March and continuing to April and May while values start to drop in October until around January and February. Such an observation is the same as what was presented by Estoque et al. [28] and Manalo et al. [65] in their framework showing the climate and seasons in the Philippines based on combined rainfall and temperature. Between March to May, the Philippines experiences a hot dry season which explains the high recorded air temperature.

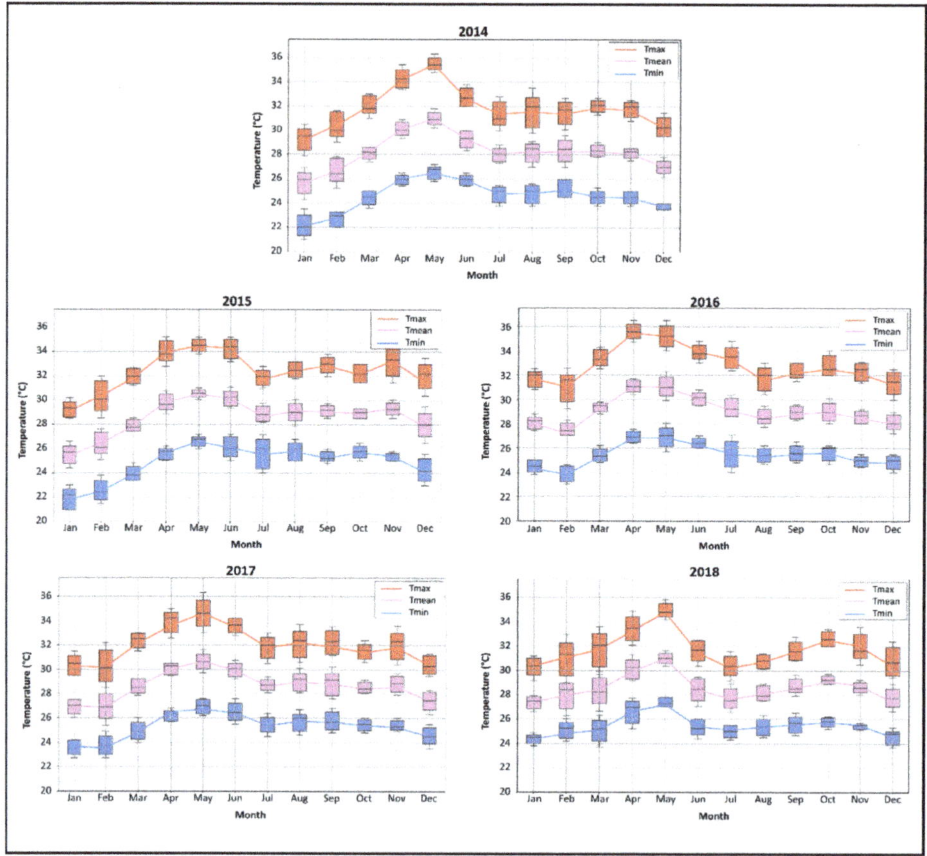

Figure 6. Monthly maximum (Tmax), mean (Tmean), and minimum (Tmin) air temperature trends from 2014 to 2018.

Additionally in our paper [45], we found a significant linear correlation between air temperature (maximum, mean, and minimum) and land surface temperature (day and night) as analyzed from available daily data shown in Table 12. On the other hand, the relative humidity shows a weak correlation with the LST data although it is shown to be significant for LST_Night.

Table 12. Corresponding interpretation of the quantitative values from the correlation analysis [45]. (* not significant).

	LST_Day	LST_Night
Tmax	moderate	strong
Tmean	moderate	strong
Tmin	moderate	strong
RH	weak *	weak

3.2.2. Outdoor Thermal Comfort Assessment

Using the same meteorological data (Tmean, Relative Humidity, and Wind Speed) taken in Port Area, Manila City, from 2014 to 2018, the Physiological Equivalent Temperature (PET) thermal index was estimated through the RayMan model. The diurnal data were computed and then averaged per month and are shown in Figure 7. Additionally, the corresponding physiological stress levels for each of the values are indicated.

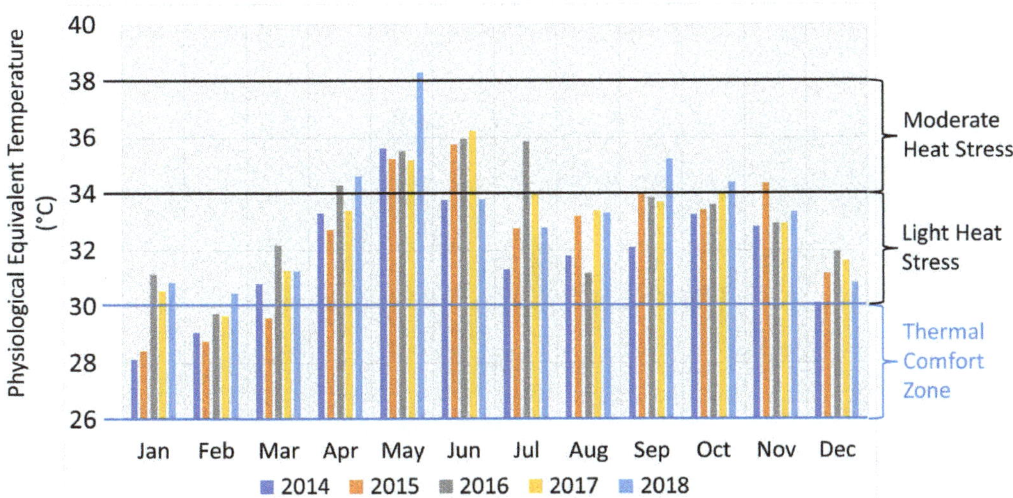

Figure 7. Monthly estimated Physiological Equivalent Temperature (PET) based on the RayMan model from 2014 to 2018.

As shown, moderate heat stress can be consistently felt in May and at some points in April and June. From July to December, light heat stress was observed, while the thermal comfort zone where there is no thermal stress only appeared in January and February. Understanding the thermal comfort in this area can also give us an idea on what is the expected outdoor thermal comfort in the other parts of Manila City. These results will be used as part of the assessment method in the latter part of the study.

3.3. LULC Indicators and Evaluation Methods
3.3.1. Multivariate Cluster Analysis

From the space-time cube generated for spectral indices (NDVI, NDWI, and NDBI) used as land use and land cover indicators and top-of-atmosphere land surface temperature (TOA_LST), the k-means clustering algorithm was used to identify the clusters within the dataset. Four groups were initialized to see a cluster for high LST (1 cluster), mid-LST (2 clusters), and low LST values (1 cluster). Standardized parameter values were plotted to clearly show the distribution of clusters, as the measurement units are not the same.

Figure 8 shows the boxplot of the result of the multivariate cluster analysis. The clustering results indicate that for the high LST cluster, values with low NDWI, moderate NDVI, and high NDBI values are clustered together. This is also expected since low NDWI correlates to low water content and high NDBI corresponds to urbanized regions. In contrast, mid-range NDVI values correspond to urbanized areas. For the low LST cluster values are clustered with high NDWI values, low NDVI values, and low NDBI values. A high NDWI refers to a high-water content, a negative NDVI to water bodies, and a low NDBI to undeveloped regions. Consequently, two mid-LST clusters were produced because of varying parameter combinations. The first set of clusters for mid-LST (orange line) is seen to be a combination of negative NDBI, high NDVI, and a higher mid-value of NDWI which translates to lowly built-up, high vegetation with a fair amount of moisture content. On the other hand, the second set of mid-LST clusters (light blue line) is composed of NDBI, NDVI, and NDWI values close to zero which can be interpreted as areas with low to no built-up and low water content.

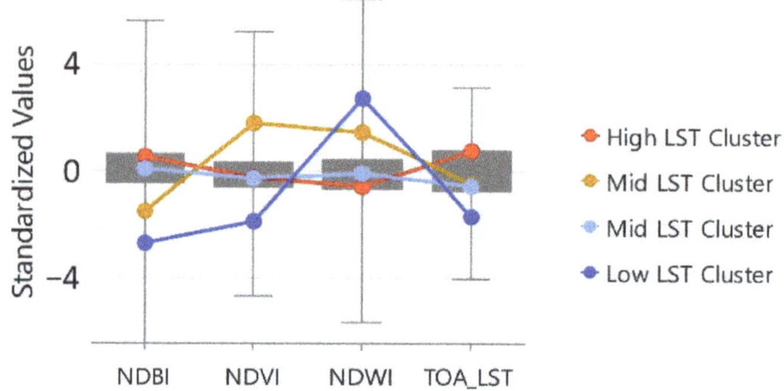

Figure 8. Boxplot of the multivariate cluster analysis result.

3.3.2. LULC Indicators and LST Correlation Analysis

The same dataset was used to see the correlation of these parameters (NDVI, NDWI, NDBI) with land surface temperature (TOA_LST). GeoDa software was used to calculate the Pearson correlation and plot the results.

Figure 9 shows the relationship between LST and LULC indicators with their corresponding slope of linear fit and frequency distribution chart while all indicators are significant at $p < 0.01$. The results show that there is a direct relationship between LST and NDBI at a $r = 0.361$ which means that highly built-up areas have high recorded temperature values. This observation agrees with the multivariate analysis. An indirect relationship is, however, observed between LST and NDVI ($r = -0.064$) and LST and NDWI ($r = -0.365$). The low Pearson correlation value between LST and NDVI indicates that both water body values and vegetation are expected to have low temperatures while mid values correspond to being built-up. With LST and NDWI, areas with high water/moisture content are more likely to have lower surface temperatures compared to areas with low water/moisture content. Based on these results, it can be inferred that the correlation values suggest that NDWI is a better indicator than NDVI for land surface temperature, which is aligned with the findings of Alexander et al. [66]. In addition, results also suggest that NDBI is a good indicator for LST.

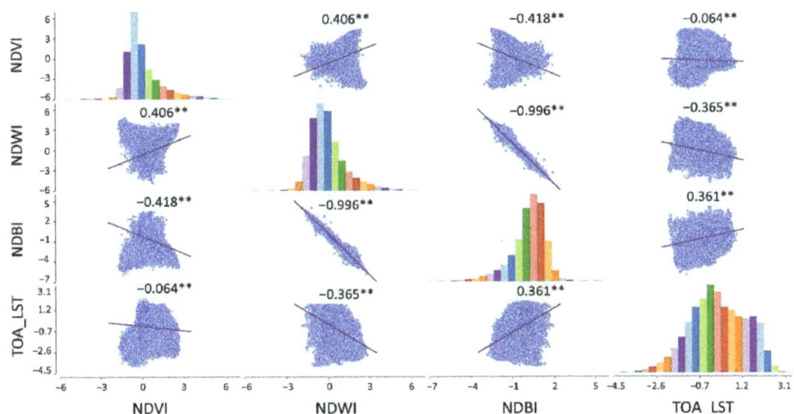

Figure 9. Relationship between LST and spectral indices with their corresponding slope of linear fit and frequency distribution chart. ** significant at $p < 0.01$.

3.4. LST Spatiotemporal Pattern Analysis

3.4.1. Emerging Hotspot Analysis

Based on the generated Emerging Hotspot Analysis (ESHA) Map, a reclassified map was also produced to indicate areas to preserve, monitor, and intervene.

As shown in Figure 10, cold spot and hot spot areas were mapped using the trend categories and a corresponding new class.

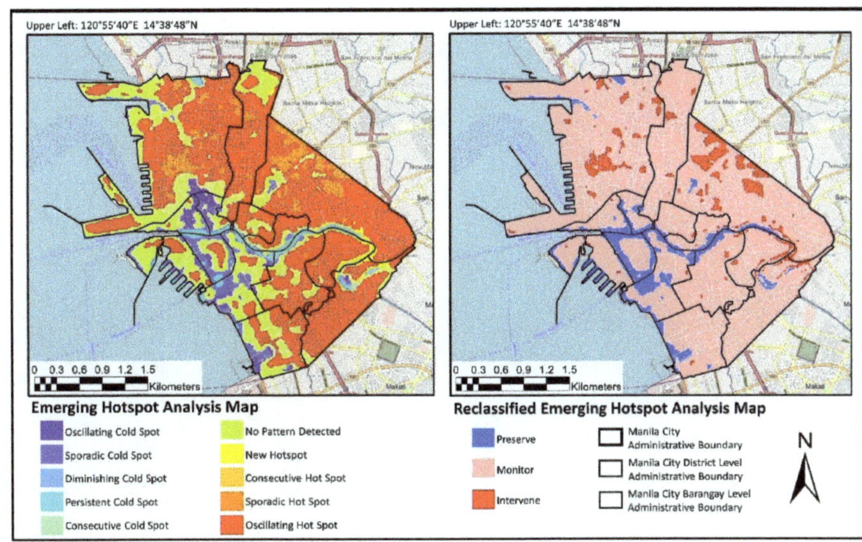

Figure 10. Emerging Hotspot Analysis Map and the reclassified map with the corresponding new class.

3.4.2. Local Outlier Analysis

Based on the generated Local Outlier Analysis (LOA) Map, a reclassified map was also produced to indicate areas to "preserve", "monitor", and "intervene". In Figure 11, the trend categories of clusters and outliers are shown on the left while the corresponding new class is also provided in the map on the right.

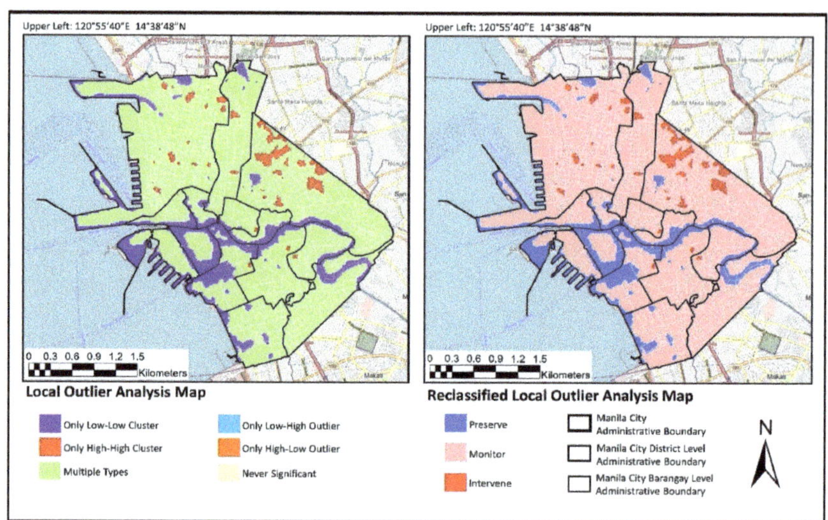

Figure 11. Local Outlier Analysis Map and the reclassified map with the corresponding new class.

3.5. Intra-Urban Island Map

Using the generated maps presented in Sections 3.2.1 and 3.2.2, a suitability analysis model was used to combine the raster maps. The suitability analysis was carried out by giving numerical equivalents for the new classification maps for emerging hotspot analysis and local outlier analysis with a common suitability scale.

Figure 12 (left) shows the resulting suitability map with suitability values per pixel. Consequently, the equivalent Intra-Urban Heat Island (IUHI) Class of Action was produced as shown in Figure 12 (right).

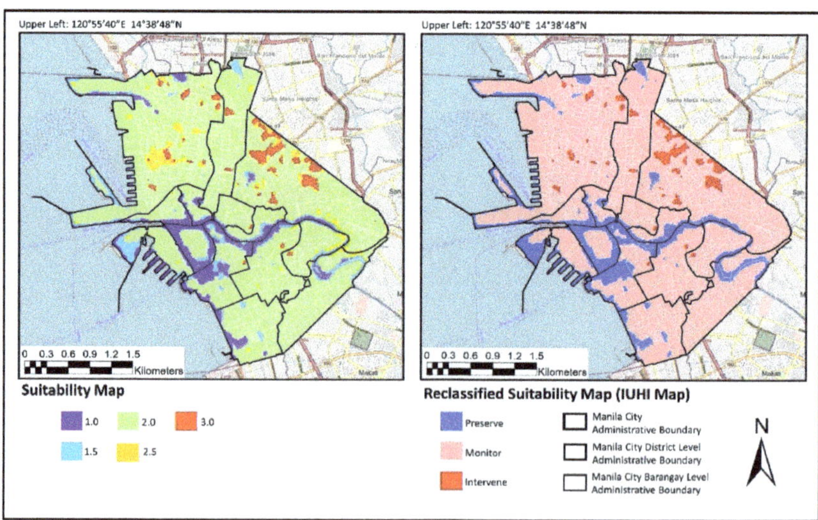

Figure 12. Suitability Map and the reclassified suitability (IUHI) map with the corresponding new class.

In Figure 13, the final Intra-Urban Heat Island (IUHI) Map of Manila City (2013–2022) was created. To keep the map as intuitive as possible, the class of action as well as the administrative boundaries at the city, district, and barangay levels were provided. This allows an easy understanding of the map while still showing the locations where areas need preservation, monitoring, and intervention.

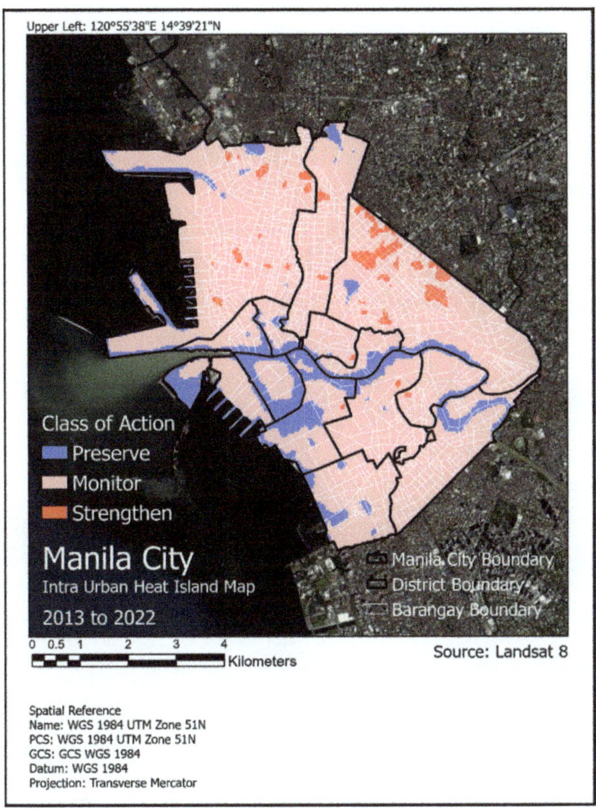

Figure 13. Intra-Urban Heat Island Map of Manila City (2013–2022).

3.6. Intra-Urban Heat Island Map Assessment and Mitigation Strategies

3.6.1. Location Assessment

Using the IUHI Map of Manila City, areas classified as "preserve" and "intervene" were examined visually using high-resolution maps from Google Earth Pro.

From the IUHI map, areas that need intervention were assessed by visually inspecting the locations to see the morphology of the areas exhibiting consistent surface temperatures during the study period. Based on the inspection, most of these areas fall within the Sampaloc district, which is part of Manila City's university belt shown in Figure 14E–H catering to Manila's academic population. The area's abundance of hotels and boarding houses makes it ideal as a dormitory and as a commuting town [36]. Moreover, there are also a few areas situated in Tondo District (A, B, and C) which is among the biggest urban poor communities in Manila City. Area D, on the other hand, mainly points toward a commercial location in Paco District.

Looking at the high-resolution satellite images, the areas shown in Figure 14 represent commonality in terms of their urban structure. It is noticeable that these areas (A, B, C, E, F, G, and H) are mostly residential and is characterized by predominantly settlement and housing locations with narrow streets and sidewalks. Although there are attempts to introduce urban soft scape via trees and vegetation, these are few and sparsely distributed within the areas of concern. In general, roads and walkways are mainly built with asphalt and concrete which might contribute to higher surface temperatures. There is also commercial space identified, such as (D), which seemed to have establishments and buildings and parking spaces made of either asphalt or concrete as well.

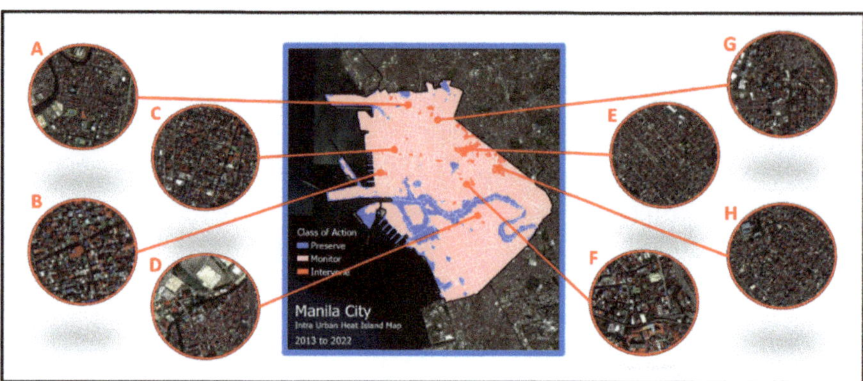

Figure 14. Some areas with the "Intervene" Class of Action. (**A–H**) are the areas highlighted to show their morphologies.

The same approach was applied in examining the areas to be preserved shown in Figure 15. Aside from the stretch of Pasig River amidst Manila City, the Intramuros district including Rizal Park Complex (part of Ermita district) as shown in (D) shows large areas with relatively lower surface temperatures. It is the historic core of Manila and is described as the "walled city" where walls surrounding the area are present until today. The Intramuros area has evident low surface temperature due to its strategic location. Aside from being situated near a body of water (Pasig River), the area is surrounded by greenery (mostly grass and some shrubs and trees) which is part of a golf range. On the other hand, the Rizal Park complex is one of the largest urban parks in Asia wherein the area is a combination of vegetation and trees, gardens water features, and shaded areas.

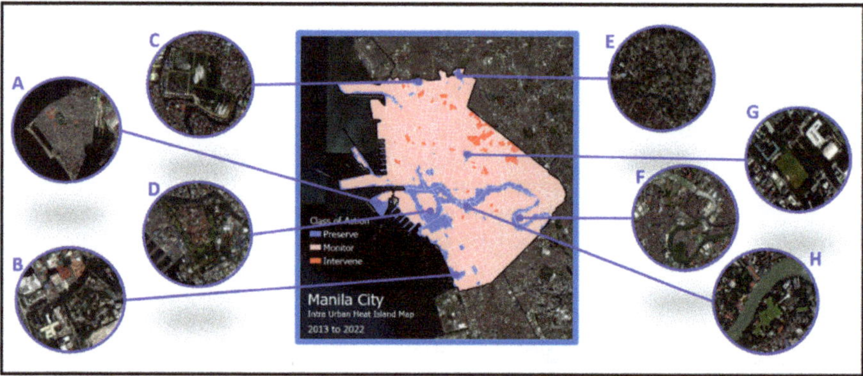

Figure 15. Some areas with the "Preserve" Class of Action. (A-H) are the areas highlighted to show their morphologies.

Predominantly, most of the areas shown in Figure 14 exhibit common morphological characteristics. For instance, areas shown in A, F, and H are either surrounded or akin to bodies of water and other water features, while areas shown in C, D, and G contain substantial vegetation and green areas. In addition, areas like B and E, although residential, also contain a decent quantity of trees spread within the area.

In this visual inspection, the two areas have distinguishable features which relate to the surface temperature in the area. Understanding the morphological characteristics of the cold spots (preserve) can help in planning the mitigation strategies needed to improve the thermal condition of the hotspots (intervene).

3.6.2. IUHI Class of Action and LULC Indicators Assessment

Overlaying the 2022 maps with the IUHI Map, the average values per class of action are shown in Table 13. It can be observed that the average NDVI values do not provide a clear distinction among the classes of action since the expected cold spots (water bodies and vegetation) have values at the extremes of the index. On the contrary, NDWI and NDBI average values convey the results. For instance, for "preserve", the average NDWI translates to higher water content while the average NDBI shows non-built-up areas. A similar remark can be drawn for "intervene" values where the average NDWI means low water content and the average NDBI falls in the built-up area category.

Table 13. Average values of LULC indicators per IUHI class of action.

Class of Action	Average NDVI	Average NDWI	Average NDBI
Preserve	0.209	0.089	−0.090
Monitor	0.190	−0.027	0.028
Intervene	0.158	−0.079	0.081

Using the same data, we also investigate how the individual index classification is distributed among the IUHI class of action to validate it with the literature. Table 14 provides the distribution of NDVI-based LULC per class of action. It can be observed that areas considered as "preserve" have a higher proportion of water bodies and vegetation while areas considered as "intervene" mostly fall into the urban built-up category.

Table 14. Distribution of LULC per IUHI class of action based on NDVI.

Class of Action	Water Body	Urban Built-Up	Vegetation	Total
Preserve	1.76%	6.11%	6.24%	14.10%
Monitor	0.21%	55.24%	27.80%	83.25%
Intervene	0.00%	2.26%	0.39%	2.65%
Total	1.96%	63.61%	34.43%	100.00%

Table 15 shows the distribution of water content category per IUHI class of action based on NDWI. Based on the proportions, most parts of the areas considered "preserve" have high water content while those for "intervene" have low water content. This shows that the water content of the area has an impact on its surface temperature.

Table 15. Distribution of Water Content category per IUHI class of action based on NDWI.

Class of Action	High Water Content	Low Water Content	Total
Preserve	10.07%	4.03%	14.10%
Monitor	26.72%	56.52%	83.25%
Intervene	0.11%	2.54%	2.65%
Total	36.91%	63.09%	100.00%

Table 16 shows the distribution of built-up categories per IUHI class of action based on NDBI. As shown about two-thirds of the "preserve" area occupy non-built-up locations while almost all parts of the "intervene" area are built up. This illustrates the effect of built-up areas such as infrastructures, roads, and buildings that contribute to higher surface temperatures in the city.

Table 16. Distribution of Built-up category per IUHI class of action based on NDBI.

Class of Action	Built-Up	Non-Built-Up	Total
Preserve	4.19%	9.91%	14.10%
Monitor	57.19%	26.06%	83.25%
Intervene	2.56%	0.10%	2.65%
Total	63.94%	36.06%	100.00%

Based on the observations above, LULC indicators allow us to assess the IUHI maps according to different aspects of the indices. By understanding such categories and how they are related to the IUHI map class of action, the areas can be quantitatively described and later can be used to incorporate mitigation strategies.

3.6.3. IUHI Class of Action and High-Resolution Settlement Layer Assessment

The high-resolution settlement layer which consists of population per pixel and settlement categories was also used to assess the IUHI map. The demographic data represent the year 2018 which is the latest available during the conduct of the study.

By superimposing the generated IUHI Class of Action Raster and High-Resolution Settlement Layer containing population per pixel and settlement class, an attribute table is generated. From this attribute table, statistics about the population data and settlement information are taken and summarized in Tables 16 and 17. An example of the attribute table is shown in Figure 16. The object ID represents the corresponding pixel where values related to the attributes are provided. In the Population/Settlement column, population per pixel is shown while those that indicate zero mean a non-settlement pixel.

Table 17. Distribution of affected population per IUHI class of action.

Class of Action	Estimated Affected Population	Population Percentage
Preserve	85,601	4.92%
Monitor	1,594,166	91.55%
Intervene	61,531	3.53%
Estimated Total Population (2018)	1,741,298	100.00%

	A	B	C	D	E	F	G	H	I
1	OBJECTID	Population/ Settlement	Class of Action	LULC	NDVI	NDWI	NDBI	NDWI_Class	NDBI_Class
2	74464	0	1	1	-0.3993	0.0681	-0.2203	High Water Content	Non Built-up
3	39913	0	1	1	-0.3760	0.1921	-0.3623	High Water Content	Non Built-up
4	31026	0	1	1	-0.3721	0.3202	-0.3606	High Water Content	Non Built-up
5	40222	0	1	1	-0.3601	0.1873	-0.2780	High Water Content	Non Built-up
6	73704	0	1	1	-0.3549	0.3502	-0.3354	High Water Content	Non Built-up
7	39668	0	1	1	-0.3535	0.2042	-0.3017	High Water Content	Non Built-up
8	29919	0	1	1	-0.3482	0.3238	-0.3515	High Water Content	Non Built-up
9	38223	42	1	1	-0.3462	0.1000	-0.3444	High Water Content	Non Built-up
10	40161	0	1	1	-0.3459	0.2682	-0.3010	High Water Content	Non Built-up
11	39852	0	1	1	-0.3459	0.3087	-0.3622	High Water Content	Non Built-up
12	28681	44	1	1	-0.3448	0.3413	-0.3581	High Water Content	Non Built-up
13	59464	0	1	1	-0.3438	0.1204	-0.2669	High Water Content	Non Built-up
14	39266	0	1	1	-0.3428	0.4137	-0.3961	High Water Content	Non Built-up
15	29781	0	1	1	-0.3373	0.3672	-0.3924	High Water Content	Non Built-up
16	39331	113	1	1	-0.3352	0.5650	-0.5293	High Water Content	Non Built-up

Figure 16. Excel Sheet of the superimposed IUHI Class with Population/Settlement Data.

In Table 17, although the percentage of "intervene" areas is small compared to the other IUHI categories, there are still about 61 thousand of the population affected by higher surface temperatures. As Manila is a densely populated city, the population despite its small percentage is still not negligible.

In Table 18, the distribution of settlement categories (from the high-resolution settlement layer data) with IUHI class of action is presented. We can see that about three-fifths (1.70%/2.65%) of the "intervention" area falls on settlement areas. This implies that most of these areas are inhabited by people, which was backed up by the visual inspection in Section 3.4.1. For the "preserve" class of action, most of the areas are non-settlement areas which are mostly vegetated locations, parks, and those near the water features.

Table 18. Distribution of settlement category per IUHI class of action.

Class of Action	Settlement	Non-Settlement	Total
Preserve	2.37%	11.73%	14.10%
Monitor	41.88%	41.37%	83.25%
Intervene	1.70%	0.95%	2.65%
Total	45.96%	54.04%	100.00%

3.6.4. IUHI Class of Action and Land Surface Temperature

To compare the variation of temperature between the cold spots (preserve) and hotspots (intervene), the yearly land surface temperature was calculated for each class of action.

A summary table of the average LST per year per class of action is shown in Table 19. As can be seen, the average difference between the warmest and coldest areas in Manila City is 6.13 °C. The difference through the years has a small deviation wherein the lowest is recorded in 2013 while the highest is in 2017. To better see the trend, a graphical representation of Table 18 is shown in Figure 17.

Table 19. Average LST (°C) per year per IUHI class of action.

	Preserve	Monitor	Intervene	Difference
2013	28.56	31.87	33.94	5.38
2014	34.32	37.74	39.47	5.15
2015	37.24	41.96	44.07	6.83
2016	38.88	43.19	44.78	5.90
2017	32.46	37.03	39.74	7.28
2018	33.84	37.49	40.23	6.39
2019	36.00	40.81	43.12	7.12
2020	33.90	37.36	39.12	5.22
2021	36.25	40.89	42.76	6.51
2022	32.91	36.79	38.39	5.48
Average LST	34.43	38.51	40.56	6.13

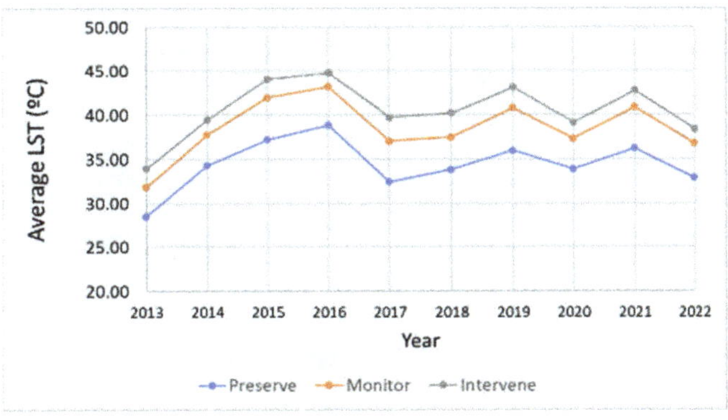

Figure 17. Average LST per year per class of action.

3.6.5. Mitigation Strategies for Areas That Need Intervention

With the assessment done in Sections 3.6.1–3.6.4., the differences in temperatures at different urban morphologies were tackled. SDG 11, with its aim to make cities and human settlements inclusive, safe, resilient, and sustainable, can only be realized by not only understanding the city's current situation but also providing means to identify vulnerable areas and implementing solutions to solve existing problems. While the assessment provides information about the presence of intra-urban heat islands in Manila City, this also offers insights into which area in the city policymakers can focus on in offering mitigation strategies. In the analysis, for example, urban settlement and residential areas with narrow streets and sidewalks, asphalted roads and walkways, and concrete commercial spaces can contribute to high surface temperatures, while areas surrounded by and near bodies of water/water features, substantial green spaces/vegetation/trees, and residential areas with decent quantities of trees are places of lower surface temperature. With this in mind, the following mitigation strategies are suggested to help ameliorate the effect of urban heat islands, some of which were adapted from the compendium of strategies by the U.S. Environmental Protection Agency [9].

As part of the local institutional mechanism to address SDG 11, the government can include the following in their priority development initiatives, especially in the identified areas for intervention:

1. Water mist/dry-mist sprayer on pavements and pedestrians. Since the provision of water features may not be possible, mist sprayers can be installed on pavements and pedestrians with the likelihood of people staying or passing by. This inhibits the heat island effect at a low cost and immediately cools the outside air directly [67].
2. Provision of shade structures. Shading can be done in multiple ways, such as with large, canopied trees (which is unlikely based on the assessment) or overhead features to reduce heat buildup in an area. Aside from heat buildup mitigation, it can also be used as protection for people under the heat of the sun.
3. Using cool materials for pavements and roofs. Cool materials are characterized by high solar reflectance and high infrared emittance which result in affecting the temperature of the surface [68]. Replacement of asphalted and concrete roads and pavements with these materials can be done while government-related projects can use cool materials for their roofs and other infrastructures.
4. Provision of cooling centers. Also known as "heat refuge", this includes libraries, community centers, commercial spaces, and other public buildings with cooling systems available to city residents during extreme heat events [69]. Manila City has these spaces already, so additional facilities and designation of such areas is the only requirement.

Additionally, the current densely populated city cannot accommodate extra large-scale trees and vegetation anymore, so the following alternatives can be employed:

5. Conversion of regular walls to green walls. Green walls are partially or completely covered with vegetation and seem lush. They are both beautiful and energizing. Consequently, they absorb warm air, reduce interior and exterior temperatures, and enhance air quality and visual appeal [70]. They are several areas in the city with empty walls but with enough space to convert them to green walls.
6. Plants in plant boxes, road isles, and indoors. One indication of urbanization is the shortage of green spaces [71], so planting in plant boxes, road isles, and indoors can help in improving the thermal landscape without planting trees. Although this cannot provide shading as with a tree canopy, the presence of plants can help in air temperature reduction and evapotranspiration [72]. Manila City still has those spaces for plant boxes and road isles and can encourage its residents to do indoor planting, which is common in the Philippines now.

These are just some of the mitigation strategies applicable to Manila City in its current state. For the attainment of SDG 11 and to address the ill effects of UHI that would result in a sustainable and livable city, a holistic approach is necessary for implementing such strategies. It should be highlighted that the local government unit including its population plays an important role in this.

4. Discussion

The result of this study shows evaluation methods using multiple sources to understand the presence of Intra-Urban Heat Islands in Manila City, Philippines. The satellite data retrieved from Landsat 8 provided distribution maps from 2013 to 2022 which include land surface temperature and LULC indicators such as NDVI, NDWI, and NDBI. More satellite data from MODIS Terra were also obtained to provide point data for land surface temperature data for both day and night. In addition, in-situ data were obtained at Port Area, Manila City, with meteorological data measurements from 2014 to 2018. Finally, raster data containing population density and urban settlement category for 2018 were acquired to represent demographics data for Manila City.

The LST and air temperature data show that beginning in March and continuing through April and May, there is an increasing tendency in the values, whereas values begin to decline in October and continue through January and February, which is similar to the observations in [28,65]. This trend is because March to May is the hot dry season in the Philippines while October to January is rainy and December to February is the cool dry season. In addition, it was found that there is a significant linear relationship between air temperature and land surface temperature based on daily data, while relative humidity shows a weak correlation with the LST data.

In terms of outdoor thermal comfort, a limited analysis was done due to limitations provided by the point measurements of meteorological data in Port Area Manila, City from 2014 to 2018. Despite these limitations, we used the meteorological parameters to estimate the Physiological Equivalent Temperature (PET) thermal index using the RayMan microclimate model. With the calculated PET thermal index values, corresponding physiological stress levels were provided to understand the outdoor thermal comfort. We observed that mild heat stress may be routinely experienced in May, and at certain times in April and June. From July through December, moderate heat stress was seen; however, the thermal comfort zone, where there is no heat stress, did not emerge until January and February. Understanding the thermal comfort in this location may also help us predict the outdoor thermal comfort in other areas of Manila City. It should be noted that the location of Port Area, Manila City is near Manila Bay, which may indicate that the meteorological parameters may not be representative of the whole of Manila City. The calculation of thermal index is calculated based on the meteorological parameters while these meteorological parameters were correlated with land surface temperature. With this, we have associated thermal comfort indirectly with the land surface temperature such that while Port Area, Manila City is not considered as an area for intervention, it still experiences heat stress. Therefore, other areas which are considered areas for intervention are more likely to experience worse thermal stress than Port Area, Manila. This observation and the generated IUHI map can be the basis for selecting additional meteorological stations in areas that may experience worse heat stress, so it can be monitored and provided by mitigation strategies in the future.

Land Use Land Cover (LULC) indicators such as NDVI, NDWI, and NDBI were very useful in understanding the morphological characteristics of Manila City, while their relationship with land surface temperature was also considered. Results of the multivariate analysis show that clusters can be generated based on combinations of these LULC indicators relative to land surface temperature. The clustering findings reveal that values with low NDWI, moderate NDVI, and high NDBI are grouped in the high LST cluster. Low NDWI corresponds to low water content, and high NDBI corresponds to urbanized zones; therefore, this is also predicted. Correlation between LULC indicators and LST shows the link between LST and LULC indicators with their respective slope of linear fit and frequency distribution chart. The data demonstrate a direct association between LST and NDBI at r = 0.361, meaning highly built-up regions have high reported temperatures. The multivariate analysis supports this finding. LST and NDVI (r = 0.064) and NDWI (r = 0.365) have indirect relationships. A Low Pearson correlation between LST and NDVI implies low temperatures for water bodies and vegetation, whereas mid values imply built-up areas. High water/moisture locations exhibit lower surface temperatures using LST and NDWI. Based on these data, it can be argued that NDWI is a better indication than NDVI for land surface temperature, which agrees with Alexander et al. [66]. NDBI is a good indication for LST, according to the data.

The creation of a space-time cube for LST made spatiotemporal pattern analysis easier. Using the space-time mining tools in ArcGIS Pro, Emerging Hotspot Analysis and Local Outlier Analysis were performed. The resulting reclassified maps of EHSA and LOA were respectively used as input to the suitability analysis model to generate an easy-to-understand Intra-Urban Heat Island (IUHI) class of action map between 2013 to 2022. Such a map contains the class of action (preserve, monitor, and intervene) as well as the administrative boundaries at the city, district, and barangay levels.

In the location assessment, the focus was given to areas to preserve and intervene. Understanding the morphology of "preserve" locations helps in the provision of mitigation strategies for the "intervene" locations. The results show that the highest temperatures are in areas with a concentration of urban settlement areas, buildings, and establishments while those with low temperatures are areas with enough vegetation and near bodies of water. Visual inspection revealed that most "intervene" areas are in the Sampaloc district and university belt. Such an area has a high concentration of universities and colleges while within it are settlement areas, establishments, and concrete roadways which are deemed contributory to the high surface temperature. Knowing this is crucial because aside from its residents, the population in this area swells due to students and employees coming from the nearby province during the daytime. Other intervention areas can be found in the Tondo district, which is home to urban poor communities, while there are also hotspots in the Paco district, which mainly points toward a commercial location. These regions are largely residential, with small streets and sidewalks and a concentration of settlements and dwelling sites. In the regions of concern, initiatives to create an urban soft scape employing trees and plants are limited and scarce. Roads and sidewalks are often constructed with asphalt and concrete, which may contribute to greater surface temperatures. There is also an identifiable commercial area, which seems to have asphalt or concrete companies, buildings, and parking spaces.

On the other hand, "preserve" areas are mostly located in Intramuros, Rizal Park, and sites near the Pasig River banks. Most of the regions have similar physical characteristics. For example, these places are either next to or resembling bodies of water and other water features, while other areas have extensive vegetation and green landscapes. Additionally, residential neighborhoods feature a significant number of trees. Noting these characteristics, mitigation strategies appropriate to the "intervene" areas can be established.

The IUHI class of action was also assessed relative to the corresponding LULC indicator values. While NDVI does not provide a clear distinction among the classes of action, NDVI and NDWI convey their results. For example, the average NDWI for "preserve" indicates a greater water content, but the average NDBI indicates undeveloped lands. Similar observations may be made for "intervene" values when the average NDWI indicates a low water content and the average NDBI falls under the category of "built-up area." Using the same data, we also investigate how the individual index classification is distributed among the IUHI class of action to validate it with the literature. It may be noticed that regions designated as "preserve" have a greater percentage of water bodies and vegetation, higher water content, and occupy non-built-up locations while regions designated as "intervene" are in urban built-up areas with lower water content.

With the high-resolution settlement layer (HRSL), the distribution of the affected population including the settlement category for 2018 was assessed. Upon superimposing the HRSL with the IUHI class of action map, about 61 thousand of the population are affected by higher surface temperatures as indicated in the "intervene" areas. Despite the small percentage of "intervene" locations compared to the entire Manila City; it is evident that such a small percentage is not negligible due to the city's dense population. In terms of the settlement category, the "intervene" locations are mostly located in settlement areas while the "preserve" locations are in non-settlement areas. Such observation is aligned with what was observed in the visual inspection of locations using high-resolution satellite images.

Summarizing the LST values per year per class of action reveals an average LST for "preserve", "monitor" and "intervene" as 34.43 °C, 38.51 °C, and 40.56 °C, respectively. The result of this study clearly shows differences in temperature within Manila City. With these data, the average difference between cold and warm areas is about 6 °C, just as in the discussion in [20]. As the LST statistics are based on the highest LST readings for each site, it should be understood that the highest LST recorded differentiates 6 °C between specific urban areas. We avoided pixel-based comparison in the overall analysis to evaluate clusters of warm and cold regions appropriate to a city viewpoint and to make the analysis more significant.

Finally, applicable mitigation strategies based on the assessment of cold spots and hotspots in the city were proposed. These strategies support the attainment of SDG 11 in making cities and human settlements inclusive, safe resilient, and sustainable. Such strategies are (1) water mist/dry-mist sprayer in pavements and pedestrians, (2) provision of shade structures, (3) using cool materials for pavements and roofs, (4) provision of cooling center, (5) conversion of regular walls to green walls, and (6) plants in plant boxes, road isles, and indoors.

5. Conclusions

This study presents the use of satellite-derived data and meteorological data to assess the presence of an intra-urban heat island in Manila City, Philippines. To address SDG 11 and provide better insights to make cities and human settlements inclusive, safe resilient, and sustainable in terms of UHI, different assessment methods were used and established. The assessment includes (a) understanding the temporal variability of air temperature measurements and outdoor thermal comfort based on meteorological data, (b) comparative and correlative analysis between common LULC indicators (NDVI, NDBI, and NDWI) to LST, (c) spatial and temporal analysis of LST using spatial statistics techniques, and (d) generation of an intra-urban heat island (IUHI) map with a recommended class of action using a suitability analysis model. Finally, the areas that need intervention are compared to the affected population, and suggestions to enhance the thermal characteristics of the city and mitigate the effects of UHI were established. Results show that there exists a clear difference between cold and warm areas within Manila City. Overall, residential areas, asphalted and concrete roads and walkways, and some commercial establishments and buildings exhibit higher surface temperatures compared to areas with vegetation and near bodies of water. Based on the results, mitigation strategies applicable to Manila City were proposed to improve the areas which need intervention.

In the future, we plan to realize these strategies by partnering with the local government unit to implement these proposed measures. We also advise providing additional meteorological stations to some of the hotspots, to understand outdoor thermal comfort in Manila City better. In addition, the methods used in this study can also be used in other cities as well as municipalities that require assessment due to the presence of intra-urban heat islands.

Author Contributions: Conceptualization, M.A.P., M.C. and T.Y.; methodology, M.A.P., M.C. and T.Y.; software, M.A.P.; validation, M.A.P., M.C. and T.Y.; formal analysis, M.A.P., M.C. and T.Y.; investigation, M.A.P.; resources, M.A.P.; data curation, M.A.P.; writing—original draft preparation, M.A.P.; writing—review and editing, M.C. and T.Y.; visualization, M.A.P.; supervision, M.C. and T.Y.; project administration, M.A.P.; funding acquisition, M.A.P. and M.C. All authors have read and agreed to the published version of the manuscript.

Remote Sens. **2022**, 14, 5573

Funding: This work was funded and supported by Adamson University, Kyushu Institute of Technology, and the Department of Science and Technology—Science Education Institution (DOST-SEI) through the STAMINA4Space Program of the University of the Philippines-Diliman.

Data Availability Statement: Not applicable.

Acknowledgments: The authors would like to thank Adamson University, Kyushu Institute of Technology, and the Department of Science and Technology—Science Education Institution (DOST-SEI) through the STAMINA4Space Program of the University of the Philippines-Diliman for supporting and funding this research. Moreover, we thank the following who one way or another helped in the realization of this work: Climate and Agrometeorological Data Section (CADS), Climatology and Agrometeorology Division (CAD), Philippine Atmospheric, Geophysical and Astronomical Services Administration (PAGASA) for the weather data; Maria Fe B. Abalos, Knowledge Management and Communications Division, Information Technology and Dissemination Service, Philippine Statistics authority for Philippine statistical data including the population for Manila City; Julius M. Judan, SSED-Ground Receiving Station, DOST-Advanced Science and Technology Institute for giving access to remote sensing data and satellite images; Joven Javier, for assisting in communicating with the government agencies such as PAG-ASA and DOST-ASTI; Joseph Ronquillo and Anna May Ramos for the statistical analysis, Ronnie Serfa Juan and Evelyn Raguindin for revision comments; and BIRDS-4 satellite project members for the support.

Conflicts of Interest: The authors declare no conflict of interest.

References

1. United Nations Department of Economic and Social Council. *World Urbanization Prospects: The 2018 Revision*; United Nations Department of Economic and Social Council: New York, NY, USA, 2019.
2. United Nations Economic and Social Council. *World Demographic Trends: Report of the Secretary General*; United Nations Economic and Social Council: New York, NY, USA, 2018; Volume E/CN.9/201.
3. World Bank. *Issue Brief for SDG 11—Sustainable Cities and Communities: Make Cities and Human Settlements Inclusive, Safe, Resilient, and Sustainable*; World Bank: Washington, DC, USA, 2017.
4. Estoque, R.C. A Review of the Sustainability Concept and the State of SDG Monitoring Using Remote Sensing. *Remote Sens.* **2020**, *12*, 1770. [CrossRef]
5. Sachs, J.; Schmidt-Traub, G.; Kroll, C.; Lafortune, G.; Fuller, G. *Sustainable Development Report 2019*; Bertelsmann Stiftung and Sustainable Development Solutions Network (SDSN): New York, NY, USA, 2019.
6. Seto, K.C.; Güneralp, B.; Hutyra, L.R. Global Forecasts of Urban Expansion to 2030 and Direct Impacts on Biodiversity and Carbon Pools. *Proc. Natl. Acad. Sci. USA* **2012**, *109*, 16083–16088. [CrossRef] [PubMed]
7. Mushore, T.D.; Odindi, J.; Dube, T.; Matongera, T.N.; Mutanga, O. Remote Sensing Applications in Monitoring Urban Growth Impacts on In-and-out Door Thermal Conditions: A Review. *Remote Sens. Appl. Soc. Environ.* **2017**, *8*, 83–93. [CrossRef]
8. Hua, L.; Zhang, X.; Nie, Q.; Sun, F.; Tang, L. The Impacts of the Expansion of Urban Impervious Surfaces on Urban Heat Islands in a Coastal City in China. *Sustainability* **2020**, *12*, 475. [CrossRef]
9. U.S. Environmental Protection Agency. *Reducing Urban Heat Islands: Compendium of Strategies*; U.S. Environmental Protection Agency: Washington, DC, USA, 2008; Volume 1, ISBN 9781509045303.
10. Oke, T.R. The Energetic Basis of the Urban Heat Island (Symons Memorial Lecture, 20 May 1980). *Q. J. R. Meteorol. Soc.* **1982**, *108*, 1–24.
11. Ulpiani, G. On the Linkage between Urban Heat Island and Urban Pollution Island: Three-Decade Literature Review towards a Conceptual Framework. *Sci. Total Environ.* **2021**, *751*, 141727. [CrossRef]
12. Fuladlu, K.; Riza, M.; İlkan, M. The Effect of Rapid Urbanization on the Physical Modification of Urban Area. In Proceedings of the 5th International Conference on Architecture and Built Environment with AWARDs, Venice, Italy, 22–24 May 2018; pp. 1–9.
13. United States Environment Protection Agency. Learn about Heat Islands. Available online: https://www.epa.gov/heatislands/learn-about-heat-islands (accessed on 19 December 2021).
14. Zhou, D.; Xiao, J.; Bonafoni, S.; Berger, C.; Deilami, K.; Zhou, Y.; Frolking, S.; Yao, R.; Qiao, Z.; Sobrino, J.A. Satellite Remote Sensing of Surface Urban Heat Islands: Progress, Challenges, and Perspectives. *Remote Sens.* **2019**, *11*, 48. [CrossRef]
15. Intra-Urban Relationship between Surface Geometry and Urban Heat Island: On JSTOR. Available online: https://www.jstor.org/stable/24868753 (accessed on 14 May 2022).
16. Simko, B.; Bruns, J.; Simko, V. Stable Hotspot Analysis for Intra-Urban Heat Islands. *GI-Forum* **2017**, *1*, 79. [CrossRef]
17. Kousis, I.; Pigliautile, I.; Pisello, A.L. Intra-Urban Microclimate Investigation in Urban Heat Island through a Novel Mobile Monitoring System. *Sci. Rep.* **2021**, *11*, 9732. [CrossRef]
18. EPA—US Environmental Protection Agency. Heat Islands and Equity. Available online: https://www.epa.gov/heatislands/heat-islands-and-equity (accessed on 14 May 2022).

19. Martin, P.; Baudouin, Y.; Gachon, P. An Alternative Method to Characterize the Surface Urban Heat Island. *Int. J. Biometeorol.* **2015**, *59*, 849–861. [CrossRef]
20. Igergård, F.; Mörtberg, U.; Törnros, T. Addressing the Urban Heat Island Effect in Stockholm: An Analysis of Its Presence and Relation to Land Cover and Urban Planning. Master's Thesis, KTH Royal Institute of Technology-School of Architecture and the Built Environment, Stockholm, Sweden, 2021.
21. Zhu, Z.; Zhou, Y.; Seto, K.C.; Stokes, E.C.; Deng, C.; Pickett, S.T.A.; Taubenböck, H. Understanding an Urbanizing Planet: Strategic Directions for Remote Sensing. *Remote Sens. Environ.* **2019**, *228*, 164–182. [CrossRef]
22. Rodrigues de Almeida, C.; Teodoro, C.A.; Gonçalves, A. Study of the Urban Heat Island (UHI) Using Remote Sensing Data/Techniques: A Systematic Review. *Environments* **2021**, *8*, 105. [CrossRef]
23. World Bank. *Philippines Urbanization Review*; World Bank: Washington, DC, USA, 2017.
24. University of the Philippines—Training Center for Applied Geodesy and Photogrammetry Project GUHeat. Available online: http://www.guheat.tcagp.upd.edu.ph/studyareas.html# (accessed on 19 December 2021).
25. Landicho, K.P.; Blanco, A.C. Intra-Urban Heat Island Detection and Trend Characterization in Metro Manila Using Surface Temperatures Derived from Multi-Temporal Landsat Data. In Proceedings of the The International Archives of the Photogrammetry, Remote Sensing and Spatial Information Sciences, Manila, Philippines, 14–15 November 2019. [CrossRef]
26. Estacio, I.; Babaan, J.; Pecson, N.J.; Blanco, A.C.; Escoto, J.E.; Alcantara, C.K. GIS-Based Mapping of Local Climate Zones Using Fuzzy Logic and Cellular Automata. *ISPRS-Int. Arch. Photogramm. Remote Sens. Spat. Inf. Sci.* **2019**, *XLII-4/W19*, 199–206. [CrossRef]
27. Alcantara, C.A.; Escoto, J.D.; Blanco, A.C.; Baloloy, A.B.; Santos, J.A.; Sta Ana, R.R. Geospatial Assessment and Modeling of Urban Heat Islands in Quezon City, Philippines Using OLS and Geographically Weighted Regression. *Int. Arch. Photogramm. Remote Sens. Spat. Inf. Sci. ISPRS Arch.* **2019**, *42*, 85–92. [CrossRef]
28. Estoque, R.C.; Ooba, M.; Seposo, X.T.; Togawa, T.; Hijioka, Y.; Takahashi, K.; Nakamura, S. Heat Health Risk Assessment in Philippine Cities Using Remotely Sensed Data and Social-Ecological Indicators. *Nat. Commun.* **2020**, *11*, 1581. [CrossRef]
29. Baloloy, A.; Sta Ana, R.R.; Cruz, J.A.; Blanco, A.C.; Lubrica, N.V.; Valdez, C.J.; Cajucom, E.P. Spatiotemporal Multi-Satellite Biophysical Data Analysis of the Effect of Urbanization on Land Surface and Air Temperature in Baguio City, Philippines. *ISPRS-Int. Arch. Photogramm. Remote Sens. Spat. Inf. Sci.* **2019**, *XLII-4/W19*, 47–54. [CrossRef]
30. Cañete, S.F.; Schaap, L.L.; Andales, R.; Otadoy, R.E.S.; Blanco, A.C.; Babaan, J.; Cruz, C. Analysis of the Impact of Vegetation Distribution, Urbanization, and Solar Radiation on the Seasonal Variation of the Urban Heat Island Effect in Cebu City Using Landsat and Global Horizontal Irradiance Data. *Int. Arch. Photogramm. Remote Sens. Spat. Inf. Sci.-ISPRS Arch.* **2019**, *42*, 93–100. [CrossRef]
31. Tinoy, M.M.; Novero, A.U.; Landicho, K.P.; Baloloy, A.B.; Blanco, A.C. Urban Effects on Land Surface Temperature in Davao City, Philippines. *ISPRS-Int. Arch. Photogramm. Remote Sens. Spat. Inf. Sci.* **2019**, *XLII-4/W19*, 433–440. [CrossRef]
32. Cruz, J.A.; Santos, J.A.; Garcia, J.J.; Blanco, A.C.; Moscoso, A.D. Spatiotemporal Analysis of the Urban Cooling Island (UCI) Effect of Water Spaces in a Highly Urbanized City: A Case Study of Iloilo River and Adjacent Wetlands. *Int. Arch. Photogramm. Remote Sens. Spat. Inf. Sci.-ISPRS Arch.* **2019**, *42*, 149–156. [CrossRef]
33. Rejuso, A.M.; Cortes, A.C.; Blanco, A.C.; Cruz, C.A.; Babaan, J.B. Spatiotemporal Analysis of Urban Heat Island in Mandaue City, Philippines. *Int. Arch. Photogramm. Remote Sens. Spat. Inf. Sci.-ISPRS Arch.* **2019**, *42*, 361–367. [CrossRef]
34. Enriquez, R.; Rodriguez, M.; Blanco, A.C.; Estacio, I.; Depositario, L.R. Spatial and Temporal Analysis of Monthly Water Consumption and Land Surface Temperature (LST) Derived Using Landsat 8 and MODIS Data. *ISPRS-Int. Arch. Photogramm. Remote Sens. Spat. Inf. Sci.* **2019**, *XLII-4/W19*, 193–198. [CrossRef]
35. Zhang, Y.; Li, D.; Liu, L.; Liang, Z.; Shen, J.; Wei, F.; Li, S. Spatiotemporal Characteristics of the Surface Urban Heat Island and Its Driving Factors Based on Local Climate Zones and Population in Beijing, China. *Atmosphere* **2021**, *12*, 1271. [CrossRef]
36. Gamboa, M.A.M.; Rivera, R.R.B.; Reyes, M.R.D. City Profile: Manila, Philippines. *Environ. Urban. Asia* **2019**, *10*, 331–358. [CrossRef]
37. Salita, D.C. "Manila". Encyclopedia Britannica. 2022. Available online: https://www.britannica.com/place/Manila (accessed on 23 January 2022).
38. Philippine Statistics Authority Highlights of the Population Density of the Philippines 2020 Census of Population and Housing (2020 CPH). 2021. Available online: https://psa.gov.ph/content/highlights-philippine-population-2020-census-population-and-housing-2020-cph (accessed on 23 January 2022).
39. Kottek, M.; Grieser, J.; Beck, C.; Rudolf, B.; Rubel, F. World Map of the Köppen-Geiger Climate Classification Updated. *Meteorol. Z.* **2006**, *15*, 259–263. [CrossRef]
40. Open Street Map Wiki Administrative Boundary. Available online: https://wiki.openstreetmap.org/wiki/Tag:boundary%3Dadministrative (accessed on 25 May 2022).
41. Philippine Statistics Authority Philippine Standard Geographic Code (PSGC) | Philippine Statistics Authority. Available online: https://psa.gov.ph/classification/psgc/ (accessed on 25 May 2022).
42. Altcoder High Resolution Philippine PSGC Administrative Boundaries. Available online: https://github.com/altcoder/philippines-psgc-shapefiles (accessed on 25 May 2022).
43. OCHA. Services Philippines—Subnational Administrative Boundaries—Humanitarian Data Exchange. Available online: https://data.humdata.org/dataset/cod-ab-phl (accessed on 25 May 2022).

44. GADM. Available online: https://gadm.org/data.html (accessed on 1 October 2020).

45. Purio, M.A.C.; Cho, M.; Yoshitake, T. A Temporal Analysis of the Relationship between Synoptic Weather Station Air Temperature Measurement and Satellite-Derived Land Surface Temperature: A Case Study in Port Area, Manila City, Philippines. *Int. Geosci. Remote Sens. Symp.* **2021**, *2021*, 4640–4643. [CrossRef]

46. Huntington, J.L.; Hegewisch, K.C.; Daudert, B.; Morton, C.G.; Abatzoglou, J.T.; McEvoy, D.J.; Erickson, T. Climate Engine: Cloud Computing and Visualization of Climate and Remote Sensing Data for Advanced Natural Resource Monitoring and Process Understanding. *Bull. Am. Meteorol. Soc.* **2017**, *98*, 2397–2409. [CrossRef]

47. Gorelick, N.; Hancher, M.; Dixon, M.; Ilyushchenko, S.; Thau, D.; Moore, R. Google Earth Engine: Planetary-Scale Geospatial Analysis for Everyone. *Remote Sens. Environ.* **2017**, *202*, 18–27. [CrossRef]

48. Land Surface Temperature—Applications—Sentinel-3 SLSTR—Sentinel Online—Sentinel. Available online: https://sentinel.esa.int/web/sentinel/user-guides/sentinel-3-slstr/overview/geophysical-measurements/land-surface-temperature (accessed on 22 January 2021).

49. U.S. Geological Survey U. What Are the Acquisition Schedules for the Landsat Satellites? Available online: https://www.usgs.gov/faqs/what-are-acquisition-schedules-landsat-satellites (accessed on 5 June 2022).

50. Deilami, K.; Kamruzzaman, M.; Liu, Y. Urban Heat Island Effect: A Systematic Review of Spatio-Temporal Factors, Data, Methods, and Mitigation Measures. *Int. J. Appl. Earth Obs. Geoinf.* **2018**, *67*, 30–42. [CrossRef]

51. Weier, J.; NASA Earth Observatory; Herring, D. Measuring Vegetation (NDVI & EVI). Available online: https://earthobservatory.nasa.gov/features/MeasuringVegetation/measuring_vegetation_2.php (accessed on 5 June 2022).

52. Gao, B.C. NDWI—A Normalized Difference Water Index for Remote Sensing of Vegetation Liquid Water from Space. *Remote Sens. Environ.* **1996**, *58*, 257–266. [CrossRef]

53. Zha, Y.; Gao, J.; Ni, S. Use of Normalized Difference Built-up Index in Automatically Mapping Urban Areas from TM Imagery. *Int. J. Remote Sens.* **2010**, *24*, 583–594. [CrossRef]

54. NDBI—ArcGIS Pro | Documentation. Available online: https://pro.arcgis.com/en/pro-app/2.8/arcpy/spatial-analyst/ndbi.htm (accessed on 5 June 2022).

55. Matzarakis, A.; Rutz, F.; Mayer, H. Modelling radiation fluxes in simple and complex environments—Application of the RayMan model. *Int. J. Biometeorol.* **2007**, *51*, 323–334. [CrossRef] [PubMed]

56. Matzarakis, A. Linking Urban Micro Scale Models—The Models RayMan and SkyHelios. In Proceedings of the 8th International Conference on Urban Climate-ICUC, Dublin, Ireland, 6–10 August 2012; pp. 10–13.

57. Mayer, H.; Höppe, P. Thermal Comfort of Man in Different Urban Environments. *Theor. Appl. Climatol.* **1987**, *38*, 43–49. [CrossRef]

58. Matzarakis, A.; Mayer, H.; Iziomon, M.G. Applications of a Universal Thermal Index: Physiological Equivalent Temperature. *Int. J. Biometeorol.* **1999**, *43*, 76–84. [CrossRef] [PubMed]

59. Lin, T.P.; Matzarakis, A. Tourism Climate and Thermal Comfort in Sun Moon Lake, Taiwan. *Int. J. Biometeorol.* **2008**, *52*, 281–290. [CrossRef]

60. Cooper, R.I.; Manly, B.F.J. *Multivariate Statistical Methods: A Primer*; CRC: Boca Raton, FL, USA, 1987; Volume 150, ISBN 9781498728966.

61. (Javatpoint) K-Means Clustering Algorithm. Available online: https://www.javatpoint.com/k-means-clustering-algorithm-in-machine-learning (accessed on 8 June 2022).

62. Environmental Systems Research Institute Inc. ArcGIS®Pro (GIS Software) Release 2.9. 2022. Available online: https://pro.arcgis.com/en/pro-app/get-started/get-started.html (accessed on 20 August 2022).

63. Ord, J.K.; Getis, A. Local Spatial Autocorrelation Statistics: Distributional Issues and an Application. *Geogr. Anal.* **1995**, *27*, 286–306. [CrossRef]

64. Hamed, K.H. Exact Distribution of the Mann–Kendall Trend Test Statistic for Persistent Data. *J. Hydrol.* **2009**, *365*, 86–94. [CrossRef]

65. Manalo, J.A.; Matsumoto, J.; Nodzu, M.I.; Olaguera, L.M.P. Diurnal Variability of Urban Heat Island Intensity: A Case Study of Metro Manila, Philippines. *Geogr. Rep. Tokyo Metrop. Univ.* **2022**, *57*, 13–22.

66. Alexander, C. Normalised Difference Spectral Indices and Urban Land Cover as Indicators of Land Surface Temperature (LST). *Int. J. Appl. Earth Obs. Geoinf.* **2020**, *86*, 102013. [CrossRef]

67. Ishii, T.; Tsujimoto, M.; Yoon, G.; Okumiya, M. Cooling System with Water Mist Sprayers for Mitigation of Heat-Island. In Proceedings of the 8th International Conference on Urban Climate-ICUC, Yokohoma, Japan, 29 June–3 July 2009; pp. 1–2.

68. Santamouris, M.; Synnefa, A.; Karlessi, T. Using Advanced Cool Materials in the Urban Built Environment to Mitigate Heat Islands and Improve Thermal Comfort Conditions. *Sol. Energy* **2011**, *85*, 3085–3102. [CrossRef]

69. Black-Ingersoll, F.; de Lange, J.; Heidari, L.; Negassa, A.; Botana, P.; Fabian, M.P.; Scammell, M.K. A Literature Review of Cooling Center, Misting Station, Cool Pavement, and Cool Roof Intervention Evaluations. *Atmosphere* **2022**, *13*, 1103. [CrossRef]

70. Yeh, Y.-P. Green Wall-The Creative Solution in Response to the Urban Heat Island Effect. *Natl. Chung-Hsing Univ. Taiwan* **2015**, *8*. Available online: http://www.nodai.ac.jp/cip/iss/english/9th_iss/fullpaper/3-1-4nchu-yupengyeh.pdf (accessed on 20 August 2022).

71. Kim, H.H. Urban Heat Island. *Int. J. Remote Sens.* **1992**, *13*, 2319–2336. [CrossRef]

72. Shareef, S. The Influence of Greenery and Landscape Design on Solar Radiation and UHI Mitigation: A Case Study of a Boulevard in a Hot Climate. *World* **2022**, *3*, 175–205. [CrossRef]

remote sensing

Article

Tropical Dry Forest Dynamics Explained by Topographic and Anthropogenic Factors: A Case Study in Mexico

Yan Gao [1,2,*], Jonathan V. Solórzano [3], Ronald C. Estoque [4] and Shiro Tsuyuzaki [2]

1 Centro de Investigaciones en Geografía Ambiental, Universidad Nacional Autónoma de México, Morelia 58190, Mexico
2 Graduate School of Environmental Earth Science, Hokkaido University, Sapporo 060-0810, Japan
3 Posgrado en Geografía, Centro de Investigaciones en Geografía Ambiental, Universidad Nacional Autónoma de México, Morelia 58190, Mexico
4 Center for Biodiversity and Climate Change, Forestry and Forest Products Research Institute, Tsukuba 305-8687, Japan
* Correspondence: ygao@ciga.unam.mx

Abstract: Tropical dry forest is one of the most threatened ecosystems, and it is disappearing at an alarming rate. Shifting cultivation is commonly cited as a driver of tropical dry forest loss, although it helps to maintain the forest coverage but with less density. We investigated tropical dry forest dynamics and their contributing factors to find out if there is an equilibrium between these two processes. We classified multi-temporal Sentinel-2A images with machine learning algorithms and used a logistic regression model to associate topographic, anthropogenic, and land tenure variables as plausible factors in the dynamics. We carried out an accuracy assessment of the detected changes in loss and gain considering the imbalance in area proportion between the change classes and the persistence classes. We estimated a 1.4% annual loss rate and a 0.7% annual gain rate in tropical dry forest and found that the topographic variable of slope and the anthropogenic variable of distance to roads helped explain the occurrence probability of both tropical forest loss and tropical forest gain. Since the area estimation yielded a wide confidence interval for both tropical forest loss and gain despite the measures that we took to counterbalance the disproportion in areas, we cannot conclude that the loss process was more intense than the gain process, but rather that there was an equilibrium in tropical dry forest dynamics under the influence of shifting cultivation.

Keywords: land use and land cover change; shifting cultivation; tropical forest gain and loss; topographic factors; distance to roads; logistic regression

Citation: Gao, Y.; Solórzano, J.V.; Estoque, R.C.; Tsuyuzaki, S. Tropical Dry Forest Dynamics Explained by Topographic and Anthropogenic Factors: A Case Study in Mexico. *Remote Sens.* **2023**, *15*, 1471. https://doi.org/10.3390/rs15051471

Academic Editor: Weiqi Zhou

Received: 31 December 2022
Revised: 24 February 2023
Accepted: 4 March 2023
Published: 6 March 2023

1. Introduction

Tropical and subtropical dry forests are one of fourteen biomes identified at the global scale [1]. In Mexico, tropical dry forests (TDFs) cover extensive areas in the western Pacific lowland from southern Sonora to northern and central Chiapas [2]. TDFs host a large variety of fauna and flora, playing an important role in biodiversity conservation and providing food and shelter for local people [3,4]. Despite having the highest level of endemism in the American continent [5], TDFs in Mexico are conventionally perceived as having less commercial value compared with temperate forests, and they are mainly designated for shifting cultivation and cattle ranching [6,7].

Land use and land cover change (LULCC) contributes to about one-tenth of the annual carbon emissions [8], in which deforestation shares more than three times that of the other LULCC categories combined [9]. In Mexico, at the national level, TDF decreased at a rate of 0.4%, about 100,000 ha every year [10]. Regional studies have reported higher deforestation rates, reaching 1.4% per year [11]. Large dry forest tracts have disappeared in recent years mainly to support agriculture and cattle ranching. About 70% of pre-Hispanic TDF has been converted to other land use types, and about 62% of the remaining TDF is in an altered and disturbed state [6]. In addition to carbon loss, the loss of TDF leads to the loss of biodiversity and soil erosion and increases the vulnerability of local people who depend on TDFs for food and shelter. The disturbance also fundamentally alters environmental conditions and constrains the forest's capacity to regenerate [12].

In previous studies, different drivers have been associated with the LULCC of tropical forests, such as the expansion of agriculture (frequently large-scale and industrialized) or livestock activities, as well as socio-economic conditions such as poverty. Especially, topographical and distance-related measures have been reported as determinant factors to explain which areas will undergo a LULCC process. For example, it was reported that the probability of an area experiencing a LULCC process is related to a poverty index, the population size of nearby settlements, topographical variables, and distance to roads [13]. Nonetheless, these can sometimes be a product of complex relationships [13,14].

Shifting cultivation is widely practiced in the global south and plays an important role in food security in Asian, African, and Latin American countries [15,16]. In Mexico, shifting cultivation is the main driver of disturbances in TDFs, especially in the southern part of the country [17]. This agriculture system includes cycles of clearing, cultivation, and fallow period. During clearing, the standing vegetation is cut down and burned to create fields and produce ash which provides nutrients for farming. The cleared parcels have an average size of 2.5 ha and crops are grown for subsistence [18]. Cultivation starts during the rainy season when maize crops are planted and harvested after six months of growth. After harvesting until the next plantation, livestock graze on crop residues. The cultivation cycle repeats for about 2–3 years and then the land is left to rest in a fallow period for about 3–8 or more years, during which natural vegetation grows as a mixture of shrubs and trees. Shifting cultivation creates a landscape with a mosaic of patches currently being cropped and patches in the fallow period with natural vegetation under various stages of regeneration. The regenerated natural vegetation keeps the area a forest, however, with less biomass density [19].

This paper aims to understand the contributing factors to the dynamics of TDFs and whether there is a balance between TDF loss and gain under the influence of shifting cultivation. We first obtained areas of TDF loss and gain by comparing multiple dates of land use land cover maps created by classifying Sentinel-2A images with a machine learning algorithm. Then, using a logistic regression model, we analyzed the plausible factors including topographic, anthropogenic, and land ownership that are associated with TDF changes. Lastly, we projected the areas of future TDF loss and gain to shifting cultivation.

2. Materials and Methods

2.1. Study Area

The study area is within the Ayuquila River watershed, in the state of Jalisco, Mexico (Figure 1). It is one of the first areas in Mexico designated as a Reduction of Emission from Deforestation and Forest Degradation (REDD+) experimental area because of its importance in biodiversity and water provision, among other ecosystem services [20]. The watershed has a wide range in topography, from 260 m to 2500 m above mean sea level (amsl). The monthly average temperature is about 18–22 degrees Celsius, and the annual average precipitation is 800–1200 mm, which occurs mainly during the rainy season, from June to October [21]. The dominant forest type is TDF, which is comprised of deciduous and semi-deciduous trees that lose their leaves during the dry season, typically from November to May. TDF covers about 24% of the watershed, and it has been intensively used for shifting cultivation, cattle grazing, and fuel wood collection. As in the rest of Mexico, most of the forests (59% to 80%) in the Ayuquila watershed are under the authority of ejidos, which is a communally managed land tenure system of rural agrarian settlements [18].

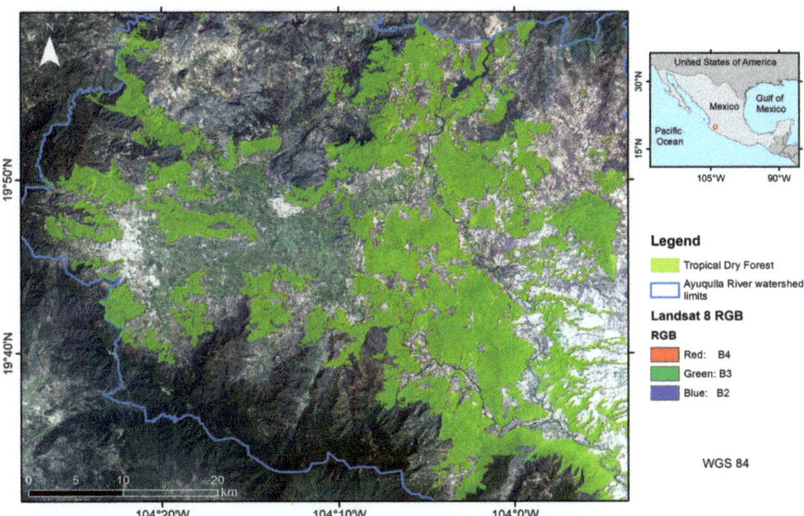

Figure 1. Location of the study area in the central west of Mexico. The background is a natural color composite of Landsat 8 imagery. The distribution of the tropical dry forest in the study area is shown in light green color.

2.2. Data

This study collected and used a variety of datasets (Table 1). Sentinel-2A images were obtained from Google Earth Engine archive. All Sentinel-2A bottom-of-atmosphere reflectance images were atmospherically corrected with Copernicus scihub using the sen2cor algorithm. In addition, elevation data (Digital Elevation Model: DEM) were obtained from the official website of the National Institute of Statistics and Geography (INEGI) of Mexico with a 15 m spatial resolution, and the topographic variables of slope and aspect were calculated using the DEM to reflect terrain changes. Data on accessibility including distance to roads and distance to agriculture were calculated using the proximity function with roads data downloaded from the INEGI and agriculture data (including both irrigated and temporal agriculture) extracted from the INEGI land use land cover and vegetation maps, series VII, at the scale of 1:250,000.

Table 1. Datasets used.

Datasets	Resolution	Date/Value Range	Description	Source
Sentinel-2A	10 m, 20 m	15 May 2019; 15 March 2022	Orthorectified, radiometrically calibrated and atmospherically corrected	https://doi.org/10.5270/S2_-znk9xsj (accessed on 12 November 2022)
Elevation	15 m	292–2132 m above sea level	Six scenes of Digital Elevation Model (DEM) (E13B12, E13B13, E13B14, E13B23, E13B24) to cover the study area, downloaded from	https://www.inegi-org.mx/app/geo2/elevacionesmex/ (accessed on 2 December 2022)
Slope	15 m	0.6–73.9 (°)	Arctan(rise/run)	Calculated using the DEM
Aspect	15 m	0.0–359.76 (°)	Represents the compass direction that the slope of the terrain faces. An aspect of 0 means that the slope is north-facing, 90 east-facing, 180 south-facing, and 270 west-facing.	Calculated using the DEM

Table 1. *Cont.*

Datasets	Resolution	Date/Value Range	Description	Source
Distance to roads	10 m	0–4235.6 (m)	Euclidean distance was used to represent distance to roads.	https://www.inegi.org.mx/temas/viascomunicacion (accessed on 2 December 2022)
Distance to agriculture	10 m	0.0–9235.4 (m)	Euclidean distance was used to represent distance to agriculture.	https://www.inegi.org.mx/temas/usosuelo (accessed on 2 December 2022)
Land ownership	10 m	1: ejido and communal. 0: other	The land ownership was categorized into two categories: both ejido and communal, and other type which cover 38.6% and 61.4% of the study area, respectively.	https://www.gob.mx/ran#709 (accessed on 2 December 2022)

2.3. Classification

Training samples were collected for the eleven land use/land cover classes (Table 2), using Sentinel-2A images and high spatial resolution images in Google Earth (GE) as reference. The number of training samples for the 2019 and 2022 imagery is presented in the last column of Table 2. Figure 2 shows the distribution of the sample classes in Sentinel-2A bands. We also referred to the INEGI land use land cover and vegetation maps, series VII (1:250,000), which was produced during the period of 2015–2017 for the spatial distribution of different types of forests and agriculture. The distribution of TDF and oak and pine forests shows a general ascending pattern following elevation. We used two classification algorithms, namely artificial neural network and random forest, to classify the Sentinel-2A images in 2019 and 2022.

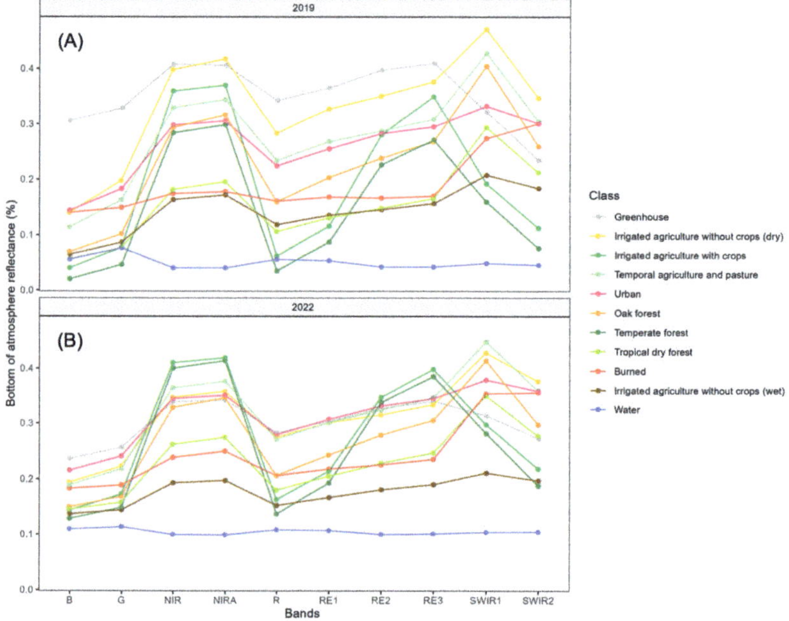

Figure 2. The distribution of the mean reflectance of the samples for the land use/land cover classes at the ten spectral bands of Sentinel-2 images for the years 2019 (**A**) and 2022 (**B**). Classes are sorted in descending order according to their mean reflectance in the NIR band in 2019. The standard deviations of the spectral reflectance range from 0.0042 to 0.142.

Table 2. Definition of land cover/land use classes (in reference to the INEGI series VII class definition).

Classes	Change Classes	Class Definition	The Number of Training Samples in Pixel for 2019/2022
Tropical dry forest	Tropical dry forest	Dense and sparse vegetation in different stages of succession	25,883/22,876
Temperate forest	Temperate forest	Coniferous and mountain cloud forest	10,504/15,296
Oak forest	Oak forest	Forest formation distributed between tropical dry forest and temperate forest	8580/20,237
Irrigated agriculture with crops		Crops in different growing stages	7307/13,581
Irrigated agriculture without crops (wet)	Agriculture	Wet agricultural fields temporarily without crops	1116/5590
Irrigated agriculture without crops (dry)		Dry agricultural fields temporarily without crops	2075/5047
Greenhouse		Agriculture fields covered by greenhouse	907/3632
Burned		Burned agriculture or forest area	314/1226
Temporal agriculture and pasture		Rain-fed annual agriculture and grassland fields	20,546/11,679
Water	Other	Waterbody	3573/6117
Urban		Urban landscape and scattered houses in rural areas	1380/833

2.3.1. Artificial Neural Network

Artificial neural network (ANN) is a machine learning algorithm that uses a network of nodes to perform supervised classifications [22,23]. Typically, each neuron in an ANN receives a series of inputs, and then performs a weighted sum of them and outputs a value of 1 if its sum is over a threshold and a value of 0 if not. Finally, the complete network can classify a different set of inputs based on the neuron's weights [24]. Training a neural network requires that the user specifies the network structure and sets the learning parameters [22]. In our case, to train the ANN, the sample data were split in the proportion of 0.7, with 70% of the data assigned as training data and the remaining 30% as test data. This algorithm applied a 5-fold cross-validation (CV) with 5 repetitions.

A 5-fold CV involves randomly dividing the training data into 5 groups, or folds, of approximately equal size [25]. The first fold is treated as a validation set, and the method is fit on the remaining 4 folds. The mean squared error, MSE, is computed with the data in the held-out fold. The process resulted in 5 estimates of the test error and the 5-fold CV estimate is computed by averaging these errors (Equation (1)).

$$\mathrm{CV}_5 = 1/5 \sum_{i=1}^{5} \mathrm{MSE}_i \tag{1}$$

2.3.2. Random Forest

Random forest (RF) is a non-parametric machine learning algorithm that generates robust predictions by creating a set of regression trees from the bootstrap sampling of the original data and improving prediction accuracy by aggregating the results [26]. RF has been shown to be resistant to problems of overfitting and noise, and it has been widely used for the supervised classification of land use and land cover [27,28]. In this study, we used RF to classify land use/land cover types of the study area in 2019 and 2022 using Sentinel-2A images. The sample was split with a proportion of 0.2, with 20% of the samples used as training data and the remaining 80% as test data. The tuning parameters were tested for the number of variables randomly sampled as candidates at each split (mtry: 2, 6, 10) and accuracy was used to select the optimal model. The final value used for the model of the 2022 image classification was mtry = 2 (Figure A1), and for the model of the 2019 image classification, the value was mtry = 6 (Figure A2). The "rf" method uses a default ntree of 500, which is a recommended value for remote sensing applications [29].

Consistent with the ANN classification algorithm, the RF algorithm also applied 5-fold cross-validation with 5 repetitions.

2.4. Change Analysis

We first reclassified the land use/land cover maps in 2019 and 2022 by grouping the classes of irrigated agriculture, temporal agriculture, greenhouse, and burned area into the class of agriculture, and by grouping water and urban into the class other (Table 2). The class burned area was grouped into the agriculture class because burning is part of the shifting cultivation and temporal agriculture field preparation practice [30]. The reclassified land use/land cover maps had five classes, namely TDF, temperate forest, oak forest, agriculture, and other.

We performed the LULCC by overlaying and comparing the reclassified maps in 2019 and 2022. To analyze the dynamics of TDF, we focused on the following classes of changes and persistence: TDF persistence (35.1% of the study area), TDF loss to agriculture (1.7%), TDF gain from agriculture (1.1%), and other changes (62.1%). To remove the noise, we applied a filter of 2 ha since the average size of shifting cultivation was recorded as 2.5 ha and applied a 4-neighbor rule (QGIS sieve function) since it obtained better results.

2.5. Accuracy Assessment

Since classification errors often propagate to the result of change analysis, especially with post-classification comparison, it is important to evaluate the accuracy of the map of change in addition to the classifications. In a map of change detection, the classes of change usually occupy small proportions in comparison to the classes of persistence, and therefore, omission errors in the classes of change classes are often exaggerated due to the imbalance in area proportions and cause big uncertainty in the estimation of accuracy and areas. To counterbalance this effect, we implemented the method that was detailed in [31]. We created a buffer area the size of 12 pixels around the classes of change, assuming changes are more prone to occur in the buffer areas, and assigned 75 points to the class of TDF persistence and 75 points to the class other persistence. Following stratified random sampling [32], and assuming the standard error of the change map as 0.01, the user's accuracy of TDF persistence 84%, TDF loss 60%, TDF gain 50%, and other 90%, and counting the 150 points from the buffer areas, the total number of random points needed for a statistically valid accuracy assessment was 1178 points. The points were distributed as follows: 331 points in TDF persistence, 300 points in TDF gain, 300 points in TDF loss, and 93 points in other persistence. The distribution of the verification points is shown in Figure 3.

We exported those points to Google Earth and interpreted them visually to obtain the ground truth data. During visual interpretation, we considered an area of 100 m^2 around each point, which is equivalent to the pixel size of Sentinel-2A images. We compared the ground truth data with the map of LULCC and verified the obtained changes and persistence. We summarized the results of the accuracy assessment in an error matrix. We incorporated the area proportion of the mapped changes and calculated the overall accuracy, producer's accuracy, and user's accuracy with confidence intervals (CIs) and estimated the weighted areas of TDF loss and gain with their respective CIs. The accuracy assessment was carried out using Open Foris tools developed by the FAO (https://github.com/openforis/accuracy-assessment/ (accessed on 15 December 2022)).

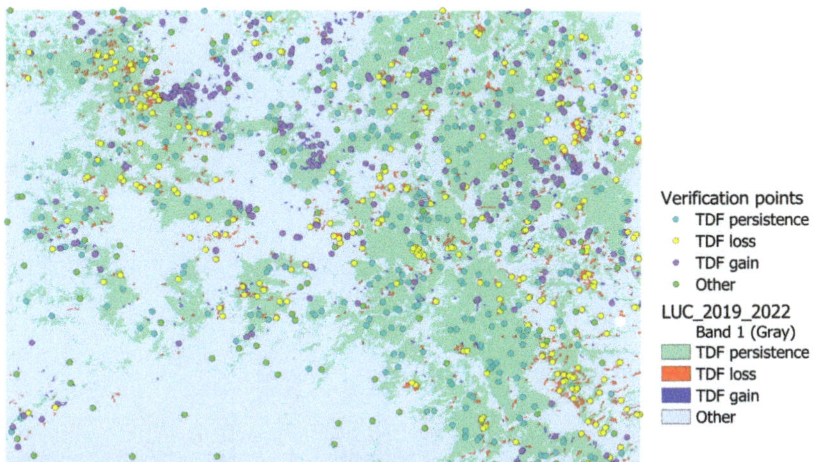

Figure 3. The map of the land use land cover change (LULCC) and the distribution of the verification points.

2.6. Examining the Spatial Variables Contributing to TDF Loss and Gain

We used the verified points of TDF change and persistence to analyze the effect of spatial variables on TDF dynamics (loss to and gain from shifting cultivation). We considered topographic variables, such as elevation, aspect, and slope, and anthropogenic variables, such as distance to roads and distance to agriculture and land tenure represented by ejido and communal lands or other land ownership (Table 1). The spatial variables were resampled to 10 m spatial resolution to be consistent with the LULCC maps. We fitted two types of logistic regression models, one for TDF loss and the other for TDF gain. In these models, the dependent variables were change (1) and persistence (0), and the independent variables are the spatial variables (Figure 4).

First, we fitted the models including all the spatial variables to detect the significant terms, and then we fitted additional models using only the significant terms. We used the Akaike information criterion (AIC) to select the best model (Equation (2)). AIC is calculated using the number of independent variables (K) and the log-likelihood estimate of the model (L). Using AIC as a criterion, the best model would explain the biggest amount of variation in the data using the smallest number of independent variables [33]. We selected the best model that had the lowest AIC and was at least two units lower than the AICs of other competing models.

$$AIC = 2K - 2\ln(L) \tag{2}$$

Before fitting the models, a Pearson correlation analysis was performed to avoid using strongly colinear variables in the models. A Pearson correlation coefficient of ≥ 0.8 was interpreted as an indicator of strongly collinear variables.

The TDF loss models had 94 verified random points for TDF loss and TDF persistence, respectively, with the explicative variables extracted to each of these 188 points. The TDF gain models had 46 verified random points for TDF gain and persistence, respectively, also with the variables extracted to each of those 92 points. Both TDF loss and TDF gain points were a subset of the verification dataset for the change analysis.

2.7. Projecting Future TDF Loss and Gain

We predicted the probability of the occurrence of TDF loss and TDF gain using the best models. We reclassified the probability maps using a threshold of 0.5 and calculated the map areas with a probability higher than 0.5 as predicted areas of TDF loss or TDF gain.

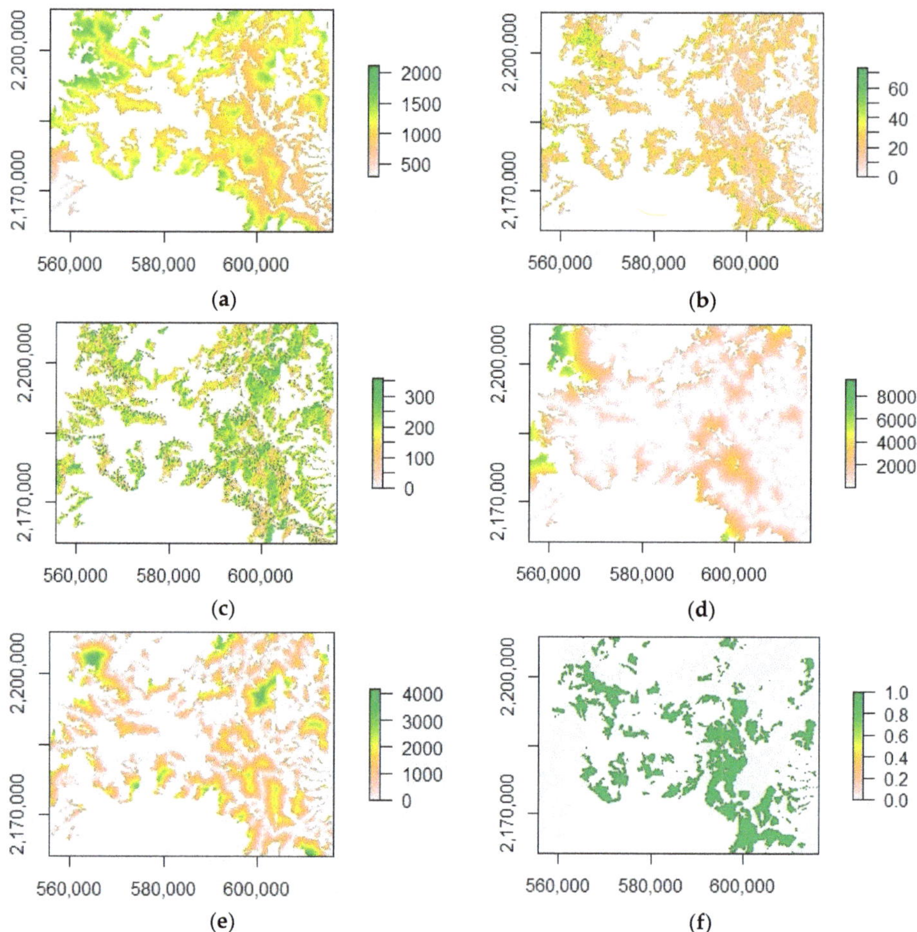

Figure 4. Explicative spatial variables: (**a**) elevation, (**b**) slope, (**c**) aspect, (**d**) distance to agriculture, (**e**) distance to roads, and (**f**) land ownership.

2.8. Comparison of Forest Loss and Gain

We compared the TDF loss and TDF gain by computing the statistics of their patch size and distribution. We assumed that smaller areas of forest loss might be related to areas under a shifting cultivation scheme (around 2.5 ha), contrary to larger areas of forest loss, which might be related to large-scale agricultural management. Thus, we expected a smaller area in TDF gain in comparison to TDF loss. We used a non-parametric Mann–Whitney test to test the difference between the median values of the areas of the TDF loss and TDF gain [34].

3. Results

This section presents the results of the classification accuracy assessment, change detection, influencing factors modeling with the logistic regression model, and future projection of TDF loss and gain.

3.1. Classification Model Validation

The accuracy of the classification result was evaluated with the test data using an error matrix (Appendix A Tables 2–4 and A1). The classification of the 2019 image using the ANN obtained an overall accuracy of 0.9714 with a 95% CI of (0.9695, 0.9732), and using the RF, the classification obtained an overall accuracy of 0.9901 with a 95% CI of (0.9894, 0.9908). The classification of the 2022 image using the ANN obtained an overall accuracy of 0.9452 with a 95% CI of (0.9423, 0.948), and using the RF, the classification obtained an overall accuracy of 0.9766 with a 95% CI of (0.9754, 0.9777).

Table 3. Error matrix with sample points. OA: overall accuracy; PA: producer's accuracy; UA: user's accuracy; CI: confidence interval; MA: map area; AP: area proportion; Buffer-TDF-P: buffer in TDF persistence, Buffer-O: buffer in other.

		TDF persistence	TDF loss	TDF gain	Other	Buffer-TDF-P	Buffer-O	MA (ha)	AP (%)	UA
					Ground Data					
Map data	TDF persistence	334	0	1	0	0	0	99,931	32.17	0.997
	TDF loss	0	151	0	0	111	38	4814	1.55	0.503
	TDF gain	0	1	58	0	113	128	3022	0.97	0.193
	Other	12	2	1	78	0	0	176,813	56.92	0.839
	Buffer-TDF-P	0	0	0	0	71	4	11,843	3.81	0.947
	Buffer-O	0	0	0	0	13	62	14,206	4.57	0.827
	Total	346	154	60	78	308	232	310,628		
	PA	0.965	0.981	0.967	1	0.231	0.267			
	Weighted PA	0.814	0.389	0.210	1	0.676	0.823			
	Estimated MA (ha) with 95% CI	122,447 ± 12,126	6235 ± 5248	2784 ± 3774	148,295 ± 13,289	16,593 ± 1402	14,274 ± 1389			
	OA with 95% CI	0.882 ± 0.018								

Table 4. Result of the logistic regression predicting tropical forest loss to agriculture with all explicative variables. '**' significant at 0.01 level, '.' significant at 0.1 level.

Variables	Estimate	Std. Error	Z Value	*p*-Value
Intercept	−0.0333	0.2062	−0.161	0.8717
Distance to roads	−0.3395	0.1894	−1.792	0.0731
Distance to agriculture	−0.0913	0.2155	−0.424	0.6718
Elevation	−0.1282	0.1960	−0.654	0.5130
Slope	−0.5004	0.1863	−2.685	0.0072 **
Aspect	0.1910	0.1578	1.210	0.2263
Tenure	0.0056	0.3178	0.018	0.9859

Null deviance: 260.62 on 187 degrees of freedom, Residual deviance: 235.37 on 181 degrees of freedom, AIC: 249.37.

Based on the accuracy assessment, both ANN and RF achieved high overall accuracy with comparable results (Tables 2–4 and A1). However, for both dates, the classified land use/land cover maps obtained using the ANN had less of a salt and pepper effect and there was less confusion between temperate forest and irrigated agriculture, especially in the southern part of the maps (Figures A3–A6). For this reason, we chose the results obtained using the ANN for the change analysis.

3.2. Verification of Detected Changes

The error matrix is presented in Table 3. The overall accuracy of the map of change was 0.882 ± 0.018. The unweighted producer's accuracy for TDF loss was 0.981 and for TDF gain was 0.967, while the user's accuracy for TDF loss was 0.503 and for TDF gain was 0.193.

3.3. Tropical Dry Forest Loss

The model for TDF loss had 188 points, 94 points for loss, and 94 points for persistence. Pearson correlation showed only a mild correlation (coefficient < 0.5) between the variables (Table A5), and, therefore, all the variables were included in the logistic regression model.

We scaled all the numeric variables using the mean and the standard deviation (STD) of the samples before running the models, i.e., by subtracting the mean and dividing the STD. The logistic regression model with all variables showed that slope was significant at the 0.05 level, and distance to roads was significant at the 0.1 level (Table 4). We built another two models, one with the two significant variables (Table 5) and the other one with only slope since it had a higher coefficient (Table 6). We selected the best model with the lowest AIC (Table 5).

Table 5. Result of the logistic regression predicting tropical forest loss to agriculture with variables of slope and distance to roads. '*' Significant at 0.05 level, '**' significant at 0.01 level.

Variables	Estimate	Std. Error	Z Value	*p*-Value
Intercept	−0.0280	0.1556	−0.180	0.8575
Slope	−0.5613	0.1752	−3.205	0.0014 **
Distance to roads	−0.4036	0.1730	−2.333	0.0197 *

Null deviance: 260.62 on 187 degrees of freedom, Residual deviance: 238.12 on 185 degrees of freedom, AIC: 244.12.

Table 6. Result of the logistic regression predicting tropical forest loss to agriculture with the variable slope. '***' Significant at 0.001 level.

Variables	Estimate	Std. Error	Z Value	*p*-Value
intercept	−0.0154	0.1529	−0.101	0.9198
Slope	−0.6454	0.1690	−3.818	0.0001 ***

Null deviance: 260.62 on 187 degrees of freedom, Residual deviance: 243.87 on 186 degrees of freedom, AIC: 247.87.

3.4. Tropical Dry Forest Gain

The model for TDF gain had 92 points, with 46 points for TDF gain and 46 points for persistence. The predictive variables were extracted to the location of each of these points. Two points for gain were deleted since they had NA values in the variable aspect. The final dataset had 90 points, including 44 points for gain and 46 points for persistence. The Pearson correlation showed a mild correlation between the variables, with the highest correlation being 0.66 between distance to agriculture and elevation (Table A6). The correlation was also found in the TDF loss dataset, with lower coefficient (0.53). We scaled all the numeric explicative variables similar to the procedure in the TDF loss models.

We tested the logistic regression model with all variables (Table 7). Both the distance to roads and slope were significant with negative coefficients, showing that the probability of TDF gain decreased with the increase in distance to roads and slope. We tested two more models, first with the two significant variables (Table 8) and then with the variable that had the highest coefficient (Table 9). When using only slope and distance to roads, the AIC of the model decreased (Table 8). We selected the best model with the lowest AIC for TDF gain.

Table 7. Results of the logistic regression model predicting tropical forest gain from agriculture using all predictive variables. '*' Significant at 0.05 level, '***' significant at 0.001 level.

Variables	Estimate	Std. Error	Z Value	*p*-Value
Intercept	−0.5174	0.4098	−1.263	0.2068
Distance to roads	−1.1118	0.4646	−2.393	0.0167 *
Distance to agriculture	−0.5781	0.4486	−1.289	0.1975
Elevation	0.0339	0.3880	0.087	0.9304
Slope	−1.2231	0.3482	−3.512	0.0004 ***
Aspect	0.3379	0.2732	1.237	0.2162
Tenure	0.3454	0.5667	0.610	0.5422

Null deviance: 124.72 on 89 degrees of freedom, Residual deviance: 81.26 on 83 degrees of freedom, AIC: 95.26.

Table 8. Results of the logistic regression model predicting tropical dry forest gain using slope and distance to roads as predictive variables. '**' Significant at 0.01 level, '***' significant at 0.001 level.

Variables	Estimate	Std. Error	Z Value	*p*-Value
intercept	−0.3025	0.2910	−1.040	0.2986
Slope	−1.3311	0.3430	−3.881	0.0001 ***
Distance to roads	−1.1214	0.3918	−2.862	0.0042 **

Null deviance: 124.72 on 89 degrees of freedom Residual deviance: 85.49 on 87 degrees of freedom AIC: 91.49.

Table 9. Results of the logistic regression model predicting tropical dry forest gain using slope as a predictive variable. '***' Significant at 0.001 level.

Variables	Estimate	Std. Error	Z Value	*p*-Value
Intercept	−0.1527	0.2427	−0.629	0.5291
Slope	−1.1712	0.3287	−3.563	0.0004 ***

Null deviance: 124.72 on 89 degrees of freedom Residual deviance: 106.11 on 88 degrees of freedom AIC: 110.11.

3.5. Probability of Future Tropical Dry Forest Loss and Gain

The probability of future TDF loss and TDF gain were predicted with the best models (Figure 5). For both TDF loss and gain, the best models included slope and distance to roads, although in the model of TDF gain, both variables had higher coefficients (Table 8).

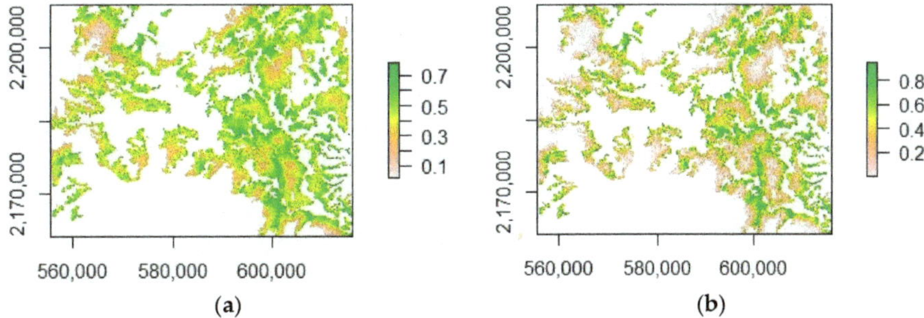

Figure 5. Probability of TDF loss (**a**) and TDF gain (**b**) predicted with the best models. (**a**) Probability of TDF loss predicted with slope and distance to roads. (**b**) Probability of TDF gain predicted with slope and distance to roads.

The potential areas of TDF loss and gain were predicted by reclassifying the probability maps with a threshold of 0.5 (Table A7). Overall, 43.1% of the total TDF area was predicted as TDF loss and 35.4% as TDF gain (Table A7), showing that TDF loss was a dominant process in the study area. This pattern coincides with the area estimates for loss and gain, although their CIs overlapped (Table 3).

3.6. Comparison of the Verified Forest Loss and Gain

The statistics for TDF loss showed a median size of 7.2 ha, an average of 14.3 ha, a minimum of 2.1, and a maximum of 130 ha. In comparison, the statistics for TDF gain showed a smaller size with a median of 5.3 ha, an average of 9 ha, a minimum of 2.2 ha, and a maximum of 62.8 ha. Despite being apparently larger in size, TDF loss in patch sizes is not significantly different from TDF gain, based on the Wilcox test for samples with a non-parametric distribution, with a *p*-value = 0.083. The size distribution of the TDF loss and TDF gain is shown with the boxplot in Figure 6.

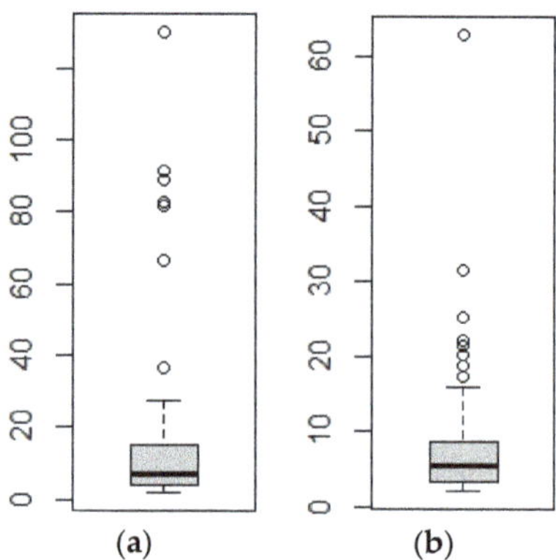

Figure 6. Boxplots of the patch size (ha) of (**a**) TDF loss patches and (**b**) TDF gain patches.

4. Discussion

4.1. TDF Dynamics

During 2019–2022, in our study area, the average TDF loss and TDF gain were estimated at 6235 ha and 2784 ha, respectively, and the average rate of TDF loss and TDF gain was estimated at 1.6% and 0.7% per year, respectively. The rate of TDF loss was higher than the national level TDF loss estimated at 0.4% per year by [10], but it was comparable with the annual TDF loss rate of 1.4% at the regional level estimated by [12]. As for TDF gain, we did not find studies at the national or regional level to compare with. Both the areas and the rate of TDF loss were much higher than TDF gain. Nonetheless, the confidence intervals of the estimated areas of both classes overlapped; thus, we cannot be sure that there is a significant difference between both estimates. According to our results, TDF gain and loss are in equilibrium in the region. However, due to the relatively large confidence intervals for both area estimates, we recommend taking this conclusion with precaution.

Having smaller patch sizes, areas of TDF gain can be more readily related to TDF regrowth during the fallow period of shifting cultivation. On the other hand, large patches of TDF loss (around or more than 80 ha) are more readily related to large-scale plantations, which might not be under a shifting cultivation scheme. According to our results, the area of TDF loss and gain were not significantly different. Therefore, although large-scale plantations are more common in the study site, most of the TDF loss areas seem to be related to small-scale areas, probably under shifting cultivation or other management that could further imply a TDF recovery.

Processes such as shifting cultivation do not affect net vegetation distribution; however, they can result in an overall decrease in vegetation density and cause carbon release into the atmosphere and affect the carbon budget [20]. For TDF regeneration, assisted natural regeneration—a practice to convert degraded forests into more productive forests with improved ecosystem services by managing regeneration rather than relying on pure natural processes—is recommended [35,36]. Although this study consists of an evaluation of TDF gain and loss in terms of area, a more detailed analysis should be made to obtain the carbon balance in the study area.

As for the contributing factors, for TDF loss, both slope and distance to roads were significant predictive variables, with slope having a higher coefficient. Both variables had negative coefficients, showing that with the increase in either slope or distance to roads, the probability of TDF being converted to agriculture decreases. For TDF gain, the same variables were significant and with negative coefficients, showing that with the increase in slope and distance to roads, the probability of TDF recovery decreases. Interestingly, in both models, slope had a higher coefficient and therefore is a more important factor to determine the probability of both TDF loss and gain. In the case of TDF loss, this pattern is probably related to the fact that frequently areas with higher slopes and farther away from established roads are not preferred to establish agricultural or livestock activities, due to the costs and difficulties associated with the transportation of products and animals to those areas. In the case of TDF gain, its dependence on TDF loss (i.e., for an area to gain TDF, first it must lose it) causes the same relation with those two variables.

4.2. Methodological Challenges and Insights

The wide range in area estimation of TDF loss and TDF gain partly comes from the fact that these two change classes occupied a small proportion of the area in comparison to classes of persistence, and therefore, (small) a number of omission errors were exaggerated due to the imbalance in area proportions. We intended to counteract this imbalance in area proportion by creating buffer zones of 12 pixels of Sentinel-2A images (120 m) around the change areas, assuming that omission errors are prone to occur near detected changes [31]. However, we did not succeed in reducing the confidence intervals. One possible reason could be that the buffer size is not wide enough to cover the points of omission errors. For future work, we will consider using different buffer sizes to find out how buffer size affects the reduction in omission errors.

The maps of predicted forest loss and gain created using our models show that forest loss and gain follow a similar spatial distribution because of their dependence on the same explicative variables: slope and distance to roads. These maps were created using a limited number of factors including slope, elevation, and distance to roads; thus, other plausible factors that might explain forest loss and gain were omitted. For example, factors related to the presence of large-scale agriculture companies or certain governmental incentives. Nonetheless, our maps can give a general idea of which areas are prone to being under shifting cultivation management.

We analyzed TDF dynamics for a rather short time window (2019–2022). Although we could capture the TDF gain and loss from shifting cultivation, an analysis with a longer time span, e.g., 10–30 years, could potentially allow us to have a more precise area estimation of TDF gain and loss (with a smaller confidence interval).

For TDF dynamics, time series analysis of climate data (air temperature, precipitation) is useful to exempt false changes introduced by climate variations. Table 10 shows the annual average temperature and annual precipitation for our study area. The precipitation data were derived from CHIRPS "Climate Hazards Group InfraRed Precipitation with Station Data" and the temperature data are from MODIS "Land surface temperature". The annual temperature was rather stable for the period of 2018–2022, while the annual precipitation showed some variation, with much higher precipitation in 2022 than in 2019 (Table 10), which could explain why the vegetation in the 2022 image was greener even in a dry season. Since we used a two-time classification comparison to analyze TDF dynamics, and we provided independent training samples for the classification at each time point, the effect of the interannual climate variations on the vegetation cover was well captured with the image classification. Since we verified the changes using visual interpretation with reference to very high spatial resolution images from Google Earth, the climate-induced effects were largely removed from contributing factor analysis of TDF dynamics.

Table 10. Annual average temperature and annual precipitation in the Ayuquila River Watershed for 2018–2022.

	Temperature (°C)	Precipitation (mm/Year)
2018	31.5	897.5
2019	31.2	885.4
2020	30.6	1179.5
2021	31.6	868.3
2022	30.5	1150.9

5. Conclusions

We used multi-temporal Sentinel-2A images and topographic, anthropogenic and land tenure factors to investigate the dynamics of tropical dry forest in terms of gain and loss and the associated factors. We estimated a TDF loss rate of 1.6% per year and a TDF gain rate of 0.7% per year. Although apparently TDF loss rate was higher than TDF gain, because of the large confidence interval in our area estimate, we cannot conclude that TDF loss was more intense but rather that the TDF loss was in equilibrium with TDF gain. In future analysis, we will assess other methods that can help reduce the confidence intervals in area estimates to obtain a clearer conclusion regarding TDF loss and gain.

As for the contributing factors, both TDF loss and gain were inversely related to slope and distance to roads; therefore, these two factors explain the probability of a TDF area in both gain and loss. In the case of TDF loss, this is related to the fact that agricultural or livestock activities generally prefer flat areas for easy access and cheap cost; for TDF gain, since it depends on TDF loss, the same relation with slope could explain the distribution of TDF gain.

Author Contributions: Conceptualization, Y.G.; methodology, Y.G.; validation, Y.G. and J.V.S.; formal analysis, Y.G.; writing—original draft preparation, Y.G.; writing—review and editing, Y.G., J.V.S., R.C.E. and S.T. All authors have read and agreed to the published version of the manuscript.

Funding: This research received no external funding.

Data Availability Statement: Data are available through request to the corresponding author.

Acknowledgments: The first author would like to thank the Direccion General de Asuntos del Personal Académico (DEGAPA) at the Universidad Nacional Autónoma de México (UNAM) for the financial support.

Conflicts of Interest: The authors declare no conflict of interest.

Appendix A

BU: burned; GH: greenhouse; IC: irrigated agriculture with crops; ID: irrigated agriculture without crops (dry); IW: irrigated agriculture without crops (wet); O: oak forest; TF: temperate forest; TA: temporal agriculture and pasture; TDF: tropical dry forest; U: urban; W: water; UA: User's accuracy; PA: Producer's accuracy; OA: Overall accuracy.

Table A1. Error matrix for accuracy assessment of the classification in 2019 using ANN.

	BU	GH	IC	ID	IW	O	TF	TA	TDF	U	W	T	UA
BU	366	0	0	0	0	0	0	1	0	0	0	367	0.997
GH	0	1100	0	0	0	0	0	1	0	18	0	1119	0.983
IC	0	0	3959	1	0	5	6	1	0	0	0	3972	0.996
ID	0	6	0	1260	0	0	0	206	0	26	0	1498	0.841
IW	0	0	0	0	1674	0	0	24	79	0	0	1777	0.942
O	0	0	0	0	0	5987	2	58	41	0	0	6088	0.983
TF	0	0	56	0	0	14	4575	0	4	0	0	4649	0.984
TA	0	0	0	174	6	24	0	3263	30	0	0	3497	0.933
TDF	0	0	0	0	11	89	1	0	6729	0	0	6830	0.985
U	0	0	0	73	0	0	0	0	0	183	0	256	0.715
W	0	3	0	0	0	0	0	0	0	0	1783	1786	0.998
Total	366	1109	4015	1508	1691	6119	4584	3554	6883	227	1783	31,839	
PA	1	0.992	0.986	0.836	0.99	0.978	0.998	0.918	0.978	0.806	1		
OA	0.970												

Table A2. Error matrix for accuracy assessment of the classification in 2019 using random forest.

	BU	GH	IC	ID	IW	O	TF	TA	TDF	U	W	T	UA
BU	951	0	0	0	0	0	0	5	0	0	0	956	0.995
G	0	2885	0	0	0	0	0	3	0	15	0	2903	0.994
IC	0	0	10,776	0	0	11	79	0	0	2	0	10,868	0.992
ID	0	0	0	3924	0	5	0	148	0	11	0	4088	0.96
IW	0	0	0	0	4423	0	12	0	27	0	0	4462	0.99
O	0	0	0	0	0	16,091	3	31	123	0	0	16,248	0.99
TF	0	0	47	0	2	12	12,148	0	4	0	0	12,213	0.995
TA	9	0	0	125	0	35	0	9166	3	42	0	9380	0.977
TDF	0	0	0	0	10	73	0	5	18,136	0	0	18,224	0.995
U	0	1	0	18	0	0	0	3	0	615	0	637	0.966
W	0	0	0	0	0	0	0	0	0	0	4914	4914	1
Total	960	2886	10,823	4067	4435	16,227	12,242	9361	18,293	685	4914	84,893	
PA	0.991	0.9996	0.996	0.965	0.997	0.992	0.992	0.979	0.991	0.898	1		
OA	0.9898												

Table A3. Error matrix for accuracy assessment of the classification in 2022 using ANN.

	BU	GH	IC	ID	IW	O	TF	TA	TDF	U	W	T	UA
BU	86	0	0	9	4	0	0	0	0	0	0	99	0.869
G	2	199	11	0	0	0	0	2	0	31	0	245	0.812
IC	0	3	2031	0	0	3	68	55	2	0	0	2162	0.939
ID	1	0	0	310	0	0	0	66	0	1	0	378	0.820
IW	2	4	8	0	320	0	0	0	0	0	9	343	0.933
O	0	0	41	0	0	2396	2	49	65	0	0	2553	0.939
TF	0	0	43	0	0	23	3094	0	0	0	0	3160	0.979
TA	0	0	70	307	0	57	0	5894	87	48	0	6463	0.912
TDF	1	0	0	0	12	93	1	61	7609	0	0	7777	0.978
U	0	64	0	1	0	0	0	45	0	343	0	453	0.757
W	0	0	0	0	0	0	1	0	0	0	1023	1024	0.999
Total	92	270	2204	627	336	2572	3166	6172	7763	423	1032	24,657	
PA	0.935	0.737	0.922	0.494	0.952	0.932	0.977	0.955	0.980	0.811	0.991		
OA	0.945												

Table A4. Error matrix for accuracy assessment of the classification in 2022 using random forest.

	BU	GH	IC	ID	IW	O	TF	TA	TDF	U	W	T	UA
BU	247	0	0	1	0	0	0	0	0	0	0	248	0.996
G	0	724	0	0	0	0	0	6	0	6	0	736	0.984
IC	0	0	5744	0	2	10	86	9	0	4	0	5855	0.981
ID	1	0	0	1377	0	0	0	28	0	11	0	1417	0.972
IW	0	0	0	0	883	0	0	0	1	0	11	895	0.987
O	0	0	0	0	0	6453	16	79	102	1	0	6651	0.970

Table A4. *Cont.*

TF	0	0	117	0	0	21	8284	0	0	0	0	8422	0.984
TA	1	1	26	294	0	123	0	16,192	59	109	0	16,805	0.964
TDF	0	0	0	0	15	297	0	50	20,506	1	0	20,869	0.983
U	2	2	0	0	0	0	0	46	0	934	0	984	0.949
W	0	0	0	0	2	0	0	0	0	0	2863	2865	0.999
Total	251	727	5887	1672	902	6904	8386	16,410	20,668	1066	2874	65,747	
PA	0.984	0.996	0.976	0.824	0.979	0.935	0.988	0.987	0.992	0.876	0.996		
OA	0.977												

Table A5. Pearson correlation between (numeric) explicative variables for TDF loss.

Explicative Variables	Distance to Roads	Distance to Agriculture	Elevation	Slope	Aspect
Distance to roads	1.00				
Distance to agriculture	0.44	1.00			
Elevation	0.45	0.53	1.00		
Slope	0.31	0.46	0.35	1.00	
Aspect	0.04	−0.05	−0.1	−0.01	1.00

Table A6. Pearson correlation between (numeric) explicative variables for TDF gain.

Explicative Variables	Distance to Roads	Distance to Agriculture	Elevation	Slope	Aspect
Distance to roads	1.00				
Distance to agriculture	0.42	1.00			
Elevation	0.48	0.66	1.00		
Slope	0.33	0.33	0.29	1.00	
Aspect	0.03	0.00	−0.02	−0.06	1.00

Table A7. Reclassified probabilities of TDF loss and gain using a threshold of 0.5.

Reclassified Probabilities	No. of Pixels	Percentage	Reclassified Probabilities	No. of Pixels	Percentage
0.5, 0.788 (TDF loss)	4,798,436	43.08%	0.5, 0.953 (TDF gain)	3,941,964	35.39%
0.017, 0.5	6,340,498	56.92%	0.00002, 0.5	7,196,970	64.61%
Total	11,138,934	1	Total	11,138,934	1

Appendix B

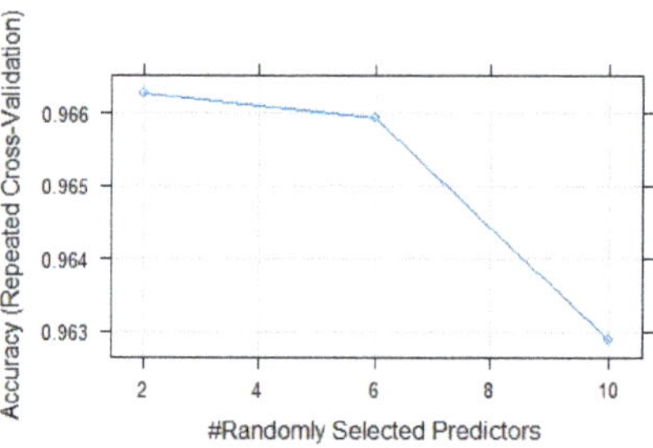

Figure A1. Cross-validation for the classification of Sentinel-2 imagery in 2022 with tested mtry = 2, 6, and 10. Mtry = 2 was selected for its highest accuracy.

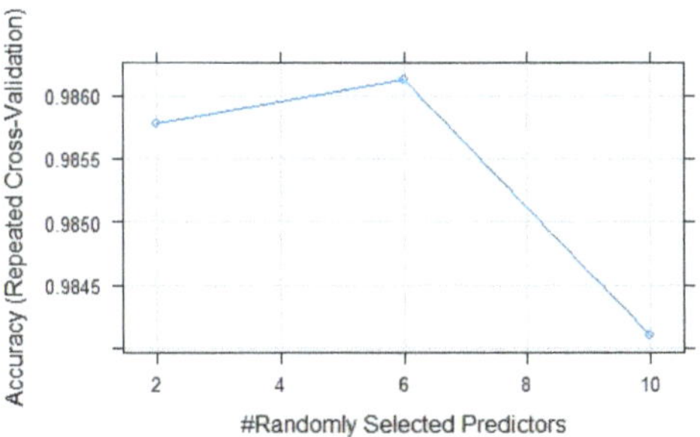

Figure A2. Cross-validation for the classification of Sentinel-2 imagery in 2019 with tested mtry = 2, 6, and 10. Mtry = 6 was selected for its highest accuracy.

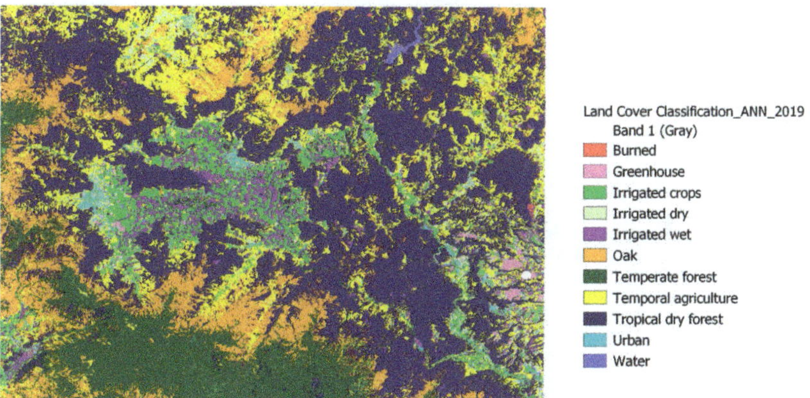

Figure A3. Land cover classification using ANN for 2019. Irrigated crops is short for Irrigated agriculture with crops; Irrigated dry is short for Irrigated agriculture without crops (dry); Irrigated wet is short for Irrigated agriculture without crops (wet); Temporal agriculture is short for Temporal agriculture with pasture; Oak is short for Oak forest.

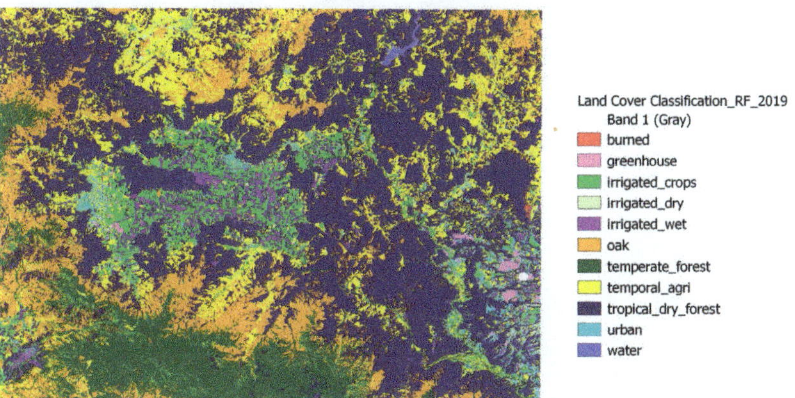

Figure A4. Land cover classification using random forest (RF) for 2019. Irrigated crops is short for Irrigated agriculture with crops; Irrigated dry is short for Irrigated agriculture without crops (dry); Irrigated wet is short for Irrigated agriculture without crops (wet); Temporal agriculture is short for Temporal agriculture with pasture; Oak is short for Oak forest.

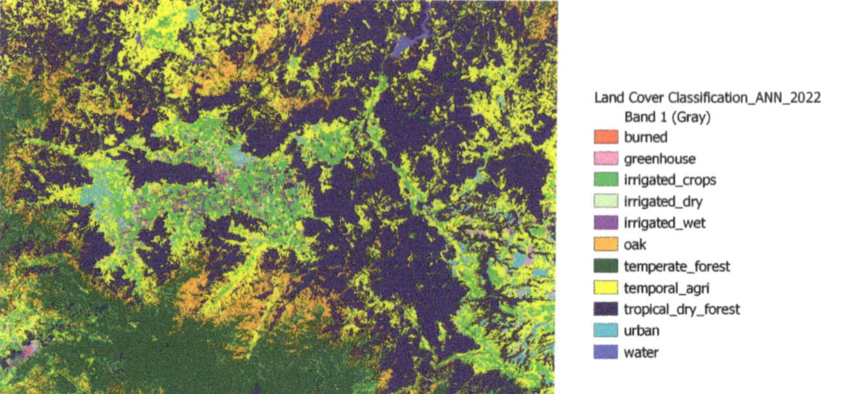

Figure A5. Land cover classification using ANN for 2022. Irrigated crops is short for Irrigated agriculture with crops; Irrigated dry is short for Irrigated agriculture without crops (dry); Irrigated wet is short for Irrigated agriculture without crops (wet); Temporal agriculture is short for Temporal agriculture with pasture; Oak is short for Oak forest.

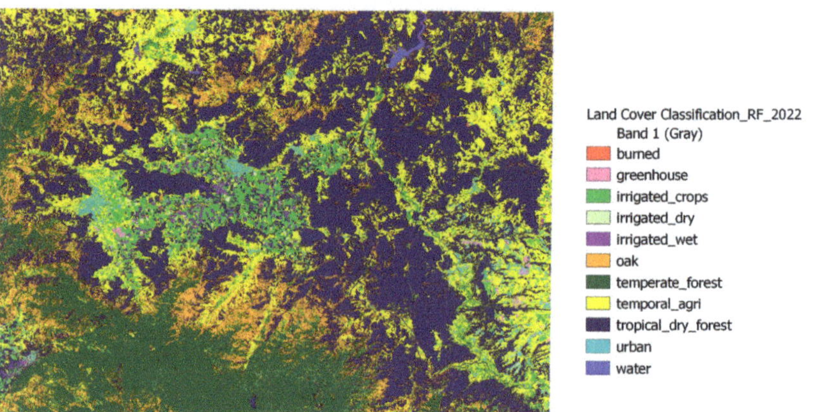

Figure A6. Land cover classification using random forest (RF) for 2022. Irrigated crops is short for Irrigated agriculture with crops; Irrigated dry is short for Irrigated agriculture without crops (dry); Irrigated wet is short for Irrigated agriculture without crops (wet); Temporal agriculture is short for Temporal agriculture with pasture; Oak is short for Oak forest.

References

1. Olson, D.M.; Dinerstein, E.; Wikramanayake, E.D.; Burgess, N.D.; Powell, G.V.N.; Underwood, E.C.; D'Amico, J.A.; Itoua, I.; Strand, H.E.; Morrison, J.C.; et al. Terrestrial ecoregions of the world: A new map of life on Earth. *BioScience* **2001**, *51*, 933–938. [CrossRef]
2. Rzedowski, J. *Vegetación de Mexico*; Edicion digital; Comisión Nacional para el Conocimiento y Uso de la Biodiversidad: Ciudad de México, México, 2006; 504p.
3. Maass, J.; Balvanera, P.; Castillo, A.; Daily, G.; Mooney, H.; Ehrlich, P.; Quesada, M.; Miranda, A.; Jaramillo, V.; García-Oliva, F.; et al. Ecosystem services of tropical dry forests: Insights from long-term ecological and social research on the Pacific Coast of Mexico. *Ecol. Soc.* **2005**, *10*, 17. Available online: http://www.ecologyandsociety.org/vol10/iss1/art17/ (accessed on 7 November 2022). [CrossRef]
4. Pennington, R.T.; Lavin, M.; Oliveira-Filho, A. Woody plant diversity, evolution, and ecology in the tropics: Perspectives from seasonally dry tropical forests. *Annu. Rev. Ecol. Evol. Syst.* **2009**, *40*, 437–457. [CrossRef]
5. Caballos, G.; Ceballos, G.; Andres Garcia, A. Conserving Neotropical Biodiversity: The Role of Dry Forests in Western Mexico. *Conserv. Biol.* **1995**, *9*, 1349–1353. Available online: https://www.jstor.org/stable/2387179 (accessed on 7 November 2022). [CrossRef]

6. Mesa-Sierre, N.; de la Peña-Domene, M.; Campo, J.; Giardina, C.P. Restoring Mexican Tropical Dry Forests: A National Review. *Sustainability* **2022**, *14*, 3937. [CrossRef]
7. Meave, J.A.; Romero-Romero, M.A.; Salas-Morales, S.H.; Pérez-García, E.A.; Gallardo-Cruz, J.A. Diversidad, amenazas y oportunidades para la conservación del bosque tropical caducifolio en el estado de Oaxaca, México. *Ecosistemas* **2012**, *21*, 85–100.
8. Le Quéré, C.; Moriarty, R.; Andrew, R.M.; Peters, G.P.; Ciais, P.; Friedlingstein, P.; Jones, S.D. Global carbon budget 2014. *Earth Syst. Sci. Data* **2015**, *7*, 47–85. [CrossRef]
9. Gasser, T.; Crepin, L.; Quilcaille, Y.; Houghton, R.A.; Ciais, P.; Obersteiner, M. Historical CO_2 emissions from land use and land cover change and their uncertainty. *Biogeosciences* **2020**, *17*, 4075–4101. [CrossRef]
10. Mas, J.F.; Velázquez, A.; Díaz-gallegos, J.R.; Mayorga-Saucedo, R.; Alcántara, C.; Bocco, G.; Castro, R.; Fernández, T.; Pérez-Vega, A. Assessing land use/cover changes: A nationwide multidate spatial database for Mexico. *Int. J. Appl. Earth. Obs. Geoinf.* **2004**, *5*, 249e261. [CrossRef]
11. Trejo, I.; Dirzo, R. Deforestation of seasonally dry tropical forest: A national and local analysis in Mexico. *Biol. Conserv.* **2000**, *94*, 133–142. [CrossRef]
12. Chazdon, R.L. Tropical forest recovery: Legacies of human impact and natural disturbances. *Perspect. Plant Ecol. Evol. Syst.* **2003**, *6*, 51–71. [CrossRef]
13. Morales-Barquero, L.; Borrego, A.; Skutsch, M.; Kleinn, C.; Healey, J.R. Identification and Quantification of Drivers of Forest Degradation in TropicalDry Forests: A Case Study in Western Mexico. *Land Use Policy* **2015**, *49*, 296–309. [CrossRef]
14. Borrego, A.; Skutsch, M. How Socio-Economic Differences betweenFarmers Affect Forest Degradation in Western Mexico. *Forests* **2019**, *10*, 893. [CrossRef]
15. Heinimann, A.; Mertz, O.; Frolking, S.; Egelund Christensen, A.; Hurni, K.; Sedano, F.; Chini, L.P.; Sahajpal, R.; Hansen, M.; Hurtt, G. A global view of shifting cultivation: Recent, current, and future extent. *PLoS ONE* **2017**, *12*, e0184479. [CrossRef]
16. Hillel, D. Soil and Water Management. In *Soil in the Environment*; Hillel, D., Ed.; Academic Press: Cambridge, MA, USA, 2008; pp. 175–195. [CrossRef]
17. Hartter, J.; Lucas, C.; Gaughan, A.E.; Lizama, L. Detecting tropical dry forest succession in a shifting cultivation mosaic of the Yucatán Peninsula, Mexico. *Appl. Geogr.* **2008**, *28*, 134–149. [CrossRef]
18. Borrego, A.; Skutsch, M. Estimating the opportunity costs of activities that cause degradation in tropical dry forest: Implications for REDD+. *Ecol. Econ.* **2014**, *101*, 1–9. [CrossRef]
19. Olofsson, J.; Hickler, T. Effects of human land-use on the global carbon cycle during the last 6000 years. *Veg. Hist. Archaeobot.* **2008**, *17*, 605–615. [CrossRef]
20. Comisión Nacional Forestal. *Visión de México Sobre REDD+: Hacia una Estrategia Nacional*; Comisión Nacional Forestal: Zapopan, México, 2010.
21. Cuevas, R.; Nuñez, N.M. El bosque tropical caducifolio en la Reserva de la Biosfera Sierra Manantlan, Jalisco-Colima, Mexico. *Bol. Inst. Bot. Univ. Guadalaj.* **1998**, *5*, 445–491.
22. Kavzoglu, T.; Mather, P.M. The use of backpropagating artificial neural networks in land cover classification. *Int. J. Remote Sens.* **2008**, *24*, 4907–4938. [CrossRef]
23. Mas, J.F.; Puig, H.; Palacio, J.L.; Sosa-Lopez, A. Modelling deforestation using GIS and artificial neural networks. *Environ. Model. Softw.* **2004**, *19*, 461–471. [CrossRef]
24. Krogh, A. What are artificial neural networks? *Nat. Biotechnol.* **2008**, *26*, 195–197. [CrossRef] [PubMed]
25. James, G.; Witten, D.; Hastie, T.; Tibshirani, R. *An Introduction to Statistical Learning, with Applications in R*, 2nd ed.; Springer: Berlin/Heidelberg, Germany, 2021.
26. Breiman, L. Random Forests. *Mach. Learn.* **2001**, *45*, 5–32. [CrossRef]
27. Gomez, C.; White, J.C.; Wulder, M.A. Optical remotely sensed time series data for land cover classification: A review. *ISPRS J. Photogramm. Remote Sens.* **2016**, *116*, 55–72. [CrossRef]
28. Solórzano, J.V.; Mas, J.F.; Gao, Y.; Gallardo, A. Land use land cover classification with U-Net: Advantages of combining Sentinel-1 and Sentinel-2 imagery. *Remote Sens.* **2021**, *13*, 3600. [CrossRef]
29. Liaw, A.; Wiener, M. Classification and Regression by randomForest. *R News* **2002**, *2*, 18–22.
30. Fearnside, P.M. Global Warming and Tropical Land-Use Change: Greenhouse Gas Emissions from Biomass Burning, Decomposition and Soils in Forest Conversion, Shifting Cultivation and Secondary Vegetation. *Clim. Chang.* **2000**, *46*, 115–158. [CrossRef]
31. Olofsson, P.; Arevalo, P.; Espejo, A.B.; Green, C.; Lindquist, E.; McRoberts, R.E.; Sanz, M.J. Mitigating the effects of omission error son área and area change estimates. *Remote Sens. Environ.* **2020**, *236*, 111492. [CrossRef]
32. Olofsson, P.; Foody, G.M.; Herold, M.; Stehman, S.V.; Woodcock, C.E. Good practices for estimating area and assessing accuracy of land change. *Remote Sens. Environ.* **2014**, *148*, 42–57. [CrossRef]
33. Burnham, K.P.; Anderson, D.R. *Model Selection and Multimodel Inference: A Practical Information-Theoretic Approach*, 2nd ed.; Springer: New York, NY, USA, 2002.
34. Wilcoxon, F. Individual comparisons by ranking methods. *Biom. Bull.* **1945**, *1*, 80–83. [CrossRef]

35. Shono, K.; Cadaweng, E.A.; Durst, P.B. Application of Assisted Natural Regeneration to Restore Degraded Tropical Forestlands. *Restor. Ecol.* **2007**, *15*, 620–626. [CrossRef]
36. Yang, Y.; Wang, L.; Yang, Z.; Xu, C.; Xie, J.; Chen, G.; Lin, C.; Guo, J.; Liu, X.; Xiong, D.; et al. Large Ecosystem Service Benefits of Assisted Natural Regeneration. *J. Geophys. Res. Biogeosci.* **2018**, *123*, 676–687. [CrossRef]

MDPI
St. Alban-Anlage 66
4052 Basel
Switzerland
www.mdpi.com

Remote Sensing Editorial Office
E-mail: remotesensing@mdpi.com
www.mdpi.com/journal/remotesensing